Transactional (Qu Microphysics

Principles and Applications

Transactional (Quantum) Microphysics
Principles and applications

Jacques Lavau

Jacques Lavau éditeur

2019

US revised edition, May 2019.

ISBN 978-2-9562312-2-6

Jacques Lavau éditeur

207 avenue Jean Jaurès

69150 Décines-Charpieu

France

jacques.lavau@free.fr

Contents

I wish to express thanks to

William Beaty, webmaster of amasci.com, who has let me know in 2003 that I was preceded on several points, by nineteen years (1979) by par Giles Henderson, by fifteen years (1983) by C. F. Bohren, H. Paul, R. Fischer and by twelve years (1986) by John G. Cramer. W. Beaty gathered and put in sight works that everybody ignored in 1998 in IN2P3, Lyon 1 University.

Bernard Chaverondier for fruitful discussions on Usenet, years 2005 to 2007.

Joël Brunet and the electronics teachers Joël Robelin and (first name?) Sainsaulieu who in 1995 asked me to explain to our pupils the huge capturing section of the carbon monoxide molecule for the precisely resonating photons. It was the first time I was in front of the proof of the convergence of this very large photon on this tiny molecule, 4.7 Å its great diameter.

Lev Lvovitch Regelson who had put in sight the paper from Schrödinger: "**Über den Comptoneffekt**", 1927 (when he had not yet the right equidistance).

Jean-Claude Coviaux who gave the initial impulse and explained the constraints in popularization.

Christiane Cesari who read the first twenty pages of that time, found out grammar mistakes, and said what was hermetic to her.

My daughter Audrey: she said where she dropped because she did not get in a too early calculus, then on page 3 (now on Appendix D).

Several contributors in researchgate.net who protested against some of my shortcuts which were impenetrable for them, and made me precise some points which lacked in the teaching they had received.

In June 2017 on Usenet, Julien Arlandis, François Guillet, and Fabrice Neyret signaled some weakness in the writing.

Any remaining mistakes are mine alone. The author would welcome any error-signaling to him by readers.

Transactional (Quantum) Microphysics, Principles and Applications

Introduction: Why transactional? And why so late?

Did you hear *Mr. Tompkins Explores the Atom* where he was told that for killing the quantic tiger, one should shoot many bullets in all directions? Or were you thrust that as owing to Richard Feynman "*Nobody understands Quantum Mechanics*", so you should not dare to understand better than him? Or that the mysteries of the Holy *Duality* are so damned subtle, that only damned math-minded initiates can master it? Or that an electron in a cathode-ray tube takes the time "*to explore beyond planet Jupiter*"? So much considering that the cat "*is*" simultaneously dead and alive, and that the submarine "*is*" *in a superposed state* between two miles further west and two miles further east, and maybe further north and further south... According to the copenhaguist mythology, as in hegemonic power for 1927, at a moment at his fantasy, the cement-maker sends a 20 tons trailer cruising at random simultaneously on all the roads. Its wave of probability spreads on all the surface of Earth, and suddenly, *Miracle!* The driver finds a customer whose hopper has room for 20 tons of cement. So is the miracle of the *collapse of the function wave* as taught by the copenhaguists... The transaction preceding the sending of the trailer to this peculiar customer is beyond their minds.

With so conceptual blurring, no wonder that charlatans praise their quacks medicine, but "*quantic huh!*": they were given all the facilities to fool you so.

It was enough with these rags and blunders, and it was high time somebody draws a much more clear and much more competent popularization of quantic physics. From experimental results through experimental results, this initiation handbook gives you benefits of advances in the Transactional Microphysics, through conversations between four characters: the reader Mr. Curious, the professors Marmot and Castle-Holder, Mr. Open-Eyes. Through annexes, the reader can restart from the beginning of the Mechanics, and verify each point. The big secret that the high priests of the mystery kept tight, is that "*quantic*", in reality, it is periodic, undulatory, and transactional, where any **individual** wave of Quantics has **one** emitter and **one** absorber, but nothing at all that could be *corpuscular*, nor *dualistic*. Nothing *corpuscular* exists in Microphysics, and only approximately in Macrophysics. Moreover, we will prove that in the frame of the transactional theory, the quantic microphysics is only a sub-domain of the general microphysics.

Zero corpuscles at the microphysical scale, zero duality, but emitters and absorbers do exist. Individually, any photon has **one** absorber and **one** emitter. So is the individual scale in Microphysics. In this handbook, you'll see the geometry of the Fermat spindle of this transfer between emitter and absorber, and you'll find the many consequences in optics, radio-electricity, radiocrystallography, electronic optics, electrotechnics, and so on.

When the solution has been found by the finder, as anybody can see how it was simple, one may question the collective disability that compelled the other researchers to find nothing, to remain stuck in tribal faults since 1925-1927. And worse, to deny tons and tons of experimental facts that embarrassed them, especially in optics: specialized too early, the academics never learned to practice horizontal transfers of technology, from one craft to another, from one branch to another. *A specialist is a person who knows very much on very little, and at the limit, all on nothing at all,* when the trans-disciplinary synthesis are made by men of synthesis.

The discipline of interdisciplinary synthesis is taught in other crafts, as the craft of geographer, or the craft of engineering, but alas it is ignored in the teaching of *"pure"* physics - a cruel lacking. Typically, an author of creativity handbook was a qualified engineer **and** doctor of psychology. An excellent example of a trans-disciplinary collective work is the FAO-Unesco Soil map of the world: The geographer who coordinates each volume is not a geologist, nor a petrologist, nor a climatologist, nor a hydrologist, nor a specialist in vegetable physiology, nor an agronomist, nor a pedologist either, but he can understand the works from all these specialists, so these necessary atlases were well carried out of press.

This handbook puts an end to numerous omissions, even censures, which seriously isolated the quantic from other branches of physics, and worse, from their experimental results. In transactional microphysics, no more need to crush the students under *"Shut up and calculate!"*, or under *"If you think you understand quantum mechanics, so you do not understand it!"*. No more need to hide from the students the resonant transparency independently discovered in 1921 by Carl Ramsauer and J. S. Townsend: that the xenon atoms become transparent to incident electrons when their energy is about 0.6 eV, it is evident when you know that each electron is his de Broglie wave, but remains incomprehensible as long as you present them as corpuscles. No more need to hide to students that the radiocrystallography, application of the physical optics laws established by Fresnel in 1819, it works with neutrons too, and though less precise because of electrostatic repulsion, with electrons too, and not only with X-rays. In radiocrystallography, no more need to hide to the students that the Scherrer law links the width of the photon to the width of the crystallites, so never any photon becomes corpuscular. In optics, no more need to hide to students that each interference proves that every involved photon is long since Young (1801) and Fresnel (1819). In optics, no more need to hide to the students that each anti-reflect coating, or each interferential color on beetles, green lizards, and many birds proves that each photon is at least several or even tens or even hundreds of wavelength wide, if enough far from its absorber and far from its emitter. In optics, no more need to hide to the students that the plane polarized light exists and that some optical devices convert all or part of a polarization into another one, nor to hide to them how some chiral molecules rotate a plane of polarization of light. In solid state physics, no more need to hide to the students that conduction electrons are each wide and long enough to interact with phonons, plasmons and polaritons, which will never become small. No more need to hide to the students the electronic properties of the dyes, nor optical properties of the crystals. No more need to hide to the students the properties nor the consequences of the Dirac equation, 1928. And so on.

In **transactional** quantic microphysics, the isolation of a pseudo-science is over, the unity of the physics is restored, ninety years later.

Jacques Lavau

P.S.

April 2019. The US edition includes exercises in Relativity, more developments in Didactic, some solved exercises on magnetic potential, and the list of the surreptitious postulates is now completed up to twenty-one, either postulates, either ferocious censures against lots of embarrassing experimental facts.

Transactional (Quantum) Microphysics, Principles and Applications

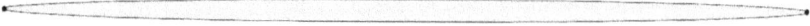

Transactional (Quantum) Microphysics, Principles and Applications

1.1. Historical position

Curious:
- Eleven years ago, it was in the New Year 2008, I had asked you to recommend a good textbook initiating on quantum mechanics. I saw you scrapping your heads, look at each other, and conclude that no such good initiation book exists, that you should roll up your sleeves and start writing this textbook. Now, what can you present to us, us the beginners, desiring to learn?

Professor Castle-Holder:
- First, we will collect some of the most spread false ideas in the heads of the laymen,

Open-Eyes:
- ... before showing that they also lay in the heads of the most beyond-suspicion academic authorities.

Professor Castle-Holder:
- Then we will tell the reader what we share in common to put him at the right level in the physics of atoms[1], electrons[2], and photons[3]. Further, he will have links to technical annexes, out of the dialog. We should defer opening the controversy that lasts publicly between us for fifteen years. First putting the reader at the proper level, debating after.

1.2. Most common howlers, a collection

(1) *An electron, it is damned small, almost a point, it is a little green ball, and it orbits around a rose nucleus. But it is true! Believe it my children!*

---

[1] **Atom**: one nucleus with its escort of electrons to balance the total charge to zero. **Atomic nucleus**: composed of Z protons and Y neutrons, it needs Z electrons to make an atom.
[2] **Electron**: The smallest part of electricity. Invented in 1891 and experimentally proved in 1897, as *cathodic rays*. Its charge is negative. It has a magnetic moment. All the electrons are indistinguishable.
[3] **Photon**: When identifiable, the smallest transferable part of electromagnetic radiation. It transfers from its emitter to its absorber one quantum of looping of Planck, **h**.

(2) *Even in the vacuum, an electron walks in random zigzags, and it patrols in all direction like a young dog. But it is true! Believe it my children!*

(3) *If the electron always keeps its electric charge, it is because it is very small, out of reach behind its damned hard walls. But it is true! Believe it my children!*

(4) *An atom is mostly vacuum between the electrons. Just like in astronomy, with only the vacuum between the planets and the stars. But it is true! Believe it my children!*

(5) *If we do not tell the beginners that an electron is a small corpuscle, they will be lost! They will not understand anything! But this is true! Believe it my children!*

(6) *At 30 CH, a homeopathic drug is still efficient[4]. But this is true! Believe it my children!*

(7) *A cat can be simultaneously dead and alive. But this is true! Believe it my children!*

(8) *The behavior of the particles, it is just a mystery of the gods, and it would be a sin of pride, to pretend to understand it. But this is true! Believe it my children!*

(9) *For killing the quantic tiger, which is a big fuzzy tiger, Mr. Tompkins must shoot many bullets in all directions. But this is true! Believe it my children!*

(10) *The light is the shock of small grains, called photons. But this is true! Believe it my children!*

(11) *Those who disagree with us, are just retired colonels of cavalry, and they want to return to classical physics. Moreover, they commit a mortal sin!*

(12) *As long as you do not master the hermitian operators on Hilbert spaces, you have nothing to do in QM!*

Be the reader reassured, all the statements above are false.

Open-Eyes:
- The howler n° 2 is written in the Landau and Lifshitz, Volume 3. Jean Bricmont and many others teach the howler n° 4. It is incompatible with the existence of electronic capture, where a nucleus catches an electron from the deepest layer, and about 500 nuclei can do so. So is the reaction: ^{57}Co (captures an internal e⁻) $\rightarrow$ $^{57}Fe^*$ + $^0\nu_e$ (used for the Mössbauer effect).

Professor Marmot:

---

[4] *The general public has not yet assimilated the Avogadro-Ampère constant, which links our macroscopic world to the atomic limit. There are six hundred thousands and two hundreds and fourteen milliards of milliards of water molecules H_2O in 18.0153 g water (one **mole**). The charlatans abuse of this public ignorance. Worse: many academic leading figures are still in difficulty with it: they confuse the individual waves of the quantic, which all have one emitter and one absorber, with macroscopic waves such as gravity waves at sea or in a water tank, that diffuse on many, many absorbers.*

- I forbid you to criticize the professor Bricmont: he is a recognized politic writer. I will denounce you as extreme rightist!

Open-Eyes:

- Jean Bricmont has no excuse: he is a professor of theoretical physics. The howler n° 9 is from George Gamow. *"Damned small, almost a point"*? And what is the *length* of a liter of milk? One can show dye molecules, where the oscillating electron, responsible for the selective absorption is more than 15 ångströms long, and more than 3 ångströms wide (0.3 nanometers). In pure metals at low temperature, a part of the conduction electrons have practically all the crystal size. *"Small"* and *"punctual"* are geometrical notions bound to our macrophysical world, but without meaning, and without coherence in microphysics, as the atomic limit exists.

Yes, the photon is the smallest unit of transferable electromagnetic radiation, in the meaning that it transfers a Planck's quantum **h** (of action per cycle) from an emitter to an absorber, but this does not make it a *small grain*: it remains electromagnetic radiation. That still comes under the laws of physical optics given in 1819 by Augustin Fresnel, and still comes under the Maxwell equations, 1873, modernized by the bosonic interaction between photons. Yes the plane polarized light exists, and we have thousands of proofs; it would not exist if the photons would be small grains. At school, we have performed experiments of light interference, but they would be impossible if photon would be small grains. Beetles and pawns, some duck feathers, some fishes have interferential colors, and they would not exist if the photons would be small grains, geometrically *"small"*.

Professor Castle-Holder:

- What is really small by our human scale, is the Planck's quantum, and often the wavelengths, at least for the visible range and beyond, in the UV.

Transactional (Quantum) Microphysics, Principles and Applications

CHAPTER 2

Explaining the stakes of the transactional microphysics

2.1. A radical revision was unavoidable

Open-Eyes:
- In these first chapters, we will precise the stakes of the transactional physics, this one which explicitly deals with **individual** waves, that each one has **one** emitter and **one** absorber. Then we will revise the key-constants of the microphysics, the basics of the atomics, and the electronic clouds around the nuclei, why spectral lines and spectrography, then we will see the chemical bonds, crystalline solids, the spectral lines, atomics, the metallic state, and plasticity of metals. We'll have to revise some parts of astronomical optics, of radio-electricity, and the interferences in optics.

Professor Castle-Holder:
- I protest! The metallic state and plasticity of metals belong to physics of solids, not to Quantic! Moreover, astronomical optics and radio-electricity are only classical physics. We do not intend to have anything to do with *classical physics*!

Open-Eyes:
- Lots of requisite knowledge do not cross the distance from one lecture room to another, on the same campus. I insist on the itinerary, and that you trust me the necessary time. Many vital pieces of knowledge from other branches of physics lack to those who presently teach *Quantum Mechanics*. They are not even conscious of the gravity of their lacunae. No way to correctly explain the photo-electric effect (Lenard 1900, Einstein 1905), without the necessary knowledge on the metallic state, that did not exist in 1905. No clear view on the photons, without taking account the experimental results in interferometry and astronomical optics, obtained from the 19th century, nor without the performances of the anti-reflect coatings, and of the birefringent quarter-wave plates, used for converting a polarization; it would be monkey business. Among the detrimental consequences, the exacted cost on the students by misunderstanding, even disgust, is exorbitant, and the yield of scientific teaching is dismaying.

Curious:

- If you already disagree about the program, then you must explain what divides you. What is the stake in your controversy?

Open-Eyes:
- Ouch! Precisely the controversy that Mr. Castle-Holder wisely tried to cool down! At least the formalism of the QM (according to their own designation: "Quantum Mechanics") is mostly correct, though futilely obscure. BUT the teachers of QM discard half of the boundary conditions, only taking account of the emitters. They are encumbered with inherited and surreptitious postulates, whose the worst are the anti-relativist, the corpuscularist, the anti-undulatory, the confusionist, and the anthropocentric ones. The students approach the QM formalism – strictly determinist and strictly undulatory – only when they paid submission to a semantics[1] rooted in the 1925-1927 years, which is unjustifiable and inexcusable. Not only this corpus of doctrine hinders the yield of the teaching of physics, alas a very low yield, but it also obstructs the heuristic[2] means the researchers should access. Their semantics, said *copenhaguist* because it was elaborated in Göttingen and adopted and defended by the Institute of Copenhagen bossed by Niels Bohr, is a knot[3] of faults in methodology, and of inexcusable absurdities. It misleads lots of researchers to swamps, where they are sucked. The repetitive faults of methodology obstruct the future of the country and its youth.

This doctrine where they reason as gunners of the first World War, who poured on zone (see George Gamow, *Mr. Tompkins explores the atom*: to kill the quantic tiger, he must shoot many bullets in all directions), is invalidated by the Ramsauer-Townsend transparency effect, by all the spectral absorptions, by all the spectra of dark lines discovered by Fraunhofer[4], and by all the spectral measures in the chemistry of gases (including the on-the-orders propaganda, and on-the-orders arguments in the on-the-orders media on the myth of "*evil greenhouse gases that will make the oceans to rise*"), by all the dyes and colorimetric methods practiced in analytic chemistry, by the technology of anti-reflect coatings in instrumental optics, by the methods of atomic absorption, by the

---

[1]**Semantics**: the study of the meanings. In physics, the semantic axioms have a structure in "*This denotes...*" followed by the experimental protocol by which you can have evident facts that will further serve as a reference. For instance, for electromagnetic waves, you could have a set of emitting dipolar antennas, rightly retarded in phase (the same general aspect as a Yagi antenna) and a dipolar detecting antenna you can move in the room of the experience. So you may have evidence of the polarization, directivity of the device, even knots and bellies if a stationary wave is organized by the operator.

[2]**Heuristics**: the art of finding.

[3]A Gordian **knot**, to be cut off by the sword.

[4]The **spectrum**, or separation of the light through a prism according to the wavelength is known from Isaac Newton. The spectroscope was invented by William Wollaston in 1802; he discovered that the solar spectrum has dark lines. The German optician Joseph von Fraunhofer performed the first spectral analysis in 1811. Fraunhofer listed 600 lines in the solar spectrum. The spectrum of the solar photosphere is now called the Fraunhofer spectrum. More than 26,000 lines in the solar spectrum are listed, where 6,000 are unique to the iron.

black color of biotite micas[5], by all the success of the Mössbauer effect, including the historic experiment from Pound and Rebka, and even by the practice of the men who mount antennas on the roofs.

Professor Castle-Holder:
- After so many unkind words, you must justify by showing pieces of evidence.

Professor Marmot:
- Moreover, I demand you to prove at once that Rudolf Mössbauer is on your side.

Open-Eyes:
- About Mössbauer, the answer is in Appendix D, including the first basis of information the reader needs as a beginner. Experience proved that if I obeyed your trick, here the vast majority of the readers would fail to keep up: they do not master this calculus – though familiar to more advanced students.

The first and enormous surreptitious and clandestine postulate you teach is the **anti-relativist** one.

2.1.1. Anti-relativistic postulate. For them, the time remains the Isaac-Newton's one, where his god could simultaneously see all. According to the formalism they teach, their time is a universal and ubiquitous parameter. However, any photon violates their postulate, as its proper time is null. The creation and annihilation reactions are simultaneous in the photon's time. However, its coherence length and duration are not null: interferences exist! Any interference demands that each participating photon has an appreciable length of coherence. Two molecules of a gas, with different speeds in different directions, do not have the same flowing of time; all the relativists know that, though all teachers of QM remain ignorant on that. Those who handle a particles accelerator know well that the accelerated electron or proton no more dwells in the laboratory time, but the teacher of QM ignores it and writes the contrary on the blackboard.

Professor Castle-Holder:
- But we had to simplify! You already complain that the formalism is heavy and so hard to master. What would be your complaints if it was relativistic?

Open-Eyes:

---

[5]The **biotite** mica is found in the most of the granites, in diorites, norites, in a vast variety of plutonic and metamorphic rocks, less often in effusive rocks. When the kids think to find gold in rocks, very often it is weathered biotite, now golden yellow.

- If you dared to have your standard output supervised by a qualitician[6] and by a didactician[7], your pride would be unpleasantly surprised. Let's go on your Newtonian time, and take an experiment in optics, such those performed at the Institut d'Optique d'Orsay. Twenty meters from the emitter to the absorber, that is a temporal shift of 66.7 ns (nanoseconds). If we trust the Lorentz transformation, then the departure and the arrival are simultaneous in the "*point of view*" of the photon, though their duration is not null in either of the two frames, about one **ns** each. Therefore there is a fault and incoherence in the Newtonian time of the laboratory, and there is plenty of this kind, for any photon emitted from anywhere to anywhere, and it is an amount for very many. And the affairs become dramatic with astronomic distances, with proved photonic short-circuits in the several milliards years range, up to fourteen-fifteen milliards years.

Curious:
- 66.7 ns? I can read it means nanoseconds. The electronics engineers know these prefixes for units, but my young daughter does not. So please give us a table of the prefixes.

Professor Castle-Holder:
- At primary school, we were introduced to deci, centi, milli, deca, hecto (a *hecatomb* is the sacrifice of a hundred oxen) and kilo. Now we have more needs in prefixes.

Decimal	abbrev.	Power of 10	Multiplic.	abbrev.	Power of 10
deci	d	-1	deca	da	1
centi	c	-2	hecto	h	2
milli	m	-3	kilo	k	3
micro	µ	-6	mega	M	6
nano	n	-9	giga	G	9
pico	p	-12	tera	T	12
femto	f	-15	peta	P	15
atto	a	-18	exa	E	18
zepto	z	-21	zetta	Z	21
yocto	y	-24	yotta	Y	24

Open-Eyes:
- The second surreptitious and clandestine postulate they carry is corpuscularist.

---

[6]**Qualitician**: an engineer in charge of quality, preferably from the beginning of the design, when the quality is still for free.
[7]**Didactician**: Teacher or researcher, devoted to coherence and sturdiness of the course the teaching assigns to pupils. The teacher is in charge of tactics, and the didactician is vowed to strategy; his/her horizon is for several years on the scholar or academic curriculum, then in the efficiency of the applying in the professional life.

2.1.2. Corpuscularist postulate. Or anti-optical postulate, as any quantitative law of optics is incompatible with the corpuscularist ideation. All the spectrography, the interferences (photonic, electronic, or neutronic), the radiocrystallography, the antennas and their directivity, the anti-reflective coatings, the interferential colors, the quarter-wave plates, the Ramsauer-Townsend transparency effect, many fine effects in polarized light, all these facts are incompatible with the myths of the corpuscularism. For an erroneous half-sentence by Albert Einstein in 1905, they were all convinced that the resurrection by Einstein of the Newtonian corpuscle, was right. However this concept is internal to macrophysics, very far from the atomic limit, and not any experiment has ever validated such extrapolation to microphysics.

Professor Marmot:
- If the highest scientists need the corpuscles, so they are necessary.

Open-Eyes:
- Remember the answer from Pierre Simon Laplace to Napoleon: *"Majesty, I did not need that hypothesis!"*. At the scale of our familiar macroscopic world, sure, some things appear so small. For instance, a grain of sand passing through the sieve of 200 μm but refused on the 160 μm looks like a good *corpuscle* by our human scale: excepted if it is of glass and cutting, or metallic and cutting or even toxic, it will pass through our bowels without harming much. Another example of *"corpuscle"*: the spore of some mold. You cannot see it in the air with your naked eyes, but it may do noticeable damages if it falls into something it can *eat* (and oxidize). But the extrapolation to microphysics, beyond the atomic limit, was never validated; it is no more than a religious dogma, droned in the lectures rooms.

Professor Marmot:
- Ah no! All the trajectories observed in bubbles chambers prove that the particles are real particles, and not your vague and muddy waves! A particle is a small place in the field where all the energy is concentrated. We have taught that all the time.

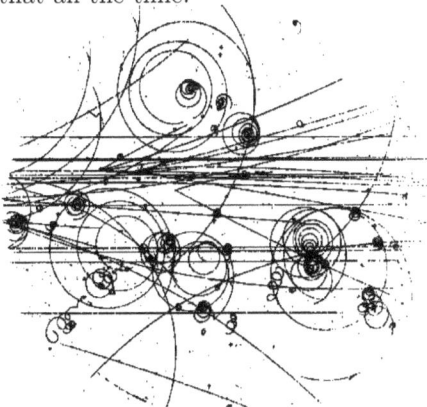

Figure 2.1.

Open-Eyes:
- I cannot encourage you too much in reading again the optics handbook, especially the chapter on the ultra-microscope; then further, to open a textbook on colloids and dispersoids, especially their optical properties. Please tell us what is the minimum diameter of the fog droplets in fog chambers, or bubbles in bubbles chambers, that is necessary to have them recorded on the photosensitive film.

Professor Marmot:
- You cheat! Photographic optics is not Quantum Mechanics! And colloids are not QM either!

Open-Eyes:
- Thanks for your confession of ignorance. These tracks are at least 0.5 μm wide, a wavelength of visible light. But you pretend that they are the proof of corpuscles that would be about hundreds of millions to milliards times smaller... Though they are historic, these tracks only prove the law of conservation of the momentum. Worse, you confuse "wave" of the quantic scale, with collective of divergent waves, and with waves in a collectivity of matter.

For many years, a multitude of dim-witted insist on confusing three classes of waves:
1 – Waves in a collectivity (of atoms or molecules). So are the gravity waves between two fluids[8], and acoustic waves, seismic too. And in microphysics, the spin waves in a ferromagnetic material, the phonons, the plasmons, and polaritons.
2 – The collectives of waves, such as light, or a beam of electrons, ions or neutrons. These collectives comprise many individual emitters and many more potential absorbers.
3 – The individual waves, for each quantic *"particle"*, photon or neutrino for instance. Each one of these individual waves converges onto one individual absorber.
For many years they deny the classes 2 and 3 and demand that all should be of class 1. Put differently, they deny the atomic limit in undulatory physics, so they could synthesize the absurdities they want to disparage – their real

---

[8]On surface **water-air**: waves, soliton of a tsunami.

Air-air: Leeward of some reliefs of simple shape, a set of stable altocumuli, each one marking the top of a wave, as the wake of the island or mountain. Cf. The wake of the Amsterdam island:

http://cache.boston.com/universal/site_graphics/blogs/bigpicture/

eobs_01_14/e13_amsterdam_tmo.jpg

Water-water: in front of the mouth of the Amazon, underwater waves of the interface salt water – fresh water; they are known by the noticeable slowing they inflict to ships.
Source: **Traité d'océanographie physique ***. Jules Rouch, Payot 1948.

tactical goal.

Professor Castle-Holder:
- Here I must intervene to emphasize that each of us was raised in ambient corpuscularism, from which we used to draw hundreds of conclusions that Mr. Open-Eyes contests. It is a matter of routine to underestimate, even cover with scorn and insults the transactional thesis sustained by Mr. Open-Eyes. But his tenacious coherence is revolutionary. Maybe it will force us to agonizing re-appraisals, but on balance, it pretends to offer considerable simplifications in the concepts.

Open-Eyes:
- Thanks! I had the luck to be learned: *"Do not stop your idea! Go further up to its ultimate end!"*. Too many people just throw a sally in the air, just for surprising the strollers, but alas avoid further deepening.
Next, we will see that everyday applications of the physical optics, like the anti-reflect coatings you see in use on the photographic lenses, bring proofs of the optic width of each photon; especially with short focal, wide fields. Other proofs, even more widespread and compelling, are the interferential colors (feathers of birds, scales of fish and reptiles), and their variations under oblique light. It was under their eyes, and they did not see.

The third and fourth surrepticious practices are censuring the plane-polarized light, and censuring the radiocrystallography.

2.1.3. Denying and censuring the existence of the plane polarized light. Though the bees, the photographers and the astronomers make good use of the plane-polarized light, denying it is necessary for the sake of the corpuscularist and anti-optical postulate listed above. Sure, when hypnotizing yourself on the mathematical formalism, you may think that two perfectly phase-opposed helicoidal photons can simulate a plane polarization. But in fact, on several kilometers, such a perfect pairing in frequency, direction, and phase is impossible to obtain. Moreover, their corpuscularist postulate forbids the quarter-wave plates to work, nor any other fraction-of-wave devices; while these devices have no kind of difficulties with the Fresnel's wave optics, 1819. The Maxwell's equations allow to photons (individual waves) any polarization intermediate between the purely plane and the purely helicoidal. So they need to hide this from their students.

2.1.4. Denying and censuring the existence and success of radiocrystallography. In mineralogy or metallurgy laboratories, we daily use radiocrystallography; preferably with X-rays, but frequently also with electrons. But the Fresnel's optics suits these interferences, and suits also their limit, the Scherrer's law. And this, for each photon, each electron. So they need to hide this from their students.

The fifth surreptitious postulate is the tribal postulate anti-Broglie and anti-Schrödinger, therefore anti-frequential.

2.1.5. Tribal postulate anti-Broglie, anti-Schrödinger, so anti-frequential.

An obligation to negate any frequential phenomenon, except those electromagnetic and mass-less. Negate the intrinsic frequencies of any particles with mass (spinorial frequency of Louis de Broglie mc^2/h, and electromagnetic frequency of Dirac-Schrödinger, $2mc^2/h$). Negate the two retrochronous components of the electron wave, brought by Dirac equation in 1928. Negate the Zitterbewegung. Hide the Ramsauer-Townsend transparency, 1921, because it proves that the electron IS its Broglian wave. And retaliations against those who do not participate in the negation of reality... If we examine the thick textbook on MQ in two volumes from Claude Cohen-Tannoudji, Bernard Diu, and Franck Laloë, the word « *frequency* » appears once on page 18, then disappears forever in page 18. Never they made explicit any value of frequency. As they remain submitted to the anti-relativist postulate, they had no chance to grab the essential point. Just one exception: Volume 2, Complément EXIII, they mention the frequency of the Mössbauer photon; and a little before, chapter XI-IIb, they mention the resonant character of the probability of transition. Only the photon frequency is evoked, but never of which frequencies it is the beat. Instead: « *Il se produit donc un phénomène de résonance lorsque la pulsation de la perturbation coïncide avec la pulsation de Bohr associée au couple d'états* $|\varphi i>$ *et* $|\varphi f>$. ». In their opinion, only the photon is frequential, and nothing else is. Obviously, they never practiced radiocrystallography[9] with neutrons or electrons, and do not know it exists. The magic adjective « *associée* » will never have a definite meaning.

Professor Castle-Holder:
- I hope you have strong evidence of these de Broglie frequencies, which are disdained by all. Are not you afraid of the retaliations? See how promptly Claude Cohen-Tannoudji brandished the war ax against Shau-Yu Lan, Pei-Chen Kuan, Brian Estey, Damon English, Justin M. Brown, Michael A. Hohensee, Holger Müller[10] because they used these de Broglie frequencies, so forbidden by the community. For your sake, you must have strong proofs!

Open-Eyes:
- We have lots of proofs mentioned above: first, each time intervene a resonance on the absorber end, then all the radiocrystallography practiced without the X-rays, but with electrons, neutrons, protons or other ions. Sure: it had needed

---

[9]**Radiocrystallography**: By the diffraction of X-rays, obtaining the equidistances of crystallographic planes, therefore finding the crystallographic structure. This invention by the father and son Bragg has revolutionized the metallurgy and the mineralogy.
[10]**A Clock Directly Linking Time to a Particle's Mass. Shau-Yu Lan, Pei-Chen Kuan, Brian Estey, Damon English, Justin M. Brown, Michael A. Hohensee, Holger Müller**. Science-2013-Lan-554-7.

a young electronics teacher to arouse my suspicions: they handle lots of oscillators and superheterodyne[11] frequency-changers. This colleague Sainsaulieu was enough to make me get rid of the blinkers; he will not be enough thanked. They think frequential and resonances, and he was right. Furthermore, the experimental proof[12] was done in 2005 at the linear accelerator in Saclay.

Professor Marmot:
- So a mere electronics teacher may be right against our highest authorities, awarded by Nobel prizes!

Open-Eyes:
- *The Spirit blows where it will*; even where nobody has access to peer-reviewed scientific publications. Sooner or later, humble people will discover that *"Midas, the King Midas, has donkey ears!"*. Another class of experiments is systematically discarded by all the authors of QM textbooks (D. Sivoukhin excepted), is the Ramsauer-Townsend transparency, discovered independently in 1921 by these two authors.
It is a fading of the diffusive obstacle of say a xenon atom, when the incoming slow electron has some energy, about 0.6 eV to 1 eV; it was accurately confirmed many times; it can be explained only if the electron is the wave imagined in 1923 by Louis de Broglie, and if it is an individual wave. This systematical censure is a *"smoking gun"*.

The sixth surreptitious and clandestine postulate is the postulate of macroscopic geometry.

2.1.6. Postulate of macroscopic geometry. The copenhaguists postulate auto-similitude of time and space at any scale, with unlimited extrapolations. Furthermore, they extrapolate to microphysics the statistical irreversibility of the macroscopic time, and the topology[13] with unlimited fineness, inherited from the mathematicians of the 19th century, down to under the atomic limit where it is no more valid at all. No, the two electrons of a helium atom are not geometrically distinguishable one from another.

Professor Castle-Holder:
- So? Do you recuse all the statements about the Planck length?

Open-Eyes:

---

[11]**Superheterodyne**: An heterodyne is a local oscillator. A superheterodyne receiver, is a type of radio receiver that uses beat by frequency mixing to convert a received signal to a **fixed** intermediate frequency (IF) which can be more conveniently processed than the original carrier frequency.
[12]http://aflb.ensmp.fr/AFLB-331/aflb331m625.pdf **Experimental observation compatible with the particle internal clock**
[13]**Topology**: Branch of mathematics devoted to the study of who is the neighbor of who, and up to how many, in which hierarchy of neighborhoods. Every metrics induces a topology, but the reciprocal is false. Some topologies exist without a metrics.

- Yes, all.

Professor Castle-Holder:
- And moreover, do you recuse the geometric calculus of the state of the helium atom?

Open-Eyes:
- The lower limits of our familiar macroscopic geometry involve some paradox, the solutions of which are not yet resolved. The paradox here is that our familiar metrics still works, though the familiar indicted topology is no more valid.
The seventh clandestine and surreptitious postulate is the geometric corollary: "*something very small*".

   2.1.7. Geometrical corollary n° 1: "Something very small". Postulate that you always can find a smaller something, enabling you to assert that something, says an electron, is "small", corpuscular, even "punctual". Bad luck: this smaller something does not exist. Though a smaller thing exists for the mold spores already mentioned: the biologists have the right microscopes, even scanning electronic microscopes if necessary.
Figure 2.2. The spores of a lycopod:

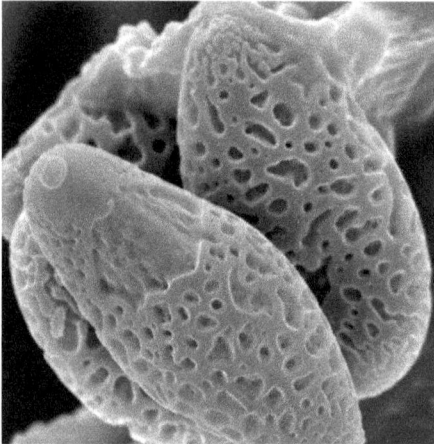

Enough accelerated (0.1 V is enough), an electron has a wavelength much smaller than the spores:
Law de Broglie : $\lambda = \frac{h}{p} = \frac{h}{mv}$ (in non-relativistic domain) $= \frac{h}{m}\sqrt{\frac{m}{2qV}} = h\sqrt{\frac{1}{2qmV}}$
where λ is the wavelength, $\mathbf{V}$ is the difference of potential for acceleration, $\mathbf{v}$ the speed of the electron, $\mathbf{m}$ its mass, $\mathbf{p}$ its momentum, $\mathbf{q}$ its charge and $\mathbf{h}$ the Planck quantum. So the wavelength under a 150 V potential is 1 Å (100 pm, a hundred picometers), 0.5 Å for 600 V, or 0.1 Å (10 pm) for 15,000 V.

Now for the needs of the microscopy, we must distinguish the width of the wave of each electron, from the width of the whole beam of electrons. A distinction that the copenhaguist QM tradition is unable to do: they have confused the statistical laws of the collectives in a beam, with the physical laws of the electron.

In transmission microscopy (light or electrons), mainly the wave width matters, and makes the resolving power; it depends on the wavelength, and on the qualities of the focusing optics. By equal dimensions and by same geometry of the electronic microscope, the wave width is tied to the wavelength, depending together on the accelerating potential. We shall give the law later, with the **geometry of the Fermat spindles**.

Curious:
- But give now the order of magnitude; we will verify later.

Open-Eyes:
- I suppose 20 cm of beam length before, and 20 cm after, and 15,000 V accelerating. It gives 1.2 μm of wave width for an electron, in transmission. However, a metallography microscope does not work by transmission, but by spectral re-emission; after careful polishing, the metallurgist uses specific chemical etchants (such as 2% nital), to enhance the contrast between grains.

Professor Castle-Holder:
- Objection! A SEM (Scanning Electron Microscope) does not work by transmission but by detection of backscattered electrons or by secondary electrons emitted by atoms excited by the electron beam. For biological targets, their surface may be made conductive by a fine coating of gold, by high-vacuum evaporation.

Open-Eyes:
- Sure! Working on backscattered electrons. So the optic width of each of these electrons on the golden surface is just the width of the reaction absorption-emission. This width is far lesser than it was at mid-journey, here one to five nanometers, that is five hundred times smaller, at least. Though the optical width of the incident beam and of the re-emitted beam is a **beam width**, containing a huge quantity of electrons, it depends on the sharpness of the emissive cathode, on the quality of the magnetic optical device, and on the precision of the scanning. However, a beam of electrons is always diverging, whichever the quality of the focusing optic device, by the electrostatic repulsion of all these negative charges. Moreover, they are all **fermions**[14], so they all must be in different states. Increasing the accelerating potential allows to restrain the diameter of the electron beam on the golden target, and to minimize the relative importance of the electrostatic repulsion.

Professor Castle-Holder:

---

[14]**Fermion**: any particle of spin 1/2, such as electron, neutron, proton, neutrino... They are ruled by the Fermi-Dirac statistics. They avoid each other; only one fermion by distinct quantic state. **Spin**: elementary angular moment, without equivalent in macrophysics.

- The attentive reader has noticed how much our colleague separates the geometry and the physics of the electron, from those of a beam of electrons. For him, only the electron is a wave, with a wavelength and a wave width. For him, the beam of electrons is not one wave, but a collective of waves. It too has a width at some distance from the source but disperses inevitably. The width of a beam is always much larger than the width of each individual electronic wave.

Open-Eyes:
- And one may doubt – understatement – that the beam could have a wavelength itself. However, this mistake is standard. We do not do the same physics as what is taught as standard.
The eighth clandestine and surreptitious postulate is the

2.1.8. Geometrical corollary n° 2, anti-absorbers. *"There are no absorbers in microphysics. Only artillery of corpuscles, exactly as in macrophysics"*.

The ninth clandestine and surreptitious postulate is:

2.1.9. Denying the acquisitions of the Solid State Physics. In solid state Physics, we deal about phonons. Sampled on tens to milliards of atoms, never the phonons can become corpuscular, nor even "very small". In metals, the conduction electrons interact with phonons. Themselves extend on tens of interatomic distances, even much more. It is incompatible with the mythology of *"something very small"*.
The reflection of light on metallic conduction electrons also forbids to them from becoming very small, considering the wavelengths of the visible and infrared waves. All this knowledge is forbidden to cross the distance separating two lecture rooms on the same campus.

2.1.10. Positivist and opportunist postulate. Systematically put a macroscopic *"observer"* in the middle of the picture, to rule the microphysical realities. Big animals with slow perceptions, instead of analyzing which may be the proper size of mesh for analysis, they deny realities, and they set the territorial comfort of the leaders above all. Slow perceptions, no common magnitudes with the frequencies involved in Microphysics. So they throw to the **Memory Hole**[15] the whole of the experimental results obtained along the 19th and 20th centuries in interferential optics with incoherent light, noticeably all the experimental facts proving the lengths and durations for each photon. We daily use these results in radiocrystallography, either with photons or with electrons or neutrons.

Even practiced by chiefs – and by definition, a chief is always right – refusing to search the right size of mesh for analysis remains a professional fault, in every

_____
[15]***Memory Hole***: George Orwell. **1984**.

craft.

The eleventh clandestine and surreptitious postulate is the

2.1.11. Anthropocentric and positivist postulate. *" The physical laws are made for satisfying the curiosity of the copenhaguist physicist, therefore to furnish him information"*. If the copenhaguist physicist cannot more know for sure the position of the submarine, then the submarine *is in a superposed state between three miles further north and three miles further south (and west and east)*. Banesh Hoffmann *scribit...* Daring to distinguish the microphysical realities from the knowledge we have on them, is said to be a heresy and relapse crime. We came about fifteen milliards years too late to dictate that the physical laws should be made for us, but they do not even notice this discrepancy.

The twelfth clandestine and surreptitious postulate is the positivo-corpuscularist and anthropocentrist, anti-Fourier corollary:

2.1.12. Anti-Fourier corollary. The properties of the Fourier transform were established one century earlier: For all wave packed, the product of its indefinition in frequency by its indefinition in length is restricted by a lower limit. Less precisely but more popularly said: the product of its length by its width on the frequency spectrum. Anyway, for each photon, its amount of electromagnetic energy in a given frame is fixed to $h\nu$ by its central frequency in the same frame, say the frame of the laboratory. The shortest the photon is, the highest is its local amplitude of field, and the shortest it is, the widest is its spectrum in frequencies. Said differently, the most its impulsion and energy are poorly definite.

All together:
- Have no pity! You must explain all!

Open-Eyes:
- There are two ways to compute the energy of a photon. The spatial one uses the density of energy, summed on the volume instantaneously occupied by the traveling photon. The volumic density of energy comes from the square of one of its specific fields, such as the electric field, the magnetic field, or even the magnetic potential $\overrightarrow{A}$ provided we choose the gauge with null at a great distance. It will be enough to compute only in the vacuum. The volume is the product of an equivalent length by an average circular area of the beam or of the Fermat spindle at this point of the propagation. We won't tell the beginner that this product of averages comes in reality from a triple integral – whose we do not know the details. If we sum this volumic density of energy on a section, we obtain a lineic density of energy, which propagates along the trajectory of the photon, and that is weaker at the ends than at the middle. So let $\Delta\mathbf{x}$ be the length of the photon, as an equivalent average.

The second way of computing is frequential, from the frequency spectrum of the photon. This spectrum is graduated in frequency ν for abscissa, and density of energy per interval of frequency $\delta\nu$: $E\nu = E/\delta\nu$. But the total area of this spectrogram is precisely the total energy of this wave pack, here an individual wave, a photon. Keeping a cautious silence on the way to obtain this sum (an integration), we say we have here an average width of the spectrum: $\Delta\nu$.

We will see farther the detail of the transformation of Fourier, and **the rule of dilation**: if you double the length of a wave packet without changing its central frequency, you define it two times better in frequency, and the width of the spectrum in frequency around its center is divided by two. In other words, the central peak of frequency is two times more elevated. To the limit, a perfectly definite frequency corresponds to a wave pack of infinite duration and length, spreading from $-\infty$ to $+\infty$. The product of the imprecisions $\Delta\nu$. Δx is constant and proportional to h. Now we convert the frequency into impulsion p, $h\nu/c$ for a photon. This constant Δpx . $\Delta x = (0.52728633 . 10^{-34}$ j.s/rad) is universal for any individual wave, photon, electron, neutron, proton, and so on. In the ideally simple case where the amplitude is a Gaussian, the spectrum is a Gaussian too, and we take the product of the average widths of these Gaussians, of known areas.

Werner Heisenberg relabeled these properties of the Fourier transform, into *Unschärfeprinzip*. Then began the fantasies in interpretations and translations. A simple and honest translation could be: « *imprecision* » or "*not sharpness*", but it became « *uncertainty* » in English, and « *incertitude* » in French. Even with « *imprecision* » would remain the original fault in methodology: Heisenberg believed in corpuscles and was upset because nature did not supply the due-because-corpuscle sharpness. Worse: infantilely overreacting to the butcheries of the World War by the obsession of "*Me! Myself and I, and My measurement and My information!*" these *Knabenphysiker* put themselves and their feelings in the middle of the microphysical image. Eugen Wigner was perhaps the most caricatural of them: Wigner, E.P., 1963. *The problem of measurement*. Am. J. Phys. 31, 6-15.
They emphasized their own personal feelings and frustration, up to "*uncertainty*", and so on: "*If I know, if I do...*", etc. Only a cruel conspiracy of Nature could hide from Heisenberg the exact position of the supposed corpuscle... So the postulate: As the properties of the Fourier transform were established one century earlier, and as Joseph Fourier was a Frenchman, and considering the emotional feelings in Germany against France in 1925, Werner Heisenberg was

perfectly right to relabel these properties into "*Principle of Cruel Uncertainty*[16] *of the Immortal Prophet*".

Niels Bohr added to the mystic by his myth of "*duality wave-corpuscle.*" These two mysticisms to conceal the properties of the transformation of Fourier (old of one century at the time) in the microphysical scale, it is as honest as the other mythologies exploited by the other clergies: that does not have any foundation, it is to throw to the trash can, like all the remaining of the corpuscularism.

Curious:

- So this *Heisenberg's Principle of Uncertainty*, you pellet it, and throw it in the wastebasket!

Open-Eyes:

- In the wastebasket, yes. When Albert Einstein came to Princeton, they asked him what office furniture he needed: "*A table, a chair, and a big and sturdy wastebasket, to throw my mistakes into.*" So now Einstein is dead, nobody has any more use of a wastebasket into which to throw his/her own mistakes, nor the mistakes we were taught? In English as in French, the vocabulary is fallacious and egocentric: nature does not have anything to do with holding us in some "*uncertainty*" on a corpuscle that has never existed. It is only a matter of indefinition, about things or "*particles*" which remain 100% undulatory: If a photon is well defined in frequency, so it is long, and its position is poorly defined. On the contrary, if it is short, then its frequency is more spread; it was known for a century, from Joseph Fourier. The trick from the hypnotist to make the suckers to believe it was new (in 1927, a century later) and deep was to be negligent, egocentric and fallacious in the vocabulary: the word "*uncertainty*" relates to us and our emotions. The indefinition below the Planck quantum is an intrinsic property of any wave pack, even for an individual wave. It is intrinsic and impersonal, so damned less exciting, not commercial enough... To sell well, you must make personal and concerned! "*Indefinition*" does not upset you enough, but "*uncertainty*" does, wow the anxiety!

The Fourier transform and its reverse (from the signal to its spectrum or from the spectrum to the signal) will be detailed later (paragraph 10.10).

---

[16]*It is a guy who suspects his wife to cheat. He hires a detective.*
The detective reports:
- Your wife followed a man in a hotel.
- And then?
- They took a room.
- And then?
- I went up in the building in front.
- And then?
- Both undressed completely.
- And then?
- They got on the bed.
- And then?
- They switched the light off, they closed the shutters, and I did not see anything more.
- Ah! **Cruel uncertainty!**

Curious:
- But? Dead or alive, the famous Schrödinger's cat?

Open-Eyes:
- That is a very good question, and I thank you for having posed it.
The thirteenth clandestine and surreptitious postulate is the

2.1.13. Wigner-Neumann animist postulate. *"Me, Myself and I, as a big macrophysical animal, self-claimed as "the observer", I am so almighty that I have the power to delay the absorption reactions and the consecutive decoherence forever, just by not observing!"*

In opposition, Erwin Schrödinger published in 1935 the humorous apologue of the *dead-alive* cat, remaining in the *dead-alive superposed state* as long as a copenhaguist physicist does not lean his august attention on the results of the experience. Overtly, Schrödinger mocked the animist postulate that the Göttingen-copenhaguists admitted; but they remain so arrogant, that eighty-three years later they still do not perceive the sarcasm.

All the marketing hype promising marvelous *"quantic computers"* is founded on the Wigner-Neumann animist postulate.

The fourteenth clandestine and surreptitious postulate is the

2.1.14. Postulate of separability and delimitablity (or postulate of triumphal laziness). As we can only write a limited system, and as anyway, we are impatient to alleviate already heavy formulas, **therefore** a quantic system is naturally delimited, naturally separated and independent of the remaining of the world. Alas, this postulate is heavily wrong.

The fifteenth clandestine and surreptitious postulate is the

2.1.15. Magic and supernatural postulate. Or if you prefer: goblin [17] and poltergeist: postulate that each quantum particle (electron, photon, neutron, proton...) is individually exempted from any physical law, but magically, in big numbers, its statistic rejoins some statistical physical laws, just as this collectivity blurs the corpuscular properties they were postulated to have. Never the copenhaguists detail the physical miracle by which the individual no-laws is transmuted into a collective law.

The sixteenth clandestine and surreptitious postulate is the

---

[17]A **goblin** is a mythological dwarf, able to appear and disappear as they like.
Poltergeist: same popular mythology, invisible strikers.

2.1.16. Anti-undulatory postulate. Though the chemists use it with success, the copenhaguists postulate that any wave under the Schrödinger equation must be fictitious, without any physical meaning, and its only use is to be elevated at the hermitian square to obtain the probability of apparition of the magical and supernatural corpuscle. This goblin and poltergeist corpuscle is allowed to explore *"beyond the planet Jupiter"* during its travel from the electrons cannon to the cathodic screen or the microchip to engrave. Feynman and Hawking have written it[18], so it must be true...

The seventeenth clandestine and surreptitious postulate is the

2.1.17. Confusionist postulate. To deny the atomic limit in undulatory, prescribe to confuse all kind of "waves", each individual wave (quantic wave) with any collective of waves, and these collectives with gravity waves or elastic waves in a collective of matter, then mathematically unify all these kinds: the individual waves (quantic ones), the collective, and the waves in a material collectivity. The Born-Heisenberg copenhaguism is founded on this trick, and it is so for ninety years. A hegemonic swindle. Tip: an individual wave has only one emitter and only one absorber. This mistake is less dangerous when applied to photons (the smallest quantity of electromagnetic transfer): They are bosons, so they attract each other and pack into herds of same frequency and habitus, especially on astronomical distances. But this mistake becomes enormous when you apply it to electrons (the smallest quantity of electric charge). On the one hand, their minus charges energetically repel each other, so the collective energetically diverges. On the other hand, they are fermions, so they avoid each other, each in a distinct quantic state.

The teaching and the popularization pull your leg on this matter: they tell you that in an interference experiment, like an Aharonov-Bohm style, the beam of electrons has **a** common phase!

Example, fig 2.3.

---

[18]http://www.physicsforums.com/showthread.php?t=513139

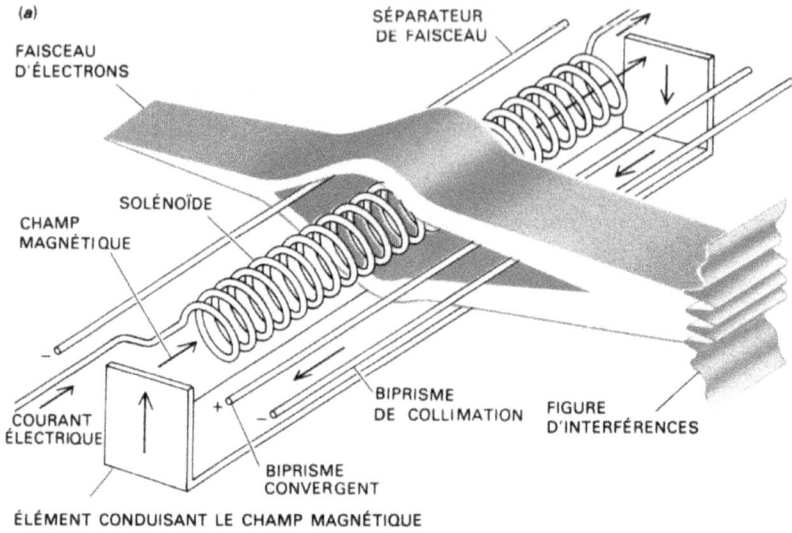

(a)

SÉPARATEUR
DE FAISCEAU

FAISCEAU
D'ÉLECTRONS

SOLÉNOÏDE
CHAMP
MAGNÉTIQUE

COURANT
ÉLECTRIQUE

BIPRISME
DE COLLIMATION

FIGURE
D'INTERFÉRENCES

BIPRISME
CONVERGENT

ÉLÉMENT CONDUISANT LE CHAMP MAGNÉTIQUE

As a fooling offering, this one is huge. See, they draw a unique and common phase for the entire beam, spreading on all the sensitive screen:

Figure 2.4.

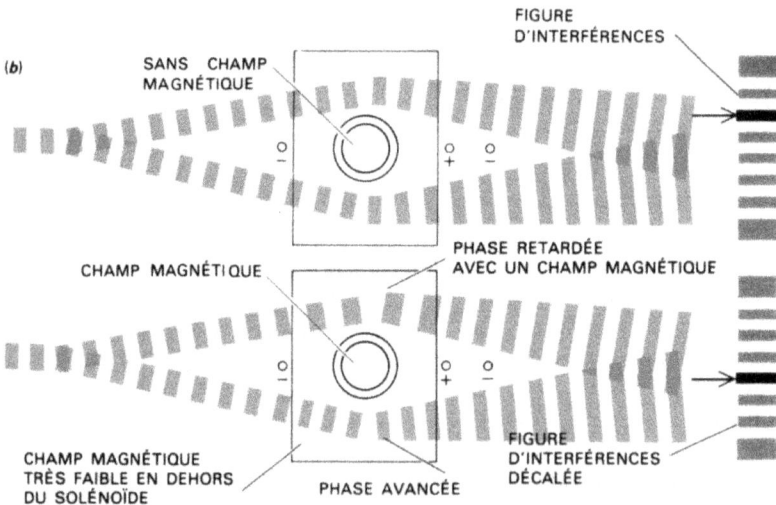

FIGURE
D'INTERFÉRENCES

(b)

SANS CHAMP
MAGNÉTIQUE

PHASE RETARDÉE
AVEC UN CHAMP MAGNÉTIQUE

CHAMP MAGNÉTIQUE

CHAMP MAGNÉTIQUE
TRÈS FAIBLE EN DEHORS
DU SOLÉNOÏDE

PHASE AVANCÉE

FIGURE
D'INTERFÉRENCES
DÉCALÉE

And from whom, these two schemes, technically well done, but physically fallacious? Herbert Bernstein and Antony Philips, in **Les particules élémentaires**, Pour la Science.

Without having ever assimilated the difference between the microphysical scale, with only one or very few quantic particles, and the macroscopic scale with enormous quantities of electrons, photons, or atoms, they still do not discern that though each electron has a phase, the bundle of electrons doesn't have any, and every electronic wave is disjointed from the other.

Curious:
- And how do you correct it, this drawing?

Open-Eyes:
- For the individual scale, I pinch the departure, and I pinch the arrival: only one electron leaves from a place as small as about ten atoms, and arrives at a small place. So the electron is diluted during the journey, and is split in two by the first met negative thread, then passing respectively on both sides of the micro-solenoid, before beginning its meeting on absorber, thanks to the two focusing threads, first positive next negative. The meeting of the two spindles of the electron (= of the electronic wave) is only complete at the arrival on the absorbing site. And yes, there was indeed a phase from the beginning to the end of every branch of the journey.

Figure 2.6.

The vertical scale of the drawing is enormously exaggerated. This drawing is very simplified, as it omits all the electrostatic splitting and focusing apparatus, that constrains the electron (the electronic wave) to split into two spindles converging on the same absorber. So for each transaction between the thermionic cathode and the sensitive screen. I did not draw the wavefronts, and you'll see why: we will compute the magnitude. We need a low accelerating difference of potential, we will take 6 V. So the wavelength is 5 Å. Did you expect to draw a 5 Å wavelength? And at which scale? Furthermore, to easily split an electron on a separation about 30 to 60 μm, we need a not too short apparatus, at minimum 1 m from the source to the screen. With an excellent shielding against any perturbation by any external electromagnetic field. Maybe cutting the rotary vane pump for suppressing its vibrations.

Curious:
- What is the interfrange we can observe?

Open-Eyes:
- Say 0.5 m between the screen and the splitting device. Admit 50 μm separation at this splitting. Developing the sine (very small) at first order only:
Interfrange $= \frac{0.5\,m * 0.5\,nm}{50\,\mu m} = 5$ μm. Five micrometers... I let you imagine how you will engrave the sensor that will resolve better than 0.2 μm. I fear that the only solution will be the fine motion (as in a microtome, by dilation) of a unique optimized sensor. You could increase the interfrange, provided you diminish the distance between the two Young slits, that implies the diameter of the micro-solenoid, and increase the distance to the screen. But beware! All that in a vacuum chamber, of excellent stiffness. And you were hoping we could

sketch all this at the scale? And for each electron...

Curious:
- No, in popularization books or magazines, there is never a magnitude computing, by fear of the fear of the readers.

Professor Castle-Holder:
- Though your so discouraging computing about the observable interfranges is correct, you lack a piece of information on the optimized experimental apparatus, which you'll find an example at the addresses:
http://iopscience.iop.org/article/10.1088/1367-2630/15/3/033018/pdf
and http://stacks.iop.org/NJP/15/033018/mmedia
They have used electrostatic lenses after the double slit - and eventual mobile mask – to lengthen the focal in a modest apparatus length, and a manageable vacuum chamber. The device magnified about 16 times, and if applied here, the observable interfrange would become 83 µm instead of 5 µm. So finding a suitable sensor, with the appropriate resolution becomes less difficult.

Professor Marmot:
- But you do it deliberately! You multiply meters by meters and divide by meters! You were never allowed to do that! One is only allowed to multiply numbers, or divide numbers, nothing else!

Open-Eyes:
- Blistering Barnacles! I forgot that you were never a physicist, just a disguised mathematician, who mistakes physical quantities for numbers, and who is apt to add geese with goats to obtain the age of the captain, or to add two dead boars and a small dog with a small Gaul and a big Gaul to pretend been attacked by *"a very outnumbering band of Gauls... They were five, as to say!"*. In their ivory tower, despising the rest of the crafts, never the pure mathematicians taught the basics of the physical quantities, nor computing in physical quantities, nor the equation to dimensions. They do not have the slightest idea of what it can be, nor what use we have for them. It is one of the most shaming secrets of the massive failure of the scientific teaching in this country.

Professor Castle-Holder:
- I suggest that you put in **Appendix I** the course on the **Physical quantities**. But here let us stay concentrated on the matter. We were on the postulate of confusion between the individual waves, according to you, and collectives of waves according to the tradition. A postulate that you criticize.

Curious:
Figure 2.7. (Public domain image)

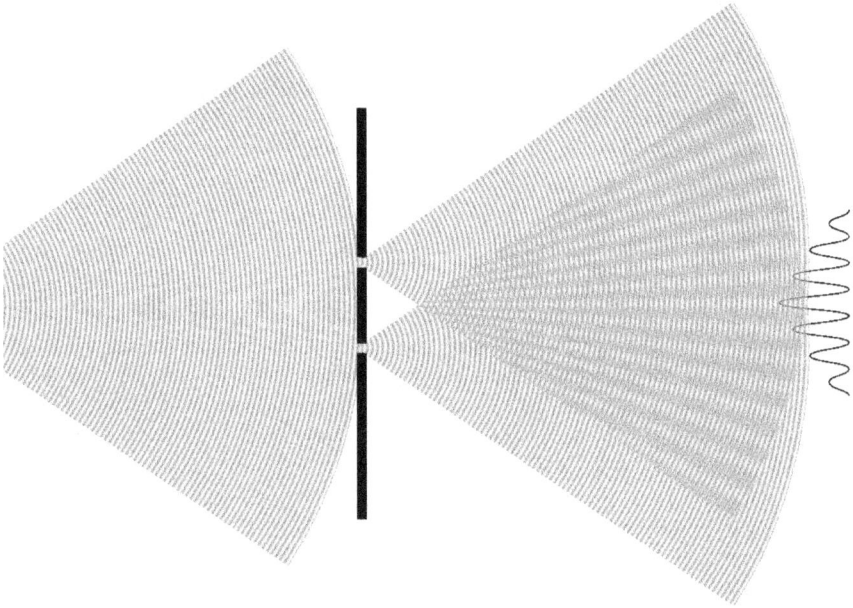

- To sum up, for the readers who never practiced interference experiments. At the Gymnasium, we had seen only macroscopic experiments, where the beam of light contained milliards of photons per second. We had no mean to distinguish the photons, and the resulting pattern of interference did not show any perceptible feature of discontinuity nor random. It is summed up by such the sketch above:

Open-Eyes:
- However, one can only ascertain the light at its arrival, so to know its path, you must intercept a part of it. And about seeing the wavelength, about 0.4 to 0.6 μm for the visible, quite impossible. This figure is largely fallacious, even on our macrophysical scale. It extrapolates from the water tanks, where the wavelengths are centimetric, and celerities very slow, about two to three centimeters per second.

Professor Castle-Holder:
- Since then, were done experiments with ultra-low intensities, where one can discern the arrival to the screen of each particle, or individual-wave-according-to-Mr.-Open-Eyes, photons or electrons. Each impact is so small that one may assimilate it to a point, and its position is unpredictable, but if the experiment last enough – several months – on the large numbers emerges the interference pattern already known in macrophysical optics, with alternating dark and bright fringes.

Open-Eyes:
- Each "*particle*", electron or photon, even neutron in radiocrystallography, even ultra-cold helium atom, has interfered only with itself, indeed passing through

the two holes or slits. So it is indeed a wave, an individual wave, and never a corpuscle. Even the helium atom... And where the dephasing is half a period, no photon nor electron arrives; the optical impedance is infinite. The calculus we did in the classroom by trigonometry remains fully valid.

The main change is that in the classroom, we handled a beam of light, that is a big collective of individual waves coming from a concentration of emitters, but in Quantic, we look down to the laws for individual waves. And the individual wave has only one emitter and only one absorber. So its maximal width at mid spindle is limited about **2z**, with $z = \sqrt{3a\lambda}/\,4$ where λ is the wavelength. **2a** is the distance from the emitter to the absorber, in a homogeneous medium. This approximation relies on the simplification with constant curvature, that is arcs of circle for the frontier of the Fermat spindle.

Curious:
- I demand a figure, a clear sketch.

Open-Eyes:
- Of course! However, no sketch in the public domain, and even in copyrighted domains may suit: They all sketched macroscopic experiments, where milliards of milliards of photons spread on milliards of milliards of milliards of potential absorbers. To illustrate the Microphysics, we must recollect old sketches from geometric optics, when they drew how through the optical device, a point of the object projected on a point in the image. Or interferences, one computed and drew the difference of optical length between the supposed source and different points of the screen.

Professor Marmot:
- What a retrograde! He is coming back to the geometric optics of the 17th and 18th centuries! We are the moderns, and he is the past!

Open-Eyes:
- The only difference with the physical optics from Augustin Fresnel, in 1819, of the quantic transactional optics of the 21st century, is that now we can evaluate the widths of each photon during its journey. Pierre de Fermat (1601 – 1665) had proved why, but could not achieve the computing, as he could not know the wavelengths in the visible range.

Curious:
- Please explain why.

Open-Eyes:
- Fermat had proved that each real optical path is minimal compared to its near neighbors. Now we know that the real condition is not so restrictive: minimum or maximum or extremum. Otherwise said: this geometrical path only differs by a second order infinitesimal from its near neighbors. So no optical path is of null width. Since then, we have computed this Fermat width, beyond which

the transmitted power is null. It will be impossible to draw at scale these tiny widths, so we only draw the axis of the Fermat spindle-beam, photon by photon, just like the astronomers of previous centuries learned to do.
Image in the public domain: Fig. 2.7.

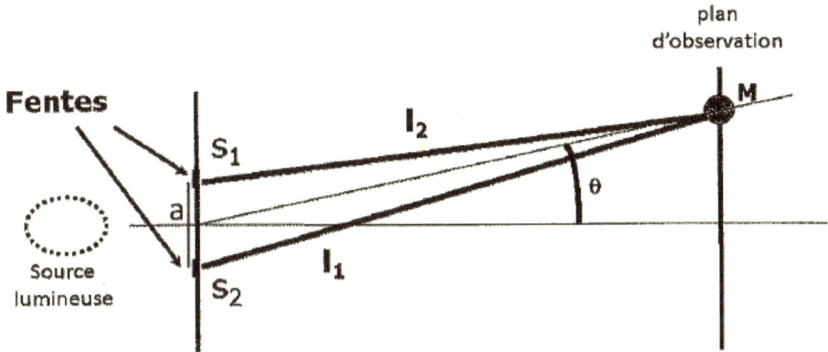

Between two fringes of maximal illumination, the difference of times of journey l_1 and l_2 must be an integer number of periods. At the minima of illumination, the difference of travel time must be an uneven number of half-periods, so the two possible electromagnetic fields coming from the two slits are in phase opposition, and their sum is null or nearly null.

Professor Castle-Holder:
- You must articulate better. You have just added the adjective "possible" to the traditional discourse, and it is not clear enough. In classical discourse, the causality flows from source to screen, so an electromagnetic field flows by one slit to the following half-space, another field from the other slit, and where they come in phase opposition, there will be no field or a very weak residual difference.

Open-Eyes:
- This classical discourse surreptitiously proceeds from the fact that the source emits milliards of milliards of photons, so at our human scale, we observe the result of a crowd of independent events. On the contrary, in transactional physics, we think and compute each transaction that evolves to a synchronous transfer from one of the emitters in the source to one of the potential absorbers of the target. So where you say "there is no field", we state that the optical impedance between this emitter and this absorber is practically infinite, so no transaction for a transfer of photon can achieve.

Curious:
- I see that you completely change the interpretation of the laws of optics and electromagnetism. You replace "There is some field" or "There is no field" and the intermediate cases according to phase shift by what, exactly? And how do this emitter and this absorber to know that the impedance is too large?

Professor Castle-Holder:
- And what does mean your expression *"synchronous transfer"*?

Open-Eyes:
- On the one hand, some emission power is offered with the given optical device, on the other hand, there are different impedances according to the positions of the potential absorbers; this determines the statistical probabilities of transfer of photons. I will return later on the impedance, and its inverse the admittance. Indeed, groping the environment by the potential emitters and potential absorbers does not proceed in our Newtonian macro-time of the laboratory. We had to revise those familiar concepts, so "above any suspicion". Synchronous transfer: during all the transfer of the photon, which may take more than a million periods, the emitter holds the absorber in frequency, in phase, and in polarization, and conversely the absorber holds the emitter; each one holds the other. The proper time of the photon is null, so in its "photonic world" absorption and emission are simultaneous – Relativity and Lorentz transform compel so – but not of null duration.

Professor Marmot:
- What nonsenses! The phase is not conservative and is not observable. So it has no physical meaning.

Open-Eyes:
- Tss tss! You pretend the contrary of what was asserted by Herbert Bernstein and Antony Philips, cited above. They wanted that the electron beam had a phase, as a collective beam. You never stop to think your *"wave of probability"* in big numbers. It shares just only one feature with the real waves of the microphysics: they share the same equation of evolution, and this equation is correct. They do not share nor the same time, nor the same space, nor the same boundary conditions, nor anything of the concrete meaning. The real waves, the individual waves, have a real phase, that your sect denies.

Professor Castle-Holder:
- And during this transfer, what produces the statistical randomness? By what do you diverge from the quantic theory we were taught in the same lecture rooms and in the same textbooks?

Open-Eyes:
- Just no! For us transactional physicist, the randomness does not take place during the transfer of a photon (or so little, if any). From the ground noise, the lapping of the de-Broglie-Dirac waves, some transactions emerge and succeed. They are three-partners transactions: the emitter, the absorber, and the optical space between them. As soon as the handshake occurs, the synchronous transfer is under deterministic laws. The subsequent questions, about how long lasts this transfer, how long last the emission and the reception of a photon, physical case by physical case, require a much more fine physics that what is

now available. A Mössbauer photon is finely defined in frequency, so it is very long (say more than 10 meters).

Curious:
- You were just in a brawl for the phase. May we observe it? Measure it?

Open-Eyes:
- We are not at the appropriate size to mount an experiment, and we never will be! We need the phase of the individual wave to build a coherent and predictive theory. It is up to us to prove it is more coherent, more economical, and more powerful than the concurrence. It is up to the concurrence the task to prove some eventual faults.

After this recall on the interferences, let us come back on the schedule. We were on comparing the fermions with the bosons about their wave behavior. I have re-diffused lots of evidence that Laue[19] and Debye[20]-Scherrer[21] diffractions have mediocre sharpness with electrons, compared with X-rays (or even neutrons). First because of electric repulsion between electrons. Next, comes the quantic but limited repulsion between fermions, which affects the neutrons too. Each neutron has a phase; no beam of neutrons may have a phase.

I return to the calculus we did on the orders of magnitude for an interference experiment, with electrons. Farther at § 10.2, we will exhibit the sketches that Richard Feynman drew for his students, in 1964 in Caltech. A small flaw, however: never Feynman computed the orders of magnitude, never; never he practiced himself the experiments he blindly cites. It is one of his flaws in methods (he had some others, alas) he used to be able of *cock-a-doodle-doo*ing "*Nobody understands Quantum Mechanics*". These corpuscularists took the right means to understand nothing. The professors Matthew Sands, Robert B. Leighton and H.V. Neher who were assisting him, were far too subdued by Feynman, to dare to criticize and correct these flaws in methods, this lack of experimental magnitudes. You, you will understand QM because this handbook is the good one, it does not tell you contradictory tales.

Curious:
- So you are at seventeen postulates, that you all throw away.

Open-Eyes:
- These seventeen postulates are hegemonic but in the clandestinity, they are surreptitious, and all are in contradiction with experimental results. But science differs from all the others systems of transmission of knowledge in that: Science is the belief in the ignorance of experts. The experts who are leading you may be wrong. We have to verify, by experiments.

As a corollary of the postulate n° 17 (the confusionist postulate), the Copenhaguists forbid to articulate one's investigation between the crowd scale and

---

[19]Max von **Laue**, German, 1879 – 1960. Nobel of physics in 1914.
[20]Peter **Debye**, 1884 – 1966. Nobel in chemistry in 1936. Born in Nederland.
[21]Paul **Scherrer**, 1890 – 1969, Swiss physicist.

the individual scale. For instance, the second principle of thermodynamics, of never-decreasing entropy, is a crowd effect. That all the gravity waves and all the acoustic waves disperse an are damped, is a crowd effect. But at the individual scale, a photon is still an electromagnetic wave, that does not diverge nor damp: it travels to an absorber, an individual absorber. The attenuation and the dispersion of a light beam are crowd effects, of which the laws of electromagnetism are not guilty.

2.1.18. Anti-Dirac and anti-Schrödinger censure. Hide from the students that among the four components of the wave equation of the electron, two of them are retrochronous.

A nineteenth postulate, not scientific but tactical, is invoked for each controversy:

2.1.19. Tactical and anti-semantics postulate. In each controversy, agglomerate the formalism and the Göttingen-København semantics, and teach they may not be separated. Then you deny any meaning to the word "semantics". This is only for a tactical purpose. Each time our common formalism wins a victory, they yelled that it proved their copenhaguist semantics, and that any other "*is just philosophical preferences without any real interest*". With its **Philosophy of Physics**, Mario Bunge had aroused many foes: he advocated that the semantic axioms should become explicit, instead of remaining clandestine. Every day we prove that one can throw away their copenhaguist semantics, without throwing the formalism – which is strictly determinist and strictly undulatory – and they immediately deny this fact.

Professor Marmot:
- All that is just philosophical fog! You ought to read books, shut up and calculate!

Open-Eyes:
- Well! Let's turn to the calculations! Later we will detail the blunders from Georges Charpak, Roland Omnès, Stephen Hawking, Leonard Mlodinow, Walter Greiner, who all recopy the blunders of Richard Feynman. There is a factor of hundred to spare on the clumsiness of some calculations inherited from the Feynman's *Integral of paths*. Feynman rediscovered the Fermat's Principle (17th century) but in far less practical, as burdened with anti-relativist and anti-undulatory presuppositions.

Professor Marmot:
- *Et voilà!* The crank who imagined to be a genius, and dares to put Nobel laureates in doubt!

Open-Eyes:

- Reminder from the same "expert": *Science differs from all the other systems of transmission of knowledge in being the irreverent belief in the ignorance of experts. The experts who are leading you may be wrong. You have to verify, by experiments.*

Avoiding to burden the rhythm of the discussion, we put off to Chapter 6, § 6.2 the details of the 20th postulate:

2.1.20. Göttingen postulate. Only "states" exist, forget the transitions. Already in 1927, coming back from Brussels in a perplexed mood, Erwin Schrödinger wrote: "*Odd physics that concentrates on the states, and omits the transitions!*". The durations and the properties of the transitions, such as the length of coherence of photons revealed by the interference phenomena described since Thomas Young, are incompatible with the corpuscularist postulate. We will detail more, later.

2.1.21. Superiority-of-the-pack postulate. *We are the modern forever. The vile objectors and non-believers are just dimwitted, brain-damaged, and retired colonels of cavalry, who try to return to **classical** physics.*
And after me, there will be no more prophets, as the new physics is complete!
All the sects and most of the packs sell prostheses of narcissism to their adherents
The more disastrous is the yield of scientific teaching, the more it *proves the superiority* of the tribe over the remaining of the world. Quotation of Niels Bohr: *If you think that you understand Quantum Mechanics, so I have not yet well explained* that nobody is allowed to understand the swindle we teach.

Curious:
- After such a ferocious indictment, which I have not yet fully understood, as there too many words that are new to me, I hope you have plots of agreement. No?

Professor Castle-Holder:
- Indeed we have. We agree on the fundamental quantities of the atomic domain, (we will expose them next). We agree on the experimental facts. We agree on the equations of evolution.

Open-Eyes:
- We disagree on the list of experimental facts that are to hide so that you could not see the dirty tricks. In my sense, they hide from you all the spectral absorptions, an extensive list. They have hidden the Ramsauer-Townsend transparency effect: it is strictly undulatory. If the electron is always undulatory, strictly undulatory, how will they keep their mystical "*wave-corpuscle duality*", which impresses so much the flabbergasted crowds?
We agree on the time-independent static Schrödinger equation, so useful to the chemists for calculating and predicting the molecules. However, on three

crucial points, we transactional physicists disagree with the copenhaguists and their heirs about the dynamical, time-dependent Schrödinger equation:
1. Their time is the Newtonian macroscopic time.
2. We explicitly write the real, intrinsic and relativistic frequencies, instead of the fictitious and unusable ones, written by the copenhaguists.
3. As soon as the problem is no more unidimensional, we explicitly part the concurrency function among the potential absorbers, from admittance function. The admittance is the inverse of the impedance. Nobody before us did this partition in optics, either photonic or electronic.

Curious:
- But what is the necessary academic level, required from me, to be apt to follow you?

Professor Castle-Holder:
- Officially in France, you tackle the quantic domain in Licence, Bac + 3.

Open-Eyes:
- Alas in an abstract and hyper-mathematized manner I do not approve. In Appendix F, you will find a glossary explaining the main words you are not so sure about, with the references to experimental facts you can verify. The cited authors are in chronological order, not alphabetical.
You must have read or seen some popularizations about the atom (or better the course of the two first years of Bachelor Degree). I will explain later their lacunae, but if you do not have these basics, you have no idea of what is the matter.
The required level of calculus, allowing you to verify you are not kidded, is the second year of Bachelor Degree. In maths, you must know the beginning of the vectorial analysis and at least the limited developments of the sine and the cosine; you have not forgotten the inscribed angles. On the experimental skills, you must know the experiments with the Snell-Descartes refraction and the Young interferences with two slits. If you discover them only today and do not know how to verify, you cannot control whether I do not fool you, with lies good enough for the kids. You must have seen the optical spectrum of a sodium and mercury lamp. In mechanics, you should master the conservations of the linear momentum and of the angular momentum – and not only when you recognize a scholar case[22]. You should be familiar with at least one oscillatory movement, such as the swinging pendulum or an elastic pendulum. You should have seen a guitar cord and waves on the water. Preferably, you have sung in the bathroom and felt the sensation that the room is coupled with your mouth, as its damping is low: I will evoke some sharp frequential resonances, and it is a novelty in the teaching of Quantics. Without these requisites, you will be reduced to believe our word, but in Science, no one believes: he or she verifies. For helping you, some annexes are put at the end of the volume, out of the dialogue. To verify and criticize the sketch on the electronic structure

---

[22]See revision at Appendix G.

of dyeing molecules, you should have half of the bachelor degree in organic chemistry; sometimes also some knowledge in analytical chemistry. But on the Dirac's bi-spinors algebra, you should have the first year of Master in maths, to be able to criticize. Otherwise, you will have to believe our word.

We will not treat on nuclear physics, nor fundamental particles. It will be enough for us with the bestiary already known in 1932: electrons, protons[23], neutrons[24], and positrons[25]. They are enough for the chemistry, and for usual radiations, those of the atomic physics. Though our body is crossed through by a cosmic ray about once per second, and though we are crossed through by innumerable neutrinos of low energy, we will not treat nor muons[26], nor neutrinos[27]. No optical experiment is feasible with muons, nor with any short-lived particles; it is also impossible with neutrinos, as only a very small part of them are detected. And here, the optics experiments are fundamental and discriminating.

Unlike the ordinary courses of atomic physics, we will use the relativistic frame, as without it, it would be monkey business. Experience has proved that a remedial course on the relativistic frame applied to microphysics is necessary; this chapter is the most difficult, and many readers will skip it first.

Professor Castle-Holder:
- Arhem! To *master the linear momentum and the angular momentum*! Maybe you ask too much. The experiment has been done in several countries: two-thirds of the students in the second year in University, replace the Newtonian mechanics by a folkloric one, rooted in the Antiquity, as soon they do not recognize a scholar case. As to say: an ice cube, laid on the table, in a plane engaged in a flat turn; they are asked to draw the run of the ice cube in two frames, one bound to the plane, the other bound to the ground. How will they master the relativistic dynamics, while 400 years after, they still do not master the Galilean relativity, and 330 years after, they still do not master the Newtonian mechanics?

Open-Eyes:

---

[23]**Proton**: heavy particle or hadron, composing the atomic nucleus. The proton has a positive electric charge, opposed to this of the electron. This hadron is composed of three quarks: u u d (and of gluons to glue them together). It has a magnetic moment.

[24]**Neutron**: heavy particle or hadron, composing the atomic nucleus, but without global electric charge. It has a magnetic moment resulting from the quarks, which have a fractional electric charge and a spin. This hadron is composed of three quarks: u d d.

[25]**Positron**: anti-particle of the electron. Its charge is positive.

[26]**Muon**: unstable and short-lived heavy electron. The muons which come to the ground were produced by collisions of cosmic rays (often protons) with some atom in the high atmosphere. The *tauon* is an even heavier and even more transient electron.

[27]**Neutrino**: ghostly and ultra-light particle, which however carries away a spin, a momentum, and some energy. Breaching the other symmetries, the neutrino has only the left helicity, though the antineutrino twists on the right. The density of low-energy neutrinos in the Cosmos is unknown, hardly perceptible, and is suspected to be huge.

- Alas! So Richard Feynman devoted two volumes in Caltech in teaching Mechanics, before the two volumes on Electricity. These two laws of mechanical conservation are only experimentally proved in University. The involution of the programs during the 20 past years does not pull toward up: the pupils have their lungs crammed with *"citizen science"* under odd dogmas and *"green energies"* (but intermittent and frisky, which plunge countries into blackouts, as in South Australia). They are toadied in the illusion that as being promoted as auxiliary militants, they will give lots of advice to their parents. In the meantime, the linear momentum and angular moment are set aside, *sine die.* The linear momentum should be mastered in the Gymnasium: they have the necessary mathematical (vectorial) background. The students may feel that the basics of Mechanics are not sexy enough, it remains that it is up to them to work the basics. But if we have to remedy their lacunae in the course and the experimental work, Ooch!

They will need to master the recoil, to attend the Compton scattering (a photon is scattered when it reacts with a conduction electron, almost free), and to distinguish the Mössbauer resonant absorptions from the other nuclear absorptions.

Curious:
- No, you will not escape so easily! If really you think that *the mass of the proletarians* (and of the bourgeoisie as well) *squelch in the mud of marsh of error*, you scientists must pull them politely out of this marsh.

Open-Eyes:
- Accepted! In Appendix G, you will find a short collection of fool traps, surreptitiously laid by the folkloric Mechanics inherited from the Antiquity, which trap so easily the general public. We will give the solutions, and the experimental means to not again fall into these traps. We begin with the basics of static mechanics, and we carry you two academic years later.

2.2. Ten transactional postulates

2.2.1. The absorbers exist. No corpuscle exist. No "*corpuscular aspects*" exist. But the properties of the absorbers exist; some of these properties may be quantic.

2.2.2. Planck and phase postulate. The unit of phase is taken into acount. The Maupertuisian action is not action-per-unit-of-phase, and Planck's constant is action-per-unit-of-phase: $h = 6.6260755 . 10^{-34}$ joule.second/cycle $= \hbar = 1.05457266 . 10^{-34}$ joule.second/radian.

2.2.3. De Broglie-Dirac postulate. If a "particle" has an intrinsic mass, the intrinsic frequencies mc^2/h and $2mc^2/h$ play each one their role. The Broglian mc^2/h for each interference of a quantic particle with itself. The Dirac-Schrödinger $2mc^2/h$ for all electromagnetic interactions, for instance, the Compton scattering.

2.2.4. Fermat-Fresnel Postulate. For each individual wave, all the real journeys come in phase to the absorber, eventually at an integer number of periods (it is then an interference). Hence the geometry of the Fermat's spindle between emitter and absorber - several spindles in case of interference on the travel. Of course, any individual wave inherits the properties of the Fourier transform.

In the case of an interference experiment, like the Aharonov-Bohm experiment, with two separate Fermat's paths:

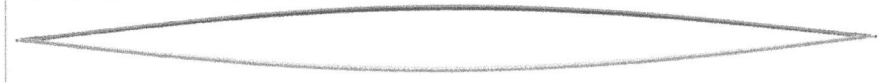

2.2.5. Every photon (= individual electromagnetic wave) has an absorber. In the cases where at least one of the absorber and the emitter are held by quantic resonance rules ("*quantic*" means dependant of the Planck's quantum **h**, via the Schrödinger's equation, or Pauli's or Dirac's one, 1928), and pass from a stationary state to another stationary state, then a photon is a successful transaction between three partners: an emitter, an absorber, and space or optical devices between them. This transaction transfers by electromagnetic means, a quantum of looping **h**, and an energy-momentum whose value depends on the respective frames of the emitter and the absorber.
The limits of this definition: One cannot quantize the acceleration of an electron in a magnetic or an electric field. So the acceleration of an electron in a vacuum tube, in a cathodic ray tube or in an electronic microscope, or in a particle accelerator, the synchrotron radiation, the *Bremsstrahlung* (braking radiation) all that escape to the quantic sub-domain: no stationary states with defined frequencies before and after. In the case where the boundary conditions are quantic, the transferred momentum is $\mathbf{h}\nu/\mathbf{c}$ in the frame where the frequency ν is considered.

Corollary. As soon as you admit that the absorbers exist, *pffuitt!* These "*Collapses of the wave-function-spreading-everywhere*" lose any interest, and are only good for the garbage can of the History.

2.2.6. The properties of a crowd of individual waves flow from the properties of the individual waves, and not the inverse. We distinguish between:
1 – Waves in a collectivity (of atoms or molecules). So are the gravity waves between two fluids, and acoustic waves, seismic too. And in microphysics, the spin waves in a ferromagnetic material, the phonons, the plasmons, and polaritons.
2 – The collectives of waves, such as light, or a beam of electrons, ions or neutrons. These collectives comprise many individual emitters and many more potential absorbers.

3 – The individual waves, for each quantic "*particle*", photon or neutrino for instance. In the quantic sub-domain, each one of these individual waves converges onto one individual absorber.

2.2.7. Macro-time $\neq$ micro-times. The god of Isaac Newton, in charge of all seeing simultaneously, does not exist. The time of Isaac Newton, a supposed universal and ubiquitous parameter, does not exist either. We distinguish the macro-time of macro-systems such as the laboratory, from the micro-times where dwell all the gropings of Broglian waves from which emerge the successful transactions. The macro-time is a statistical emergence, and it flows the same way as the entropy, a statistical emergence too. It has no causal properties in microphysics. No nucleus ages.

Corollary:
We cease to presuppose that the causal irreversibility proved in macrophysics for the macro-time could be extrapolated to the micro-times the transactions emerge from. This irreversibility is a crowd effect. We cease to disdain and censure the two retrochronous solutions resulting from the Dirac equation for the electron, or any fermion.

2.2.8. Kirchhoff's Principle of retrosymmetry. For the 17th century, it is known that in our low gravity, far from a Schwarzschild horizon, every real optical path is reversible. In 1859, Gustav Kirchhoff proved that the Fraunhofer dark spectral lines from a cold gas or vapor correspond to bright spectral lines of the same elements in a hotter gas. So the spectral emission of a photon is exactly the same physical phenomenon as the absorption. Generalization: the retrosymmetry applies to the low energies of all the atomic physics, the molecular spectroscopy, and all the solid state physics.

Limitations: No experimental proof of retrosymmetry exists in the domain of nuclear physics, nor in the nucleosynthesis in the implosion of a supernova, nor in high energies. Where a neutrino is emitted, no retrosymmetrical experiment is feasible.

2.2.9. No, it is impossible to isolate a quantic system. No, it is impossible to isolate a quantic system as we isolate its equations at the blackboard: No mean exists to shield the Dirac-de-Broglie noise. It is impossible to predict which transaction will emerge from this lapping, nor when. Only the admittances may be computed and predicted. The implied de Broglie frequencies are inaccessible from our human scale, and the theorem of the requisite variety, from William Ross Ashby, is here to ruin all our fantasies of panoptical omniscience. Moreover, the innumerable involved micro-times are bi-directional: orthochronous and retrochronous.

Then the moral principle: we refrain from censuring the experimental results that embarrass the doctrine in power.

2.2.10. Moral principle: Hiding so many experimental facts to the students is wrong, and it violates the scientific deontology. When a theory is, at last, correct, it does not have exceptions.

Many experimental results embarrass the Göttingen-copenhaguists: All the spectral absorptions (they are frequential, while their theory is anti-frequential), all the interferences such as all the radio-crystallography, the anti-reflective coatings, the quarter-wave plates, the interferential colors, the Goos-Hänchen effect in plane polarization, Imbert-Fedorov in circular polarization, all proofs of the non-negligible width of each photon. A very long list! They hid from you the transparency effect Ramsauer-Townsend (1921), which is strictly undulatory. But if the electron is strictly undulatory, how will they continue to impress the naive public, by their mystic *"wave-corpuscle dualism"*? Many other everyday experimental results are incompatible with the corpuscular ideation of the Göttingen-Københavnists.

2.2.11. The economy of postulates and concepts is on our side. No more need to erect the properties of the Fourier transform as a new postulate: they are simply inherited. The magical concepts of *"superposition of states"*, *"intrication* (of supposed theoretical and corpuscular states), measurement, *psychism and consciousness of the observer"*, all that is dropped: **Majesty, I did not need that hypothesis!**
So is the transactional microphysics we shall develop in the body of this handbook.

2.3. The protagonists of this popularization

The protagonists of this popularization are a curious amateur, wishful to understand, and a transactional physicist, that is one of the physicists who independently re-discovered that the transactional re-reading of the quantic experimental facts was unavoidable. He is baptized "Open-Eyes". We know that a specialist is a man who knows very much on very little, and at the limit, all on nothing at all. On the contrary, our "Open-Eyes" science-venturer practiced several disciplines, rarely brought together in the same head; this facilitates to him the confrontations of facts and the synthesis that the too much-specialized person cannot achieve.

The professor Marmot is frozen in the same theoretical position since the Solvay meeting in October 1927. Sometimes physicist, sometimes not a physicist at all, the Professor Marmot is an anti-transactionist as he is a conformist animal of-the- pack. He always intervenes for the worst. Sometimes condensed, sometimes literal, his interventions were written on Usenet and on he Internet, some in English. In conformity with the violent traditions of his pack, this anti-transactionist still may heap abuses on the science-venturer who co-discovered the transactional microphysics, but what is wholly new is that here he cannot ban nor erase the innovator. His pseudonym comes from "*La famille Fenouillard*" by Christophe[28], as the doctor aboard the cargo noted that Mr. Fenouillard slept in the way of the bears, Mrs. Fenouillard in the ways of the marmots, and their daughters in the way of the dormice[29]. But a pseudonym like "*Dogmatix & Idéfix*" could do well, too.

The professor Marmot is a composite assembly of quotations from several persons, so do not be surprised if he contradicts from places to places. Please forgive some of them who are honest persons, but are prisoners of a perverted system.

It would be unfair to leave alone on the stage this violent and dishonest character, so the professor Castle-Holder is in charge of the honest pedagogical tasks, where he prefers scientific spirit to pack spirit. Even when his tribal tribe induced him in error when he was a young student.

Professor Marmot:
- Please be more respectful! We are the official science, anyway! And you are only a little shrimp of conspirationist!

Open-Eyes:
- Please clear up us from doubt: Your argument above, is it a scientific argument, according to you? Or a communautarist and tribal argument?

Professor Marmot:

---

[28]**Georges Colomb**, 1846-1945. Signing as « Christophe », is the author of **L'idée fixe du savant Cosinus, Les facéties du sapeur Camember, La famille Fenouillard, Les malices de Plick et Plock**. Professor of natural sciences and author of handbooks in biology.

[29]http://aulas.pierre.free.fr/chr_fen_05.html

- Your strong expressions undoubtedly belong to the "*classical world*": this is the classical reaction of one who has no arguments to respond. Let me repeat Wittgenstein's maxima: "*Limits of my language are limits of my world*". Available from:
https://www.researchgate.net/post/Is_there_classical_counterpart_for_Plancks_constant_and_State_Vector.

Open-Eyes:
- The institutions may use successively four tactic tricks against the man who dares to innovate:
1. First suppressing the troublemaker, at least by bureaucratic means. So Dan Schechtman was kicked out of his laboratory—the discoverer of the quasi-crystals.
2. If the physical or bureaucratic suppression does not work, disqualify him: "*This is new, so this is false!*"
3. Next: "*OK, it is true, but it is not new! We already knew that.*"
4. Next: "*Hm well, it is new, and it is true, but it is not him who discovered it, it must be somebody else!*". According to professor Jean Bernard, the three last tactics were used against Jean Dausset, and his discovery of tissue groups HLA.

Curious:
- Say! It is the war among the scientists!

Open-Eyes:
- The physicists are territorial animals like the others: rats and dishonest like the others as long as they do not feel supervised. The institutional Superego of deontology of knowledge only intervenes in rare cases, only when they fear the regard of the general public whose taxes pay their wages and their laboratories.

Transactional (Quantum) Microphysics, Principles and Applications

CHAPTER 3

The atomic limit, its fundamental quantities

Professor Castle-Holder:
- First matter: the atomic limit, of which I will recall some fundamental quantities.
The **Avogadro-Ampère constant**: six hundred and two thousands two hundreds and fourteen of milliards of milliards of molecular units in a mole (we had called it molecule-gram, in the past). For instance six hundred and two thousands two hundreds and fourteen of milliards of milliards of water molecules in 18.0153 g water. I have dropped some decimals.

The **quantum of action per cycle**, discovered by Max Planck (1858 – 1947) in December 1900:
$h = 6.6260755 \cdot 10^{-34}$ joule.second/cycle $= \hbar = 1.05457266 \cdot 10^{-34}$ joule. second/radian.
10^{-34}, it is a hundred of milliardth of milliardth of milliardth of milliardth (here, of joule.second/radian).

Professor Marmot:
- Either you are a traitor, either a donkey, estimated colleague! You have just written that $h = \hbar$, when we all teach our students that they are very different!

Professor Castle-Holder:
- It seems that you did not notice that the difference is the unit of phase or cycle. The Dirac's form uses the radian, though the Planck's form uses the whole cycle, but h or $\hbar$ remains the same physical constant. The physicist pays attention to the physical units, when the mathematician does not even know they exist.

Open-Eyes:
- For the sake of clarity, you may replace the last unit **J.s/rad** by the unit of angular moment: **m ^ kg . m / s**, where the caret ^ denotes the multiplication of two perpendicular lengths; the first length is the lever arm of the linear momentum, otherwise said, what matters is the outer projection of the lever arm on the linear momentum. It is an outer product, maximized when the two factors are perpendicular vectors. So h is a quantum of looping, though the Maupertuisian action (Pierre Moreau de Maupertuis 1698-1759) is a circulation of a momentum, or sum along the path, of the inner product of two vectors

(maximum when they are colinear, when one is exactly its inner projection on the other). An irreducible difference in nature.

The symbol caret ⌢ denotes an outer product. Alas, the french readers of french books are mislead by the french tradition: to improperly use the caret to mean a *cross product*. Alas, this traditional *cross product* violates both the mathematical coherence and physics. The outer product of two vectorial quantities is an antisymmetric tensor, of rank 2. For pedagogical needs for the workshops, I have translated "*antisymmetric tensor of rank 2*" into "**gyror**".

Harasser marmot:

- Ignorant! Heretic! This *crank* deludes himself that two quarter turns make a half turn! Let's all league to ban him from all Usenet!

Open-Eyes:

- I recall that this harassing marmot, frenzied invader, is not a physicist, and ignores all about the use of dimensional analysis in physics. Despotic, ignorant and presumptuous, he deludes himself in believing that physics is just a lean-to, annex of the maths he learned years ago. We do not adhere to these erring, alas traditional. Here are some physical quantities which are not vectorial but gyratorial: an (angular) moment of force, an angular moment, a spin[1], an angular speed, a magnetic field $\check{B}$, a magnetic moment... The bulk of the course for the geometric syntax of the physics is on the wiki: http://deontologic.org/geom_syntax_gyr

The quantities about the atomic limit, continuation.

Professor Castle-Holder:

- Light celerity: $c = 299{,}792{,}580$ m/s.

Orders of magnitude: The Moon is at 1.28 light-second from the Earth, the Sun is at most at 499 light-second from us, that is between eight minutes and eight minutes and twenty seconds. Jupiter is at an average distance of 2595 light-seconds, or 43 light-minutes.

Charge of the proton: $1.60219 \cdot 10^{-19}$ C.

« **C** » denotes the coulomb, unit of electric charge. When I was a kid, the definitions of the ampere and the coulomb were electrochemical, based on the weight of silver electro-plated on the cathode: one coulomb plates 1.118 mg of silver. So it needs six hundred and two thousands two hundreds and fourteen of milliards of milliards of silver cations, or any other monovalent ion, to carry one coulomb.

Mass of the proton: $1.67265 \cdot 10^{-27}$ kg.

Mass of the electron: $9.1093897 \cdot 10^{-31}$ kg

Conversion into relativist energy: $m_e.c^2 = 511$ keV (kiloelectron-volt) $= 8{,}187 \cdot 10^{-14}$ J (joules)

---

[1]Spin: elementary angular moment, without equivalent in macrophysics.

Practical application: when you have a PET-scan (positron emission tomography), typically for evaluating a cancerous tumor, 511 keV is the energy of each
of the two gamma photons, emitted in opposite directions by the annihilation
electron-positron. All around you are disposed gamma detectors. The software
considers that the emission was on the line joining the two coinciding sensors.
As there was far more than one emission, the approximate intersections of all
these lines give the location of the emissions.

Atomic nucleus: Composed of Z protons and Y neutrons, it holds Z electrons
to form an atom.
Atom: Only one nucleus, escorted with the right number of electrons to balance
the electric charge of the nucleus.
The **atomics** deals with the electronic cloud around the nucleus.
The **nuclear physics** deals with the nuclei of the atoms.
Molecule: several nuclei, bound by electrons engaged in covalent bonds, that
is sharing of pairs of electrons of opposite spin, in the most far orbitals of the
atoms, plus eventually some mixed bonds or hydrogen bonds.
Ion: atom or molecule where the number of electrons does not balance the
number of protons.
Anion: negative ion, more electrons than protons. Examples: OH^-, $SO4^{--}$,
HCO_3^-, Cl^-...
Cation: positive ion, fewer electrons than protons. Examples: Na^+, Ca^{++},
H_3O^+. Electrolyte: a solution containing anions and cations, so able to conduct electric current. The most common electrolytes are aqueous solutions but
other solvents exist, and molten salts are electrolytes too.

Open-Eyes:
- Now the reader has all the means to calculate the intrinsic frequency of the
lectron, discovered by Louis de Broglie in 1923:
$v_e = m_e c^2$ / h = 9,1093897 . 10^{-31} kg * (299 792 580 m/s)^2 / 6,6260755 . 10^{-34}
joule.second/cycle = 1,235 59 . 10^{20} Hz (1 Hz = 1 cycle per second).
This intrinsic frequency and its phase wave are the foundation of the quantic
part of microphysics.
De Broglie Wavelength corresponding to the mass of the electron (said *Compton
length* by the American): $\lambda = \frac{h}{mc}$ = 2.42631058 . 10^{-12} m
Its inverse the electronic wave number, that is 4,121,483,900 cm^{-1}, or more elaborated, the electronic **wave vector** intervene in the reactions between electrons
and gamma photons.

Professor Castle-Holder:
- When we only say wavelength, we discard the direction of propagation. Implicitly, the implied unit of phase is the cycle or 2 π radians. The wavenumber is
used by the spectroscopists. Another craft prefers the wave vectors, but change
the unit of phase, preferring the radian. It is not a matter of principle, but of
local habits. The diameter of a hydrogen atom: about 0.11 nm (a hundred and

ten picometers, or 1.1 ångström[2]). It is intrinsically fuzzy.

Open-Eyes:
- Here is an image of the density of the electron in its fundamental state around a nucleus of hydrogen: Fig. 3.1. (image from the public domain).

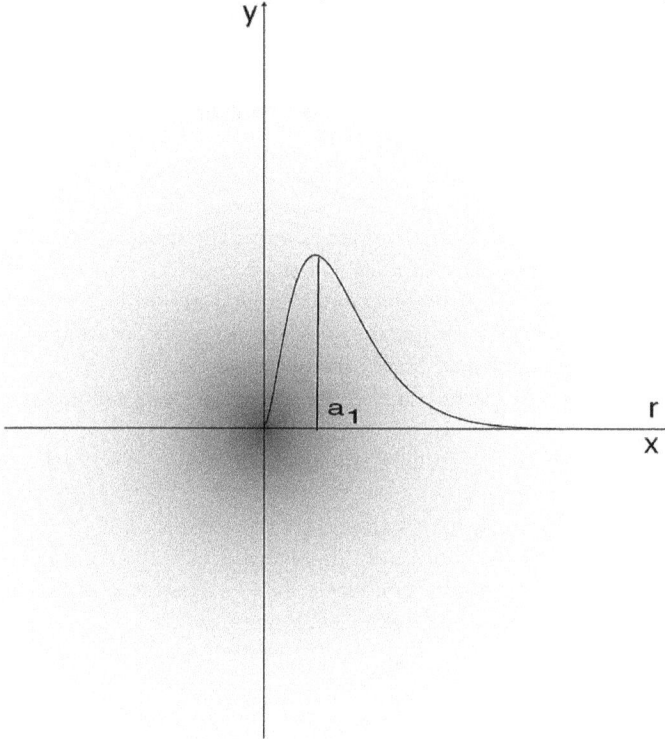

This electronic density is the square of the amplitude of the stationary wave of the Schrödinger equation. Squared or not, it remains an exponential function of the radial position: $\exp(-r/r_0)$, or $\exp(-2r/r_0)$ for the square, with its maximum at the null distance, that is just on the nucleus.

The reader has noticed the spherical symmetry, the fuzziness and the large size of this electron. This contrasts with familiar popularization by images or videos, which presents you a small planet orbiting around a central star. However, the trajectory of a planet is a spire, which has none of the symmetries of a sphere, nor worse, of a fuzzy sphere. This is a major contradiction in the corpuscularist ideation. Seen from the outside, the chemical properties of a hydrogen atom are those of its electron, captive of a central H^+ ion. Notice that you can find atomic hydrogen only in intersidereal space. In the pressure bottles, or in the laboratory you only find the dihydrogen molecule H_2 in the gaseous state, and it is much more difficult to describe and draw. We will see later the electronic

---

[2]Ångström, (ɔŋ.'strøm/), is 100 picometers, 10^{-10}m.

map of a dinitrogen molecule N_2.

Professor Marmot:
- Shame on the perfidiousness of this weird science-venturer! Instead of telling as he was taught to *"Probability of finding the electron"*, he says *"density of the electron"*! Instead of denying the wave, as he was taught to, he reinforces it, and he disavows the corpuscular aspects! What a scandal, as we had physically defeated Erwin Schrödinger and his wave in December 1926[3] and October 1927! What a shame, and what a weird mind!

Curious:
- Let's apply your scientific discipline: from what experimental protocol does this image result? How did you measure that?

Professor Castle-Holder:
- Not any direct experiment is possible. In macroscopic electrostatics, it would be possible to place a small trial-charge at a varied distance from the central charge, and to measure the local potential, and even the field. At the Palais de la Découverte, they can demonstrate it to the public. But it is impossible to set a *small test-body* around the proton as we could in macroscopic electrostatics, neither for measuring the potential, neither for measuring the local electronic

---

[3]http://citoyens.deontolog.org/index.php/topic,1141.0.html : **Les procédés employés par Niels Bohr pour vaincre Schrödinger**: The telling is by Werner Heisenberg himself, though he had much to earn in defeating Erwin Schrödinger, either by a loyal or by a disloyal fight.
Source: Franco Selleri. *Le grand débat de la physique quantique*. Champs Flammarion, Paris 1986. Page 96.
This telling is confirmed from second hand by Emilio Segrè, in *Les physiciens modernes et leurs découvertes*. Fayard, Paris 1984 for the french translation.
Selleri quotes the original source with the letter from Heisenberg:
S. Rozenthal, ed. Niels Bohr. North-Holland, Amsterdam, 1968.
Quotation: Schrödinger had to face a difficult fight in Copenhagen. Bohr invited him to give a lecture at the end of 1926 *"and asked him to, not only give a lecture on his undulatory mechanics, but also to stay in Copenhagen enough time for discussing of interpretation of Quantum Mechanics"*.
Heisenberg describes so the intensity of the fight:
"Though Bohr was really an obliging and attentive person, he was able in such discussions about the epistemological questions he considered as absolutely vital, to insist in a fanatic manner, and with a terrifying inflexibility on the complete clarity of all the arguments. After hours of wrestling, he still could not resign himself in front of Schrödinger, to admit that his interpretation was insufficient, and even unable to explain the law of Planck. Each attempt form Schrödinger to evoke this embarrassing point was refuted, slowly, point by point, in painstaking and endless discussions. Probably because of this overworking, after some days Schrödinger fell ill and had to stay in bed in Bohr house. But even then it was difficult to keep Bohr far from the bed of Schrödinger..."
And Heisenberg concludes: *"Finally, Schrödinger went out of Copenhagen rather demoralized, when at the Bohr institute we felt we got rid of the interpretation given by Schrödinger to the quantic theory, which hastily used classical undulatory theories as models."*
So were treated the fundamental questions, and so a small-group became hegemonic: by sheer violence.

density. Much worse than a practical impossibility, it is in principle impossible: no *small test-body* smaller than a whole atom may exist. Nor we ever will have a micro-mover, able to set around this mythical test-body. We only can give the calculated model of the bound electron, by resolving the Schrödinger equation. In this simplest case, we can carry the computing up to the end. Here is the solution for the lower and stable state, where the distance from the nucleus is noted ρ: $R_{10} = \left(\frac{Z}{a}\right)^{\frac{3}{2}}.2e^{-\frac{\rho}{2}}$

But we have many indirect proofs of the validity of this kind of electronic density images in the atoms or molecules or crystals: the compressibility of crystals, the thermal dilation, the frequencies of the vibrating gaseous molecules able to couple with an infrared photon, to absorb it or to emit it, predicting the optical properties of the dye molecules, etc. So concerning the principle, we have not more any chance to be wrong.

Curious:
- So? You only have indirect proofs?

Professor Castle-Holder:
- Indirect indeed. We do not have any other way.

Curious:
- How annoying! In microphysics, I can no more use my senses nor my muscular proprioception, when they are still usable in macroscopic mechanics.

Open-Eyes:
- And even provided that the mathematization would be not deceptive nor stupid! Which alas is not always guaranteed. Your proprioception is flouted when a teacher pretends: "*Then one has an angular speed vector that climbs along the axis*". They do not help, these inexcusable traditions, that are repeated from generation to generation... It is one of the more violent abuses, where the rules of the honest and scientifically validated unsensorialization are openly breached, precisely where nothing justifies any unsensorialization of the tools for electromagnetism and mechanics, in denial of reality.

Curious:
- So? How to verify you are not telling "*just so stories*", in the fashion of the frauds from Sigmund Freud and his church? The records do not encourage blind trust.

Open-Eyes:
- Demand lots of experimental reality-testings. Even the rationale Mechanics is denied by the bulk of people, included the students, without hesitation nor reflection, as soon as they do not recognize a scholar case; it is a part of the civil war against the learned ones. In Appendix G, we give you many experimental situations you can verify in a common life frame. But in microphysics, no pity: we are not more in our familiar sensorial world, and we had to build largely

unsensorialized tools. Here are some physical quantities used in microphysics, which have a correspondent in our familiar macroscopic world: the masses, the linear momenta, the energies, the angular momenta, the potentials, the fields, the electric charges and the magnetic moments are of the same nature as they are in our familiar macroscopic world. The only one solution for you laymen is to demand experimental reality-testings. Alas, the custom of the popularization magazines is not up to their duties: like African griots[4], they emphasize the dramatization: "*This one wins on the other! Let's swear fidelity to the brave general Tapioca[5], as he is the master of the situation!*" then they multiply the fantastical interpretations, so toady to the suckers. Example: they cheekily repeat that the famous Schrödinger cat "***is** in a superposed state, neither dead, neither alive*", when of course, they remain unable to exhibit any experimental proof. They remain intoxicated by the hegemonic fairy tales.

A problem more: nobody knows how to put you in front of an experiment which demonstrates clearly "*The Maupertuisian action, or the Hamiltonian action, it is that*". Indeed we can achieve this result for the angular momentum, but linking the two is more awkward that they make you believe.

Professor Castle-Holder:
- The atom next in complexity is the helium atom: two electrons, and in the nucleus two protons and two neutrons. Fig 3.2. (public domain):

---

[4]Griot: African singer or narrator, paid to praise the prestigious genealogy of the powerful chief of the day.
[5]***General Tapioca, general Alcazar***. Cf. *L'oreille cassée*. By Hergé.

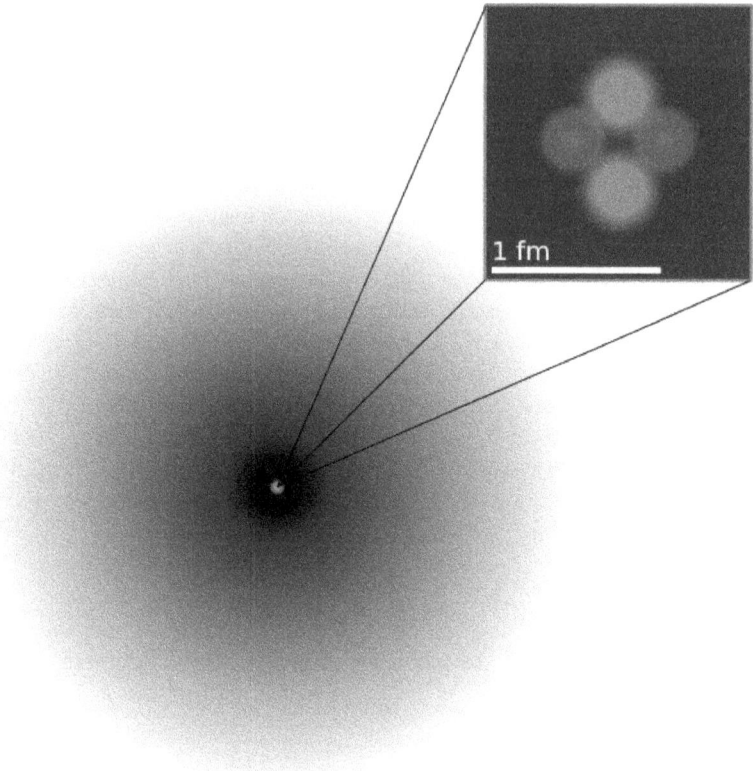

1 fm

1 Å = 100,000 fm

This atom is represented in its standard and stable state, which is again of
spherical symmetry. Read the prefix: femto_, from the Swedish *femton* =
fifteen, denotes the power -15 of 10, that is the millionth of the milliardth.
The pico_ is the power -12 of 10. The Å or ångström (ɔŋ.'strøm), is 100
picometers, 10^{-10}m. Helium does not chemically bonds in any molecule but can
slip as interstitial in many metals and rocks. Its two electrons are in the same
lower state, of spherical symmetry, they only differ by the spins, which are antiparallel, the two opposite spins make a pair, of null total angular momentum.
Again an artifact induced by the constraints on the drawing: the helium nucleus is drawn flat, with colors. Nothing such exists, nothing that is flat, with
colors. The reality of an α nucleus is beyond the means of a draughtsman. So
are the objects the Quantum Mechanics handles.

Open-Eyes:
- Are Quantic Mechanics and Microphysics synonyms? No, the relation is an
inclusion: the quantic microphysics is the subdomain of microphysics, where
stationary states exist, which de Broglie and Schrödinger had proved that they

are constrained by autoquantizations: Either the phase waves loop exactly, either the "state" is not existing, it is not a state.

Professor Castle-Holder:
- There is hardly one kind of exceptions where the QM overflows in the macro-physical domain: when herds of bosons intervene (particles with an integer spin = 0 or 1, apt to pack together, ruled by the Bose-Einstein statistics). Bosons are apt to pack together in the same quantic state. Applications: the lasers, the masers, the superconductivity[6], the super-fluidity of helium 4, the interferential astronomy, even with a large base.

Open-Eyes:
- Another half-exception: all the electric contacts, as you have so many around you, in your car, in your house, can only work by "tunnel effect", that is that the wavelength of each conduction electrons[7] as it is at the Fermi energy[8], is larger than the gap between the to nearing conductors, approximately in quasi-contact, and larger than the thin oxide layer (not seeable at naked eyes) that covers each. Ah yes, each conduction electron is much larger than the inter-atomic distance of copper or gold or aluminum of the conducting contact. No one popularizer at the TeeVee never told you that. But how a mediatic sorcerer could dazzle you with their kakarakamouchems[9] with a so common phenomenon as an electric contact? So the mediatic sorcerers keep mute on that.

Professor Castle-Holder:
- You have said too much, or too little! The tunnel effect gives a probabilistic result: some particles ought to never pass, but sometimes pass.

Open-Eyes:
- Developing: the interposed non-conductive layer – combined oxygen, adsorbed[10] dinitrogen, adsorbed water – may shoot back electrons, as also do

---

[6]In a **superconductor**, below a critical temperature, the electric resistivity falls to null. Alas, provided that the magnetic field is not too strong... Mechanism: the electrons associate by pairs – said Cooper's pairs – with one or two phonons, and their wave travels through the entire crystal in the way of a boson, of integer spin.
A **phonon** is a unit of vibration in a crystal. Further explanation at § 4.2.
[7]**Conduction electron**: at the top of the energies occupied by the electrons of the metal. Only a small minority of metals have the conduction band half-filled, with only one conduction electron by atom: copper, silver, gold, aluminum, gallium, indium, plus the alkalines. These are the only good conductors of electricity. Most metals have the conduction already fulled, with two electrons per atom. Their electrical conduction, and consequently their thermal conduction is less good, and much more difficult to explain: by recoverings of the Brillouin zones according to crystallographic directions. It is an anisotropic conduction in these metals. An **anisotropic** property differs according to the direction.
[8]Fermi energy in a metal: the highest level of electron energy occupied by an electron, at the temperature of the absolute zero. The zero is taken at the level of the most bond electrons.
[9]Molière, **Le bourgeois gentilhomme**. « *What? Kakarakamouchem means my sweet darling?* How it is admirable! What an admirable language is that Turkic! ».
[10]**Adsorption**: a weak bond of a molecule on the surface of a solid. If it retakes its liberty in gaseous form, it is said desorbed.

phonons, impurities and dislocations. If no generator would be behind to impose its rule, the accumulation of electrons on the negative side of the junction should slow the other electrons coming from the upstream. So the resistance[11] of the circuit increases; in electrokinetics[12], the contact-junction produces a supplementary drop of potential[13]. What you wanted to translate into probabilities remains a mere increase of the global resistance of the circuit: the contact resistance is noticeable. The electricians easily live without knowing that they handle quantic phenomena each time they switch a circuit on.

Another exception, which permits all the electrotechnics, the electric motors, the alternators, the transformers, and the Earth magnetism too: the ferromagnetism[14]. A similar phenomenon is the ferrimagnetism which permits the ferrite bars in your radio receptor, and the miniature transforms in the switch-mode power supplies which now are everywhere in your houses. In ferromagnetism, all the spins that are unpaired in each atom, have the goodwill to organize in herds, all in the same equiplane[15] and sign of rotation, in Weiss domains[16] enlarged enough to have macrophysical effects.

Professor Castle-Holder:
- Nobody knows why the constants mentioned above are such, and not something else. Nobody knows why the electrons have all the same charge, and why this one. Nobody knows why the Planck quantum of action is $6.6260755 \cdot 10^{-34}$ joule.second, but we have acquired the certitude that it is indeed a universal and fundamental constant.

Open-Eyes:
- Planck quantum of **looping**: $6.6260755 \cdot 10^{-34}$ joule.second/**cycle**. The initial definition of action by Pierre Moreau de Maupertuis was an inner product of two vectorial quantities: the product of the linear momentum by the way made, or circulation of the linear momentum. Whereas the angular momentum is an outer product of a lever arm by the linear momentum, that is why a unit of angle, or of cycle, or of phase intervenes in it. So is not the action of Maupertuis, neither of Hamilton and Jacobi.

---

[11]**Electrical resistance** of a circuit: the quotient of the difference of potential by the intensity of the current. Unit: Ohm = volts/amperes.
Intensity of a current: how many electrons per second. The macrophysical unit is the ampere. One ampere = six milliards and two hundreds forty-two millions of milliards electrons per second (many decimals omitted).
[12]**Electrokinetics** as opposed to electrostatics where no charge move: in electrokinetics, the currents exist, the charges are moving.
[13]**Electrical potential**: similar to the height of water in hydraulic. If you hoist a charge + to a higher potential, you increase its potential energy.
[14]The ferromagnetism is known in macrophysics as the magnetism of the iron and some alloys. The ferrimagnetism is in some oxides. For instance the magnetite Fe_3O_4.
[15]**Equiplane**: the equivalence class of all the planes having the same orientation of plane.
[16]**Weiss domain**: in a crystal, a domain where the magnetization has the same direction of plane, and the same sign of rotation.

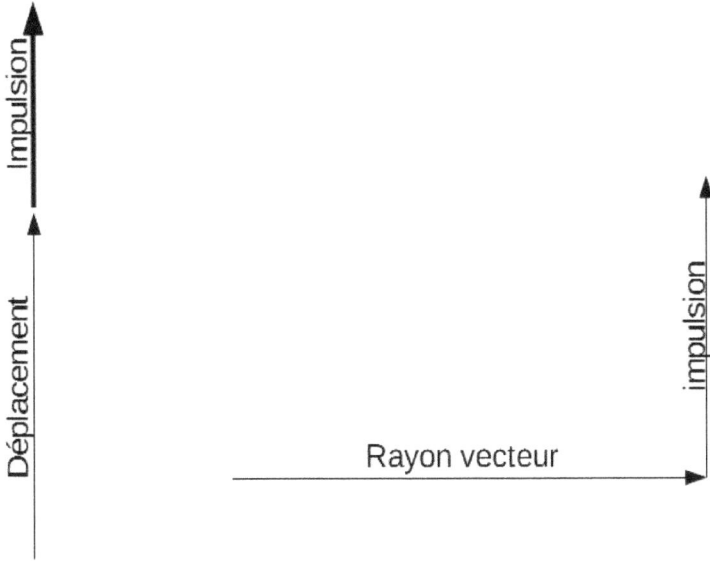

L'action est un produit intérieur de vecteurs.

Le moment angulaire est un produit extérieur de vecteurs

To do an **inner product** of two vectors, one begins by an inner projection of one vector on the other. And the cosine is an even function. The result has no orientation in the space.

To do an **outer product** of two vectors, one begins by an outer projection. And the sine is an uneven function. The order of the operation matters, it always gives a **direction of rotation** in the plane direction of the equiplane given by the two vectors.

So this is a fundamental problem, on which only a few paid attention. The twist of this story is that in 1924, Louis de Broglie had applied the first definition to the wave, along with the Bohr orbit, and two years later it allowed Erwin Schrödinger to prove that this Bohr orbit does not exist. We still have some work before us, to put all that upright.

Professor Marmot:
- What a scandal! This science-venturer from the gutter never stops to say the contrary of what we all teach! Let's slaughter him with a grenade!

Curious:
- Objection! Again here you have written lots of new words: "**spin, ferrite, ferrimagnetism, ferromagnetism, photon, tunnel effect, Fermi energy**"... You should explain all.

Open-Eyes:

- Acceptable objections. Do not forget to look at the footnotes, even when they are short. You will find a chronological glossary in Appendix F. I invite you to have more patience, as I see we have to spend some more time on atomics[17]. There it is manifest that we do not more do the same physics that the anti-transactionist copenhaguists do, and that with the same laws of stationarity of the stationary electronic wave around the nucleus, we do not present the same diagrams.

Professor Marmot:
- All that fuss is just futile and unimportant philosophical preferences! What matters is that you lower the head and calculate!

Open-Eyes:
- I intervene before the calculations: I intervene on the semantic axioms and the physical axioms, with the criterion they must be economical - and those of yours are not economical at all – but also that they adhere closely to the reality; where you are terribly failing.

Professor Castle-Holder:
- Nevertheless, it is time to shove the natural laziness of the readers, by giving them exercises, like in all serious handbooks.
Knowing that the mass of the proton is $1.67265 \cdot 10^{-27}$ kg, deduct its intrinsic de Broglie frequency. We accelerate it under a potential difference of 500 V. Calculate its wavelength in the frame of the laboratory. Compare this wavelength to some known inter-atomic distances, for instance in aluminum, 286 pm. Same questions with a bullet for an assault rifle, weighing 5 g: its Broglian frequency as a bulk, then its wavelength in the frame of the laboratory, when its speed is 800 m/s. How will you mount the experiment to show up and measure this wavelength?

The key is in Appendix H. So it will be evident to you that the mechanics are undulatory in practice only for minimal masses: electrons, protons, neutrons, etc.

---

[17]**Atomics**: a branch of physical chemistry, studying and predicting the chemical properties of the atoms according to their place in the Mendeleev periodic table, i.e. according to the structure of their electronic bevies.

CHAPTER 4

The energy levels in an atom, a molecule, a solid

4.1. Why these definite energy levels?

What was made evident by the spectroscopists along the 19th century. And exploitation of it.

4.1.1. Energy levels in an atom. Open-Eyes:

- Louis de Broglie was the first to give a physical explanation in 1924, alas incomplete: The wave of phase of the electron may loop around a nucleus as a stable and stationary wave only if it has a precisely defined frequency and a phase celerity. It can loop a turn in one cycle: principal quantum number $n = 1$. It can loop a turn in two cycles: $n = 2$. Three cycles: $n = 3$, etc.

The complete explanation was given two years later by Erwin Schrö-dinger in 1926: the wave equation he wrote had constant solutions for an atom only for electrons having a lesser energy than the one of the free electron, and only a discrete[1] suite of stable states exists. There occurs an auto-quantification, by the proper values of the equation, depending on the boundary conditions. The lower state of energy still has $n = 1$ as the principal quantum number. But from de Broglie to Schrödinger a revolution was done: the fundamental state has no more the symmetries of a turn, but of a sphere; and in that state said **S**, the electron has no more orbital angular momentum. No, the electron does not "*turn*", it does not *orbit*, but plainly **it is**, and is continuously spread around the nucleus.

Curious:
- That's odd: what you say is the contrary to what I had been told by videos.

Open-Eyes:
- I recognize which ones of the *great ancestors* did correctly their work. I sort it out, and I *throw in the garbage cans of the History* what is not worth a crap. When you read the original papers from Schrödinger, you see that what he wrote is the contrary to what his enemies, the winners of 1927, pretend in his back to be his views: The Schrödinger wave describes the electron, the stationary wave around a nucleus, and not "*the probability to see the apparition of the corpuscle*". The fact that this wave is stationary is binding. What the

---

[1] Opposed to a continuous set or subset, a set or subset is said **discrete** if it has a finite or countable number of elements, and is not dense. Example: the suite of the inverses of integers $1/2$, $1/3$, $1/4$ is a countable infinite set, but discrete. **Counter-example**: The set of the **rational numbers** Q is not discrete, as dense everywhere in the set of the **real numbers**.

spectroscopists, then next the astrophysicists discovered for the 19th century, is that these stationary states of the atoms and molecules are universal, are the same in the whole universe, with a fine definition. The Planck constant which binds the intrinsic frequency to the mass is the same everywhere. During an atomic transition, either emitting a photon or receiving it, the atom (or the molecule) oscillate between the initial and the final state, and this transition takes some time. The frequency of the emitted photon is the **beat** between the initial Broglian frequency of the atom, and the new one. Sure this difference is extremely small in relative value, but it is what the spectroscopist accesses to. The same difference at the absorption, except that the most energetic and quick state is the final state.

Curious:
- Do the spectroscopy and these spectra of lines exhaust all that we have to know about emission and absorption of light?

Professor Castle-Holder:
- Surely not! On the one hand, the thermal agitation in a gas broaden all the spectra lines, up to rendering them indiscernible it the temperature at the surface of the star is high enough. The tungsten thread of our incandescent lamps does not emit a spectrum of lines, but a continuous spectrum depending on the temperature, according to the law of Planck. On the other hand, many are the absorption mechanisms that are not especially spectral, for instance, all that which phononic; So, our solar water heaters are not spectral absorbers.

Open-Eyes:
- Beyond the phononic emissions of photons, surely two other emissive mechanisms are not bound to any stationary state, so not so spectral: the braking radiation or *Bremsstrahlung* and the synchrotron radiation.
The braking radiation is observed around nuclear reactors when they are immersed at the bottom of a pool; it is a bluish light. It is produced when high energy particles coming from the reactor have a higher speed than the celerity of light in this material. The braking radiation or *Bremsstrahlung* also occurs from the high energy electrons incurring on the anode of an X-ray tube. For a radiocrystallographist either this *Bremsstrahlung* is favorable if he intends to do a diffractogram of spots, according to the protocol from von Laue (1912), so he needs a polychromatic radiation; either it is an annoying parasite if he wants to do a diagram of powder, according to the protocol invented by Peter Debye and Paul Scherrer in 1916; then he must use filters to stop the *Bremsstrahlung*, and to let pass only the K$_\alpha$ ray of the anticathode metal. A gigantic facility for synchrotron radiation, where you must reserve your beam time a year before, with a detailed and precise file for your project of experiment, is the ESRF in Grenoble. At each bending imposed by the electromagnets to the beam of electrons at 6 GeV, a narrow beam of X-rays and gamma rays escapes tangentially. It is an extremely bright and narrow source, but widely polychromatic. It is up to the user to have mounted all the monochromators he eventually needs.

Not any kind of stationary state of the emitting electrons nor before nor after the bending.

Professor Castle-Holder:
- Let us begin with a diagram and a table of functions about which we do not yet diverge, and that our curious reader does not yet know. Here are the energy levels of the unbound hydrogen atom (not bound in a molecule, not bound in dihydrogen). On the left is the scale of the energies, in electron-volts from the fundamental level; on the right, the scale is in wavenumber in cm^{-1}, and the origin is inversed: from the free electron. Multiply by **c**, the celerity of the light, and you obtain a frequency. The oblique lines figure the spectral lines observed in the spectrum of the atomic hydrogen, which are *authorized* transitions (authorized by the transition rules) between stationary levels of the electronic wave. These stationary levels are figured by horizontal lines. This diagram comes from Edouard Chpolski, **Physique atomique**, tome 2, Ed. Mir 1978.

Figure 4.1.

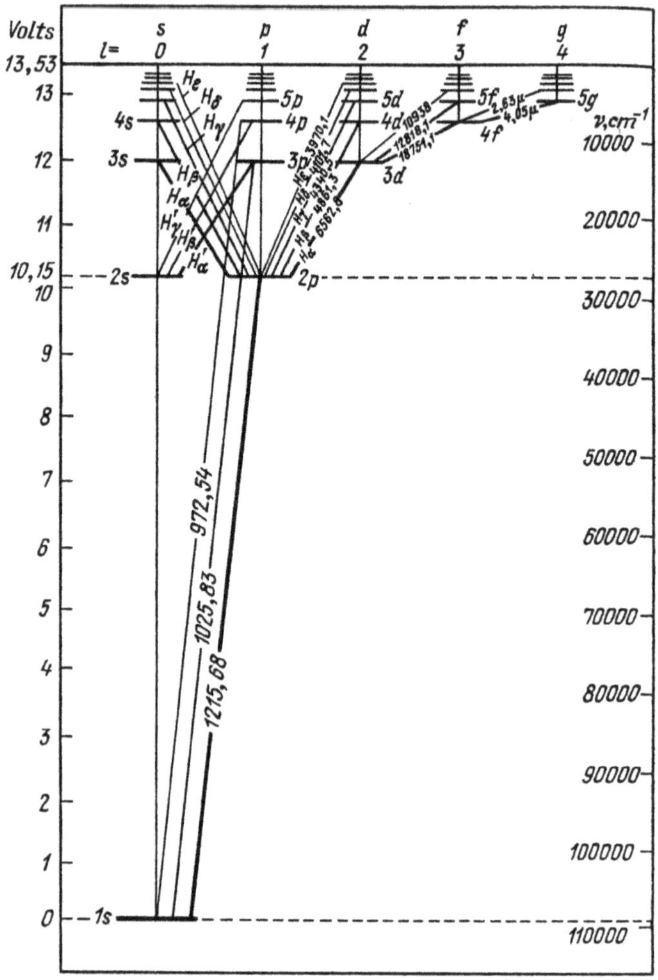

Fig. 26. Niveaux d'énergie de l'atome d'hydrogène (l'épaisseur des lignes corresp
à la probabilité de transition)

The level 1s is the fundamental state, with spherical symmetry, already figured
sooner. The levels 2s, 3s, 4s also have spherical symmetry, but with respectively
2, 3, 4 disjoint zones, so 1, 2 or 3 spherical surfaces of transition of phase, with
null intensity and null density. The level 2p has no more the spherical symmetry but instead has a plane of frontier of phase. The level 3p has both a plane of
frontier of phase and a spherical frontier. The level 3d has two planes as frontiers of phase. I let you complete for the levels 4p, 4d, 4f, 5p, etc. As the levels
step up from the fundamental, their energetic differences decrease, and anyway,
the levels have the ceiling of the free electron, 13.53 eV above the fundamental
1s, the most bound state. Here you have the table of the exact mathematical
functions for the stationary electronic waves of hydrogen, following the quantic

numbers **n**, **l**, **m**. The spin variable is omitted here. Same source: E. Chpolski. The radius is represented by the greek letter σ. All along with this presentation of the atomics, we keep the Schrödinger equation of 1926, not relativist, which is qualitatively enough for this task and remains quantitatively very good.

States of a hydrogen atom

n	l	m	$\psi(\text{normed}) = R(r)\,\Theta_{l,m}e^{\pm im\varphi}$	State
1	0	0	$\frac{1}{\sqrt{\pi}}\cdot\left(\frac{Z}{a_1}\right)^{3/2}e^{-\sigma}$	1s
2	0	0	$\frac{1}{4\sqrt{2\pi}}\cdot\left(\frac{Z}{a_1}\right)^{3/2}(2-\sigma)\,e^{-\sigma/2}$	2s
2	1	0	$\frac{1}{4\sqrt{2\pi}}\cdot\left(\frac{Z}{a_1}\right)^{3/2}\sigma e^{-\sigma/2}.\cos\theta$	2p
2	1	+1	$\frac{1}{8\sqrt{\pi}}\cdot\left(\frac{Z}{a_1}\right)^{3/2}\sigma e^{-\sigma/2}.\sin\theta.e^{i\phi}$	2p
2	1	-1	$\frac{1}{8\sqrt{\pi}}\cdot\left(\frac{Z}{a_1}\right)^{3/2}\sigma e^{-\sigma/2}.\sin\theta.e^{-i\phi}$	2p
3	0	0	$\frac{1}{81\sqrt{3\pi}}\cdot\left(\frac{Z}{a_1}\right)^{3/2}(21-18\sigma+\sigma^2)\,e^{-\sigma/3}$	3s
3	1	0	$\frac{1}{81\sqrt{\pi}}\cdot\left(\frac{Z}{a_1}\right)^{3/2}(6-\sigma)\,\sigma.e^{-\sigma/3}.\cos(\theta)$	3p
3	1	+1	$\frac{1}{81\sqrt{\pi}}\cdot\left(\frac{Z}{a_1}\right)^{3/2}(6-\sigma)\,\sigma.e^{-\sigma/3}.\sin(\theta)\,e^{i\varphi}$	3p
3	1	-1	$\frac{1}{81\sqrt{\pi}}\cdot\left(\frac{Z}{a_1}\right)^{3/2}(6-\sigma)\,\sigma.e^{-\sigma/3}.\sin(\theta)\,e^{-i\varphi}$	3p
3	2	0	$\frac{1}{81\sqrt{6\pi}}\cdot\left(\frac{Z}{a_1}\right)^{3/2}\sigma^2.e^{-\sigma/3}.(3\cos^2(\theta)-1)$	3d
3	2	+1	$\frac{\sqrt{2}}{81\sqrt{\pi}}\cdot\left(\frac{Z}{a_1}\right)^{3/2}\sigma^2.e^{-\sigma/3}.\sin(\theta).\cos(\theta).e^{i\varphi}$	3d
3	2	-1	$\frac{\sqrt{2}}{81\sqrt{\pi}}\cdot\left(\frac{Z}{a_1}\right)^{3/2}\sigma^2.e^{-\sigma/3}.\sin(\theta).\cos(\theta).e^{-i\varphi}$	3d
3	2	+2	$\frac{1}{81\sqrt{2\pi}}\cdot\left(\frac{Z}{a_1}\right)^{3/2}\sigma^2.e^{-\sigma/3}.\sin^2(\theta).e^{i\varphi}$	3d
3	2	-2	$\frac{1}{81\sqrt{2\pi}}\cdot\left(\frac{Z}{a_1}\right)^{3/2}\sigma^2.e^{-\sigma/3}.\sin^2(\theta).e^{-i\varphi}$	3d

The binomial **2 - σ** has one zero, the trinomial **21 - 18 σ + 2 σ^2** has two zeros, etc.

Open-Eyes:
- But with the next diagram, things get worse. The majority, not to say the hegemonic authors have coded the corpusculist theory of Born and Heisenberg with the soldering iron in the teaching handbooks. They graph the solutions functions only after elevating them to the square. So we have to remake all the graphs.

Figure 4.3.

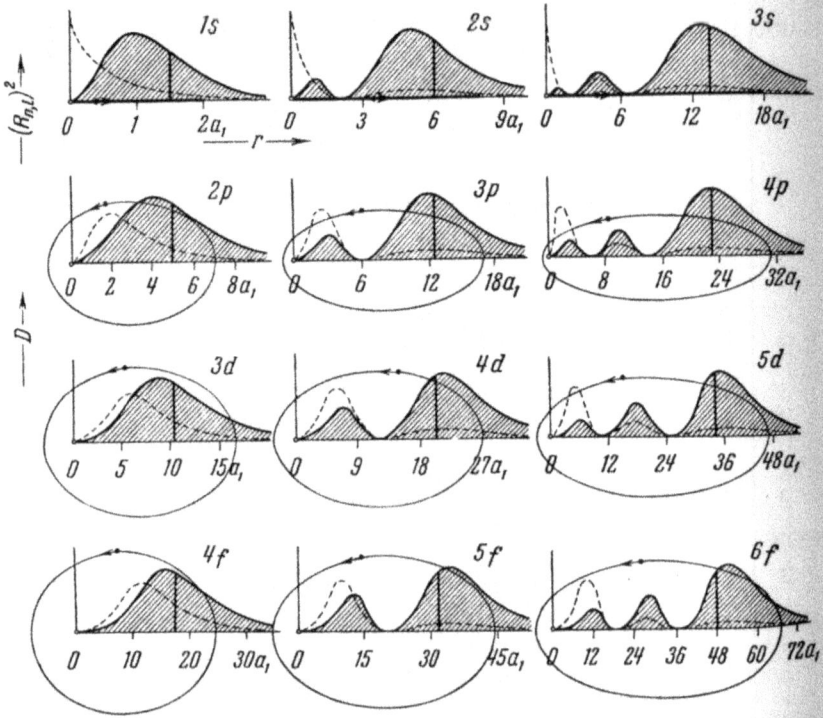

Fig. 27. Courbes de la composante radiale de la densité de probabilité $D = 4\pi r^2 (R_{n,l})^2$ pour différents états de l'atome hydrogénoïde. En pointillé est donnée la variation de la fonction $(R_{n,l})^2$. La position de la grandeur moyenne $\overline{R^2_{n,l}}$ est marquée par un gros trait vertical

OK, we will redo the necessary graphs, with the wave itself, and not its square, for the states 2s, 3s, 3p, (4p, 4d, 5d, 5f, 6f later). In blue at the screen, the punctual spatial density; in red it is multiplied by the square of the radius, giving the total presence of the electron at a distance represented by the absciss. You will notice how large are the higher states, compared to the fundamental. Figure 4.4.

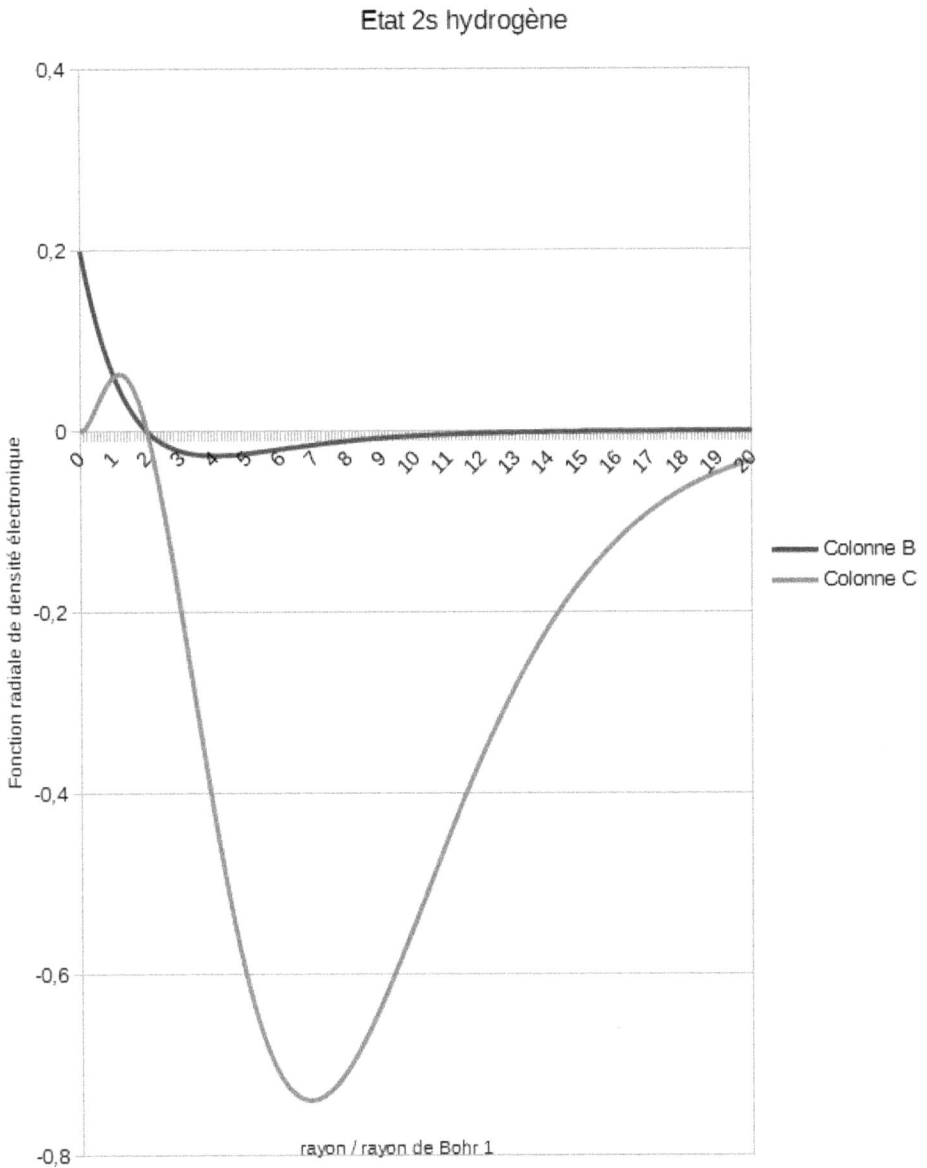

Figure: Etat 2s hydrogène

See that for the 2s state, the density has a null at twice the Bohr radius. The phase changes of sign. 2s ==> two phase-zones, one frontier of phase.

Figure 4.5.

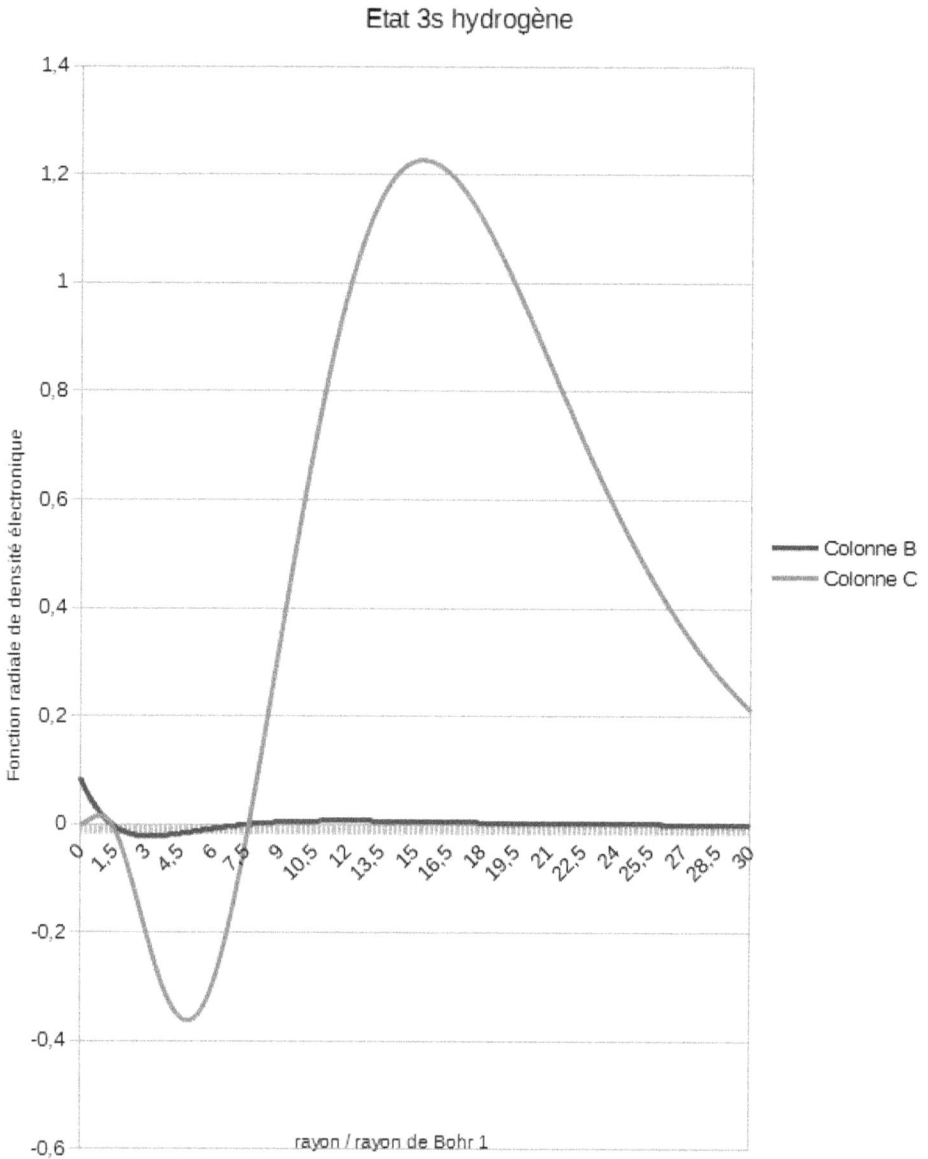

Etat 3s hydrogène

In the state 3s, the density has two zeros at 1.3775 and 7.6225 times the Bohr radius. 3s ==> three phase zones, and two frontiers of phase.

Figure 4.6.

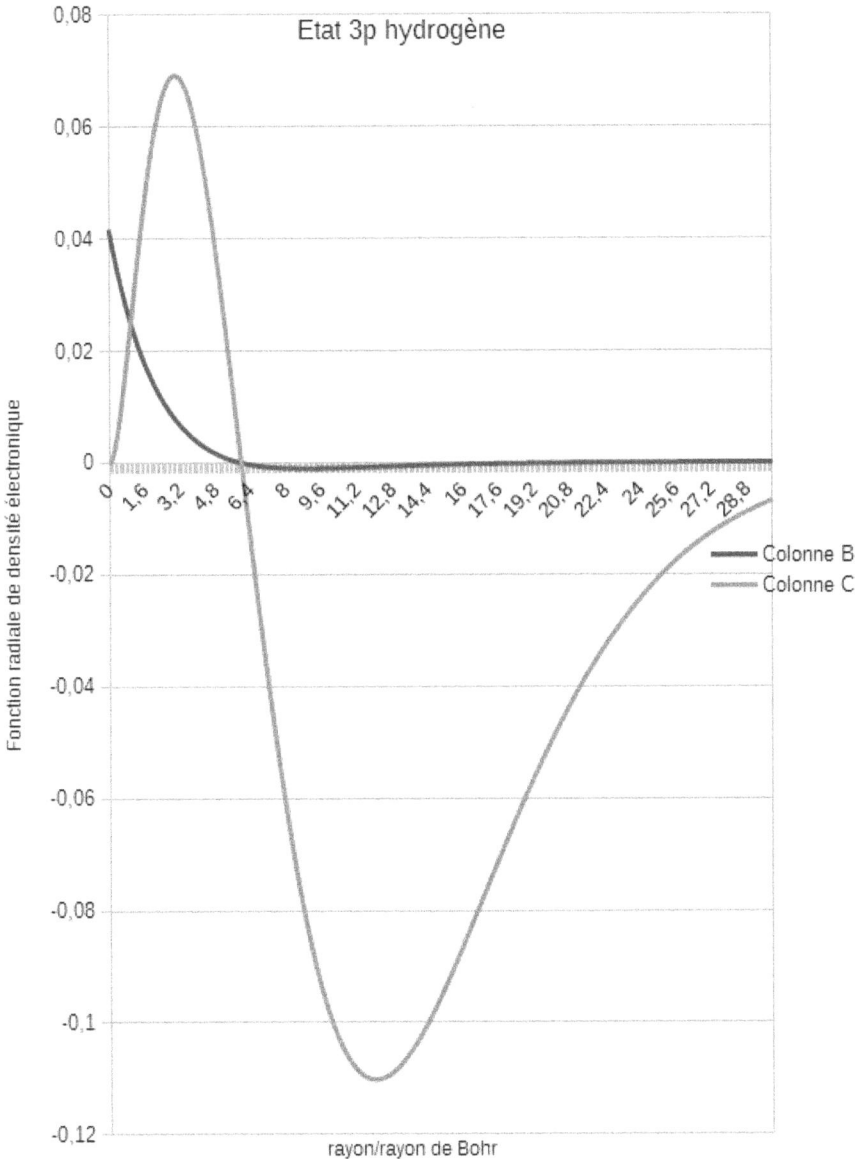

In the state 3p, the zero is at six times the Bohr radius. Moreover, the density is divided in angle in two lobes, separated by a null density plane. 3p ==> four phase zones, separated by two perpendicular surfaces of phase frontier. In University they carefully hid all that from you, as these stationary waves so precisely defined, it does not adjust to their "*probability of apparition of the gremlin and poltergeist corpuscle*". A warning is useful: Here we have solved the stationary and static Schrödinger equation, independent of time. So the flowing of time could be the one of the god of Isaac Newton. There is no propagation of the phase from one zone to the other in the electron; all is ideally synchronous. So is the simplified theory in 1926, before P.A.M. Dirac uses the relativistic frame.

Professor Castle-Holder:
- To complete this basis in atomics, you have to learn that the most the atoms are heavy, the most are electrically charged the nuclei, so the most the fundamental states dwell near the nucleus. So the remaining of the world does not notice them, and mainly discerns the most peripheral electrons. Very soon, the emission rays concerning the deep layers quit the visible domain, and belong to the UV and X domains. So is the Moseley[2] law: $\sqrt{\nu} = k_1.(Z-k_2)$ published in 1913, linking the atomic number to the frequencies of the rays K_α (and so the rays L_α and K_β too) for each element. k_1 and k_2 are constants depending on the kind of ray.

Now let us proceed to sketch the rules for building the Mendeleev table. Hydrogen and helium: one layer with only one sub-layer, which may receive only two electrons. Why two? Because the spin may have only two values, relative to the nucleus of the atom, and because two electrons never occupy the same quantic state. Each further sub-layer may receive four electrons more than the preceding, as it has new angular resources, so they are 2, 6, 10, 14, 18 and we do not go further as the heavy enough nuclei do not exist. So the second layer of electrons has a ceiling at 8 in two sub-layers: Lithium, Beryllium, Boron, Carbon, Nitrogen, Oxygen, Fluorine, and Neon. The third layer begins with the eight electrons in the two first sub-layers: Sodium, Magnesium, Aluminum, Silicon, Phosphorus, Sulfur, Chlorine, and Argon. Next comes a surprise, as instead of the third sub-layer, the first sub-layer of the fourth layer (spherical with max two electrons) begins to fill up, this gives an alkali, the Potassium and an alkaline-earth, the Calcium. Then only fills up the third sub-layer of the third layer, this gives the ten *"transition"* metals: Scandium, Titanium, Vanadium, Chromium, Manganese, Iron, Cobalt, Nickel, Copper and Zinc.

Only then resumes the normal filling of the fourth layer: Gallium (under the Aluminum in the table), Germanium (under the Silicon), Arsenic, Selenium, Bromine, Krypton. They count well as six elements. And the odd surprise comes again with the first sub-layer of the layer 5: Rubidium, Strontium, then ten transition metals, etc. Next in the 5th row, at n° 58 begin fourteen lanthanides. ... Why? According to the cost of energy for filling up such theoretical sub-layer.

Open-Eyes:
- We have reported in Appendix A a recapitulative table of the occupations of the electronic layers and sub-layers in the Mendeleev table, for the elements in their stable state, in non-ionized vapor. The table stops after the last element with at least one stable isotope[3]: the Bismuth. Beyond the Bismuth, no more

---

[2]Henry Moseley engaged in the World War, and was killed in 1915 at the Dardanelles.
[3]**Isotopes**: atoms where the nuclei have the same number of protons, but not the same number of neutrons. They do not differ by their chemical properties, only by the atomic weight (maybe by the nuclear stability).

elements exist with stable isotopes, only with more or less unstable isotopes. The reader has noticed that as we stopped so, the columns 5_3, 6_2 and 7_0 are still empty. The sub-layer 7s or 7_0 begins with the Francium (n° 87), which is an alkali; the 6d or 6_2 begins at the Actinium, transition metal as the Scandium and the Yttrium; and the sub-layer 5f or 5_3 begins at Protactinium. The series of 14 lanthanides begins with the Cerium and ends with the Lutetium. The series of 14 actinides begins with the Thorium and ends with the Lawrencium. The discussion is not yet clear cut to put the Lanthanum in the 15 lanthanides, and to put Actinium in the 15 actinides. The sub-layer with 18 positions of electrons 5g or 5_4 was not found in the stable electronic states, but it receives excited states and occurs in rays that the spectroscopists have observed. This table is copied from Hume-Rothery and Raynor, *The structure of metals and alloys*. The Institute of metals. Pp 14-16. One has noticed that the three elements giving good conductors, the group 1b, copper, silver, and gold, have only one electron in the peripheral layer. This property is still there in a metallic state: only one electron per atom in the conduction band; so this conduction band is only half-filled, that is the ideal condition for a good electric conductivity (and thermal conductivity, by the way).

Curious:
- Say! It is tedious to read, your table!

Open-Eyes:
- This is why chemistry is a so difficult and complicated science. Moreover, this table of orbitals, only valid for an atom alone, relates to a primitive state of the theory, and cannot take into account the hybridizations between orbitals of neighbor atoms in a molecule or a crystal; though only these hybridizations can predict the tetrahedral symmetry of carbon, silicon, and germanium when they are bound in a crystal, a quasi-crystal, a glass, or a molecule. On the tunnel effect, I had said the essential: in an electric contact, the electron crosses some air and the isolating oxides because they are very thin, hardly wider than the interatomic distances in the metal, and largely smaller than the phase[4] wavelength of conduction electrons; the electronic wave easily spans over such thin obstacles. Why this word "tunnel"? Like as if a tunnel was bored under this mountain, allowing some crossing through the potential barrier. Sure, a poor electric contact rises the impedance of the circuit, and consequently heats more than the remaining; there is a potential difference to pay to cross the contact for a given intensity.

Professor Castle-Holder:
- We are stuck. We must include a whole periodic table. The book is small, so we must split it into three parts.

---

[4]**Phase wave**: We owe to Louis de Broglie (1892-1987) in the years 1923-1924, to have distinguished the group velocity v of the electron, always lower than c (celerity of the light), from the phase velocity, always supraluminal. Their product is c^2. If the electron moves, its phase wavelength is h/mv, at non-relativistic speeds. Already calculated at § 2.1.5.

Group → 1   2   3   4   5   6   7
↓ Period

Period	1	2	3		4	5	6	7
1	1 H							
2	3 Li	4 Be						
3	11 Na	12 Mg						
4	19 K	20 Ca	21 Sc		22 Ti	23 V	24 Cr	25 Mn
5	37 Rb	38 Sr	39 Y		40 Zr	41 Nb	42 Mo	43 Tc
6	55 Cs	56 Ba	57 La	*	72 Hf	73 Ta	74 W	75 Re
7	87 Fr	88 Ra	89 Ac	**	104 Rf	105 Db	106 Sg	107 Bh

*	58 Ce	59 Pr	60 Nd	61 Pm
**	90 Th	91 Pa	92 U	93 Np

7	8	9	10	11	12	13
						5 B
						13 Al
25 Mn	26 Fe	27 Co	28 Ni	29 Cu	30 Zn	31 Ga
43 Tc	44 Ru	45 Rh	46 Pd	47 Ag	48 Cd	49 In
75 Re	76 Os	77 Ir	78 Pt	79 Au	80 Hg	81 Tl
107 Bh	108 Hs	109 Mt	110 Ds	111 Rg	112 Cn	113 Nh
61 Pm	62 Sm	63 Eu	64 Gd	65 Tb	66 Dy	67 Ho
93 Np	94 Pu	95 Am	96 Cm	97 Bk	98 Cf	99 Es

13	14	15	16	17	18
					2 He
5 B	6 C	7 N	8 O	9 F	10 Ne
13 Al	14 Si	15 P	16 S	17 Cl	18 Ar
31 Ga	32 Ge	33 As	34 Se	35 Br	36 Kr
49 In	50 Sn	51 Sb	52 Te	53 I	54 Xe
81 Tl	82 Pb	83 Bi	84 Po	85 At	86 Rn
113 Nh	114 Fl	115 Mc	116 Lv	117 Ts	118 Og

67 Ho	68 Er	69 Tm	70 Yb	71 Lu
99 Es	100 Fm	101 Md	102 No	103 Lr

4.1.2. Energy levels defined in a solid
. **Professor Castle-Holder**:
- Our Curious still has no idea of the pattern of electronic energies in a solid in *"permitted bands"*, divided by *"forbidden bands"*. Instead of few extremely fine levels as in the pure elements in vapor, the solid metals, and all the crystalline solids have these levels widely enlarged by coupling into **energy bands**. A dielectric (isolating) crystal has its last occupied band separated from the next permitted band by a big gap of energy. A semiconductor as the high-purity silicon has a forbidden band, a gap tight enough that the thermal agitation from the room temperature may be enough to throw some electrons over, into the next permitted band. Germanium has a narrower gap, so its conduction is more dependent on the temperature. So germanium is given up by the industry of the diodes and transistors – though it had begun with germanium. A true metal either has no gap as its conduction band is only half-filled – so are copper and aluminum – either has a narrow gap, depending on the overlapping of the bands according to the crystallographic directions. In this last case, the gap may be overcome by slight zigzags; this case is the most frequent. Examples: iron, calcium, magnesium, etc.
As stated above on the influence of atomic number and charge of the nucleus in the same column of the periodic table, the more we step down into heavy atomic numbers, the stronger is the metallic character: the gaps shrink. In column 4, the diamond is an excellent insulating material. The high-purity silicon is still insulating, but by adding a very few impurities, bringing either some electrons more – doping type **n** – either some deficit in electrons – doping type **p** – we make it a semiconductor. The germanium is definitely a more metallic element, and its thermal racing comes too easily. Heavier, the tin crystallizes preferentially in a metallic state than in diamond cubic. Some half-metals only conduct well the electric current in one direction. So is bismuth in monocrystal, or some antimony-bismuth alloys with less 5% antimony.

Open-Eyes:
- Thanks for this recalling. On may add that graphite carbon only conducts well in its base plane where its valence electrons are widely delocalized, but is insulating in the perpendicular direction, between dense planes. The patterns of energy bands depend on the crystallographic directions. The polycrystalline graphite is half-conducting: each grain conducts in its base plane, and the brain boundaries transmit from one crystal to another. Hence the carbon brushes, the carbon contact of the pantographs, the polycrystalline graphite electrodes for electrolyzing alumina.

Curious:
- You overdo! Indeed I already knew that the LEDs, the electroluminescent diodes that surround us and give us light rely on gaps of some electron-volts – 1.3 V to 3 V according to the emitted color – and that they light only when fed in the passing direction. I also know that none is made of silicon, but of more

unexpected semiconductors, still in diamond-type lattice crystals. Here are the most usual makings:

Color	Wavelength (nm)	Threshold (V)	Semiconductor
Infrared	$\lambda > 760$	$\Delta V < 1.63$	gallium-aluminum arsenide (AlGaAs)
Red	$610 < \lambda < 760$	$1.63 < \Delta V < 2.03$	gallium-aluminum arsenide (AlGaAs)
			gallium phospho-arsenide (GaAsP)
Orange	$590 < \lambda < 610$	$2.03 < \Delta V < 2.10$	gallium phospho-arsenide (GaAsP)
Yellow	$570 < \lambda < 590$	$2.10 < \Delta V < 2.18$	gallium phospho-arsenide (GaAsP)
Green	$500 < \lambda < 570$	$2.18 < \Delta V < 2.48$	gallium nitride (GaN)
			gallium phosphide (GaP)
Blue	$450 < \lambda < 500$	$2.48 < \Delta V < 2.76$	gallium-indium nitride (InGaN)
			zinc selenide (ZnSe)
			silicon carbide (SiC)
Purple	$400 < \lambda < 450$	$2{,}76 < \Delta V < 3.1$	
UV	$\lambda < 400$	$\Delta V > 3.1$	Diamond (C) aluminum nitride (AlN)
			aluminum-gallium nitride (AlGaN)
White	*Hot to cold*	$\Delta V = 3.5$	

Open-Eyes:
- Thanks for being so active! The electronic specialists also noticed that even for emitting in the infrared, the LEDs have a tension threshold well above the one of the germanium, 0.3 V, and even of the silicon, 0.7 V. So, at least one craft is warned of occupations of solids by electrons, in permitted bands separated by forbidden bands. "*Band*" is here far from any geometric meaning: it is only related to the Cartesian graph energy x occupation.

Professor Castle-Holder:
- I correct Mr. Curious: When there are two or more alternating elements, as in the gallium phosphide GaP, the geometry of the diamond cubic lattice is re-named into "*sphalerite cubic*". The mineral sphalerite is zinc sulfide ZnS. Moreover, not all the crystal lattices of this family are suitable to make a light-emitting diode: the energy of jumping must be emitted via a photon, not internally dispersed by phonons (elastic modes of vibrations). For more details about this physics, see the thesis from Daniel Ochoa:
http://daniel.ochoa.free.fr/These/PageThese.htm
and especially http://daniel.ochoa.free.fr/These/AnnexA.PDF.
Remember that a direct gap is necessary; « *direct* » means in the same crystallographic direction, while the silicon has only an indirect gap, with changing of crystallographic direction, so cannot emit a photon.

Open-Eyes:
- I am disappointed that this author (Ochoa) does not clearly precise the said crystallographic directions of the direct jump. Perhaps finding elsewhere?

Slightly less deceiving is the Ashcroft & Mermin[5] who quote the original publication from Hogarth, 1965, to which to report for a more detailed legend of the figures. The direction [0 0 0], I still do not understand...

Professor Castle-Holder:
- That is because of your reasoning as a pure crystallographer! The families of straight line directions [1 1 1] or [1 1 0] are familiar to you, but here it is the wave vector of the electron.

Open-Eyes:
- It would imply with [0 0 0] a null wave vector, a motionless electron. It does not fit with what we will see later: Fermi energies and Fermi speeds.

Curious:
- I do not follow you anymore. You have forgotten me.

Open-Eyes:
- Your objection is justified. We were just reaching the Fermi level of the electrons in a solid.
In this domain of the physics of the semiconductors, there is no difference between the calculations by the transactional microphysics and by the standard MQ: the calculation is undulatory. However, the semantics still strongly differs. Example: "*As the impurity is very localized in space, according to the Heisenberg principle of uncertainty, the uncertainty on the momentum of the trapped electron is very high*", page 263 in the thesis of Daniel Ochoa. Now I correct him: As the impurity is very localized in space, the properties of the Fourier transform induce a high indefinition of the momentum of the trapped electron.

Open-Eyes:
- Another surprise for the general public: in crystallography as in electromagnetism, I have used concepts and notations which are not those taught as standard: **outer projection**, **inner projection**, **inner product**, **outer product**, **equiplane**. I will not re-do the course here; I direct you to the course online: **http://deontologic.org/geom_syntax_gyr** . An **equiplane** is the equivalence class of the planes with the same plane direction. As for now, remember that neither the magnetic field $\check{B}$ nor the moment of a force, nor a spin are vectorial quantities, but are gyratorial quantities. The vectorial quantities have a straight line direction, and a sign of displacement along this line direction; though the gyratorial quantities have a plane direction (an equiplane) and a sign of rotation in this equiplane. Their symmetries are opposite, and their dimensional behaviors oppose them too. As a result, no "*magnetic monopoles*" may exist, the *hedge-hogs theorem* forbids it: it is impossible to comb a hairy sphere entirely.

---

[5]Neil W. Ashcroft, N. David Mermin. **Solid State Physics**. Harcourt Brace College Publishers, Saunders College Publishing, 1976.

Similarly, we define an **equiline** as the equivalence class of the straight lines sharing the same direction of straight line.

Fermi energy levels: The Fermi level is the uppermost of the occupied positions by the electrons of a metal. It is at 7 eV (above the most bound level) in copper, at 5.48 eV in silver, at 11.63 eV in aluminum. For the aluminum, the Fermi speed, which is the group velocity of each conduction electron in the metal, is 2.02 . 10^6 m/s or 2,020 km/s, measured at the room temperature. 6.7 thousandths of **c**, it is not usually considered as a relativistic speed, while the phase velocity is highly supraluminal.

Curious:
- Something is wrong in your affair: I have practiced electrotechnics, and I know that at the densities of current we handle, an electron does not progress more quickly than a ten of micrometers per second!

Open-Eyes:
- Indeed. It is the average drift, bouncing back from obstacle to obstacle. Be conscious that in metals, the conduction electrons are those with the highest energies, the less bound to each metallic ions of the crystal.

Curious:
- And what are these obstacles that make electrons bounce back like a flipper ball?

Open-Eyes:
- I fear it will drive us very far. Too far?

Curious:
- You are already so far into the too far...

Professor Castle-Holder:
- Before turning the page, let's give the reader an exercise to do. In any popularization videos, the general public has seen a planet-electron orbiting around a central star-nucleus. So it would have the symmetry of a turn, with a plane of turn, and an orbital moment in this plane. However, solving the Schrödinger equation has given for the fundamental solution a fuzzy sphere, without any orbital moment. When you dare to question them, the high priests of the corpuscularism reply that this orbital plane does not stop to change, so that in the long term, no trace of it subsists. Your mission, if you accept it, consists to conjecture the physics of this ever-changing plane. Good courage!

And you are not over, as at best so you may obtain a neat sphere, like a ping-pong ball. But to obtain the radial density given above by exactly solving the Schrödinger equation – which is widely confirmed by experimental results - your mission is indeed impossible.

4.2. The metals and their crystalline arrangement

Open-Eyes:

- Here we will see the impurity atoms which disturb the regular arrangement of the atoms in a perfect crystal, the dislocations, the grain boundaries, and the phonons, these quantified thermal agitations of the crystal, according to the temperature. We will not see the plasmons and polaritons here; only much later, when we will revise the photoelectric effect in chapter 11.

The general public has not precise ideas on the crystalline arrangements, so here is an image of a compact hexagonal, drawn as an assembly of hard balls; it is the lattice of zinc, cadmium, beryllium, magnesium, titanium:
Figure 4.7.

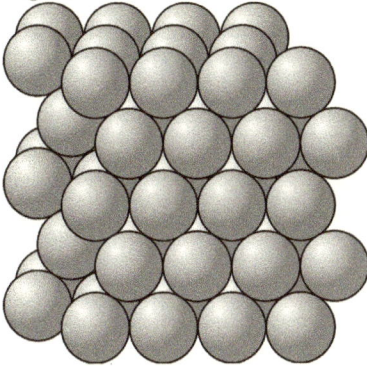

The plane of the sheet of paper is a compact plane: it represents the hexagonal plane where the "balls" occupy the plan at best. The horizontal and at 60° rows are compact straight directions. You may observe a similar compact plane on market stalls, for presenting the oranges. In compact hexagonal, the successive planes are in the sequence ababababab, etc. But many other metals have another compact sequence, with higher symmetry, said face-centered cubic or fcc, where the sequence of the compact planes is abcabcabc, etc. So is the case of all the coinage metals and good conductors, gold, silver, copper, aluminum, but also nickel, platinum, strontium, and even iron between 914°C and 1360°C, and austenitic[6] stainless steels containing nickel, the Hadfield steel containing manganese, used in the point rods, crossing frogs, and rivet setters.

Figure 4.8.

---

[6]**Austenite** or gamma iron is face-centered cubic. This lattice may be stabilized at room temperature by additions of carbon, nickel, manganese, nitrogen, copper, zinc, said gammagen elements, which increase the stability domain of the austenite.

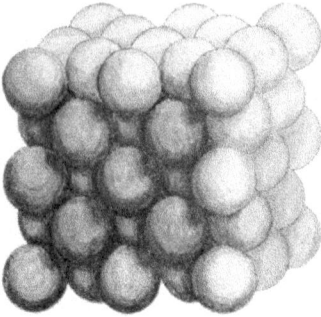

Fig. 35.

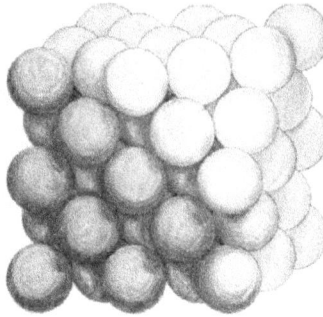

Fig. 36.

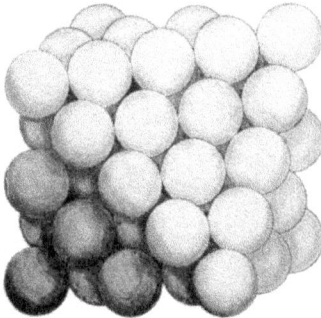

Fig. 37.

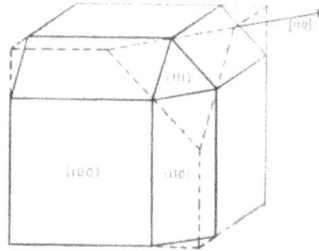

Fig. 38.

Figs. 35–38.—Unit Cell of Face-Centred Cubic Structure.

Certainly, these drawings are due to William Hume-Rothery, as they were in
The Structure of Metals and Alloys published by The Institute of Metals.
Cobalt is usually undecided between hexagonal and cubic stacking and may
present any variants mixing **ababab** with **abcabc** stacking orders. Otherwise,
many metals crystallize in a less compact lattice, said body-centered cubic
(bcc): all the alkalis, the iron above 1400°C or under 914°C (but it is unique:
because it is ferromagnetic at low temperature), vanadium, niobium, tantalum,
chromium, molybdenum, tungsten, barium.

Figure 4.9.

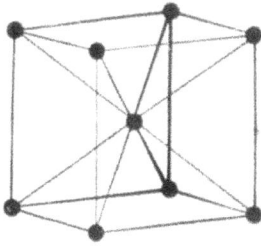

Fig. 39.

Fig. 40.

Fig. 41.

Figs. 39-41.—Unit Cell of Body-Centred Cubic Structure.

All these images with hard balls sacrifice to the ease of drawing. In the real world each atom reshapes and attracts the electronic clouds of its neighbors, so these electronic clouds compose a three-dimensional continuum. However, the image they give of the interstitial spaces is rather faithful. The interstitial sites in fcc are much larger than in bcc.

Curious:
- And again lots of new words! Dislocations, phonons, grain boundaries?

Professor Castle-Holder:

- First, we will make up a lateness: the ferromagnetism is a state where all single electrons unite to align their spins in the same equiplane in a magnetic domain. In ferrimagnetism, there is a majority population of spins and a minority. On the bulk, the majority wins, by large. The ferromagnetism occurs in iron and some alloys. The ferrimagnetism occurs in oxides and ceramics, whose the first known was the magnetite Fe_3O_4.

Open-Eyes:
- From the metallurgist point of view, a point is of importance: at room temperature, the iron is totally magnetized, but in small looping and closed domains which compensate each other in bulk. There are directions of preferred magnetization in the crystal, and ones have proved by electronic micrographies by transmission stripped patterns, by alternation between two preferred orientation, each approximating the externally imposed field. In the machinery of electrotechnics, which are naturally macroscopic, the boundaries of the magnetic domains move when externally solicited. Less compact, the body-centered cubic lattice is a high-temperature allotropic form. Only the ferromagnetism explains the unexpected stability of the body-centered cubic iron at low temperatures.

Curious:
- However, The Curie point of pure iron is 770°C, far below the transition of gamma iron to alpha iron, that occurs at equilibrium at 914°C. A 144° difference! How do you explain such a lacuna in your explanation?

Open-Eyes:
- The Curie point is from our macroscopic point of view, and is a mass effect: The Bloch boundaries between magnetic domains do not spontaneously move under the Curie temperature. Between 770°C and 914°C they are moving and fluctuating, at a rhythm far beyond the human experimental means, up to now; so we cannot obtain any macroscopic magnetic polarization; however this is enough to stabilize the alpha phase. Anyway, at cooling, the kinetics of the transformation of gamma iron to alpha iron is slow, as governed by the migration of the interstitial carbon atoms. This slowness allows the quenching of the steel, in numerous variants.

Curious:
- Dislocations? Phonons? Grain boundaries?

Open-Eyes:
- First, let us talk impurities: when you buy copper for its electrical conductivity, you pay it more dearly than copper for water or gas pipes, as you need high purity, with very, very little interstitial oxygen. This copper for conduction is carefully deoxidized. If you asked me the detail of the drawbacks of oxygen and different impurities for conductivity (and mechanical properties) in copper, alas I could not answer. But it is more complex than *"it obstructs the channels"*.

The conduction electrons are the less bound, the freer. In pure copper, their mean free path is about a hundred interatomic distances, and it lengthens at low temperatures. Add nickel, and the mean free path decreases: this atom has not the same diameter, nor the same electronic affinity, nor the same electronic charge. Few metals are good conductors, only those with one electron per atom in the conduction band: gold, silver, copper, aluminum. The alkalis, lithium, sodium, potassium, are very soft and flammable, so not usable.

The other imperfections in the metal are unfavorable to conduction too. The grain boundaries are the contact surfaces between crystals; in fact, our metals for technical usage are polycrystalline. The exception is some blades of turbo-reactors, which are monocrystalline to avoid the intergranular creep at high temperature.

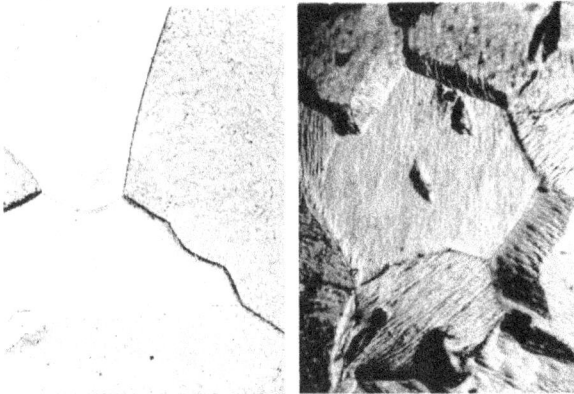

Fig. 192a.   × 25. Fig. 192b.   × 1000

Fer très pur résultant de la fusion par zone;   Attaque poussée d'un fer industriel.
   impuretés totales   quelques p.p.m.
               (comparer les grosseurs de grain : × 25 d'une part et × 1000 d'autre part).

Figure 4.10.
Only the relative scales are conserved by the reproduction. Extracted from **Traité de métallurgie structurale**, by DE SY and VIDTS. They are optical micrographies, on a metallography microscope.

The most often, the grains, or elementary crystals, have a size in the one micrometer to ten micrometers range, so often under the separating power of an optical microscope. By annealing, one can make the grain grow: the most perfect grain grows at the expense of the less perfect ones, as the *good crystal* has a lower internal energy than the *poor crystal*, and more than the grain boundary. Next, we have an electronic micrography by transmission through a thin plate.

The arrows point on the dislocations which cross the thickness of the thin plate.

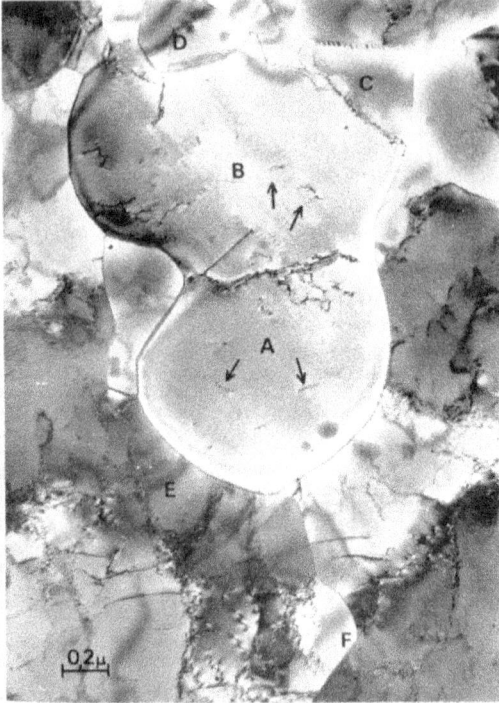

FIG. 246.   Micrographie électronique d'une lame mince de cuivre électrolytique laminé de 99 %
et recuit 7 heures à 78 C. Les blocs A, B, C, D, d'orientations très voisines sont en cours de
coalescence; certaines dislocations provenant de la désagrégation de la limite entre A et B, se
déplacent vers la frontière externe du nouveau bloc (d'après Mme Bourelier)

Figure 4.11. Extract from **Métallurgie Générale**, by Bénard, Michel, Philibert, and Talbot.

High purity aluminum, this one which has good resistance to corrosion by the quality of its adherent alumina layer, but is too soft for most technical uses, has grains about half a millimeter, so you see them all buckle differently (according to their crystalline orientations of easy slip) when you proceed a traction proof. And finally, some metallic crystals are so large as you can see them with the naked eye: those of a thin covering of zinc on electrogalvanized mobile barriers or on the poles for urban lights, or for street signals. This photo, anyone can do it:

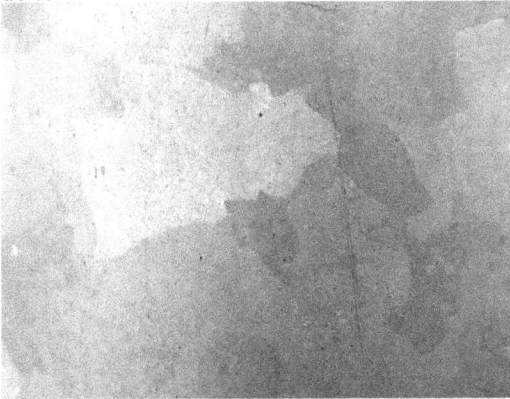

Figure 4.12.

When you solicit a metal in the plastic domain, it buckles by creating and circulating dislocations, which are linear defects in the metal. Here you have a modelization with rolls:

Fig. 235.— Representation by Means of a Series of Rollers of Dislocations Travelling Across a Crystal Plane.   [After E. N. da C. Andrade.]

Figure 4.13.

Next comes an electronic micrography by transmission in an aluminum alloy of type duralumin. The dislocations are the dark spirals (dark as less transmitting the incoming electrons, maybe dispersing them).

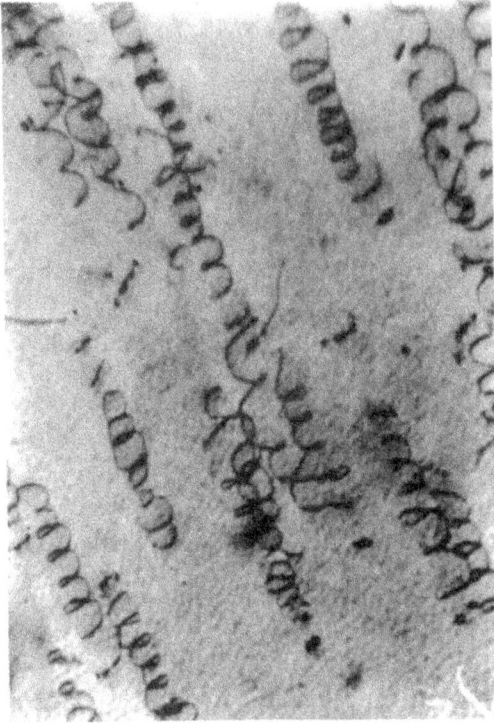

Fig. 279.—Aluminium 3% copper alloy
quenched from 550° C. into water at
room temperature. × 20,000.
(*Westmacott, Hull, Smallman, and Barnes.*)

Figure 4.14.

The edge dislocations are the boundary of a half-plane, existing at the right, and not at the left (if you prefer, under but not above).

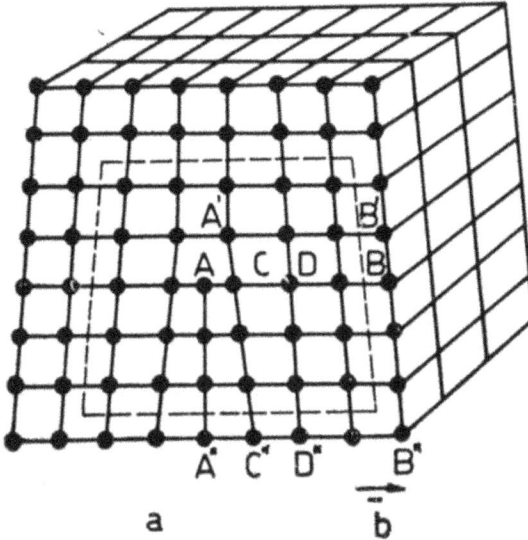

Figure 4.15.

The screw dislocations are like a helix staircase.

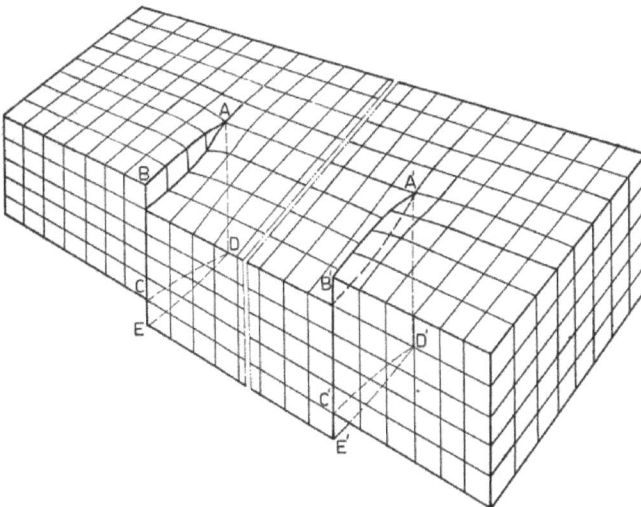

Fig. 30. Dislocation vis.

Fig 4.16.

Cold-formed metal, having been deformed by plastic work without heating, are less conducting than an annealed metal. Unlike what happens when you crush granite, the crystals of a polycrystalline metal remain bound together in the plastic deformation; the metallic bond and the metallic state remain through the grain boundaries. The metallic bonds are very little oriented.

Curious:
- Please do as if we did not know what a metallic bond is. Please recall that!

Open-Eyes:
- There are three kinds of strong chemical bonds and two weak ones.
1. The **electrostatic bond** is the most simple to understand. It bonds the Cl^- anions to the Na^+ cations in the kitchen salt. Or it bonds the cation Ca^{++} to the anion SO_4^{--} in the neutral molecule $CaSO_4$, a minority not negligible neutral ion in the sea water, strong ionic solution.
2. The **covalent bond**, or the homopolar linkage as in the diamond, and in all the carbon-carbon bonds in the organic molecules of our body. It is due to the sharing of a pair of two electrons, with opposite spins, and so with opposite magnetic moments. Similarly, the gaseous molecules of our atmosphere dinitrogen, dioxygen, or the molecule dihydrogen, the macromolecules of the molten sulfur, etc. These linkages are space-oriented, see for instance the four tetrahedral bonds for the carbon in the organic molecules.
3. The **metallic bond** is not or almost not space-oriented. The metallic atoms loosely hold their peripheral electrons, those with the highest energy, so they are shared by the whole solid, crystalline or amorphous. The metallic ions pack together in an arrangement as compact as possible, preferably compact hexagonal or face-centered cubic, but maybe body-centered cubic, according to the available orbitals and valencies. The electronic gas is mobile enough to conduct the electric current, conduct the heat, reflect the light, etc.
4. Mostly occurring in organic molecules and macromolecules, the **hydrogen bond** is the sharing of a proton H+ between two electronic clouds. From the hydrogen bond come the mechanical resistance of the collagen of our skin and bones, of the leather, and of the polyamides. It is also the basis of the "abnormal" properties of liquid water and ice, and of the liquid ammonia.
5. The weakest of all is the Van der **Waals bond**. It is of electrostatic nature, but quadrupolar, so only at very short distances and dependent on the flabbiness and plasticity of the macromolecules, apt to tightly model together. It is responsible for the resistance of the polypropylene and of the polyethylene, and anybody knows how they are very modest. One knows less that their efficiency depends much on the molecular weight, which must be high. So to have a useful material, you must obtain micelles as large as 400 Å or higher for their long axis.

Now the typology is set, most of the real cases are hybrids. In the silica, or in the alumino-silicic skeleton of the feldspars, the bond is partly covalent and partly electrostatic. In the silicon, the bond is partly metallic and partly covalent, and more metallic in the more conducting germanium; both crystallize in the diamond cubic lattice. In molecules with differing atoms, like CO and CO_2, the bond is not purely covalent, but partly electrostatic – so these molecules may couple their vibrations with resonating infrared photons. In all the at-the-disposal-of-the-oligarchy press, you endure lots of bombarding by tall stories and political extravagances on the capture of two infrared frequencies by the molecules of the atmospheric CO_2, to obtain your terror, and to submit you in full panic at the disposal of the bankster oligarchy.

About the phonons, they are individual waves of vibration or thermal agitation. They also exist in minerals, glass, and ceramics. The waves undulating on the crops by the wind give a good idea of them. Think of the Irish song: *The wind that shakes the barlow*. And why are they quantized? Because they turn around all the crystal, back and forth. In these mineral materials, the thermal conductivity and the thermal capacity are due only to phonons. In the metals, the conduction electrons intervene more and are dominating. At room temperature in copper, the mean free path of an electron before it is blocked and thrown back by a phonon is about 200 Å. The most you cool a metal and the most its conductivity increases: there are fewer phonons, and the mean free path lengthens. Moreover, some materials have the property of superconductivity, if the temperature is low enough: a wave associating two electrons with a tuned phonon runs freely without seeing any obstacle, and the resistivity drops to null. A phonon is always sampled on several atoms, many atoms, and never can become small. It is one of the reasons which compelled us to understand that a conduction electron never may become small; each one is several interatomic distances large, even several tens of interatomic distances. Even wider at low temperatures.

Curious:
- Unbelievable! It seems impossible. So your electrons are individually as big as a thousand atoms? Each?

Open-Eyes:
- Some more precisions for the conduction electrons in the copper at room temperature: the Fermi group velocity is 1,570 km/s, corresponding to 7 eV, the phase velocity is 57.2 . 10^9 m/s. On a free path of 200 Å (20 nm), so a duration of 12.7 fs (femtoseconds), the phase wave runs 0.738 mm; while the wavelength of this electronic phase is 4.6 Å. The spatial extension of this electron is about the same magnitude: few tens to several hundred ångströms. At each end of this propagator, there is an interaction with the more frequently a phonon, or a distortion of the lattice such as dislocations, lacunae[7], impurities[8]. Each electron occupies several tens of interatomic distances.

So they are not balls, they are not solids, nothing common with our familiar objects around us. Each one remains a wave, an electronic wave. Nothing holds them from occupying, simultaneously many, the same space which does not have the same features than our familiar macroscopic space (where only one solid occupies its own space). In the real microphysical world, so little familiar

---

[7]**Lacuna**: an atom is lacking in the crystalline lattice.

[8]An **impurity** in a crystal may be an atom substituted in the lattice, for instance, a zinc instead of a copper atom, or an insertion, like carbon or nitrogen in the interstitial sites between the regular atoms (used in carbo-nitridation of the steels for gears teeth). Frequently they concentrate on the grain boundaries.

to the layman, the only constraint on the electrons is that they do no occupy the same quantic state, as stated by the exclusion principle from Pauli: the electrons have a $\frac{1}{2}$ spin, so ruled by the Fermi-Dirac statistics.

And worse: it has been proved that obviously composed objects, such as protons, neutrons, helium atoms, fullerenes, and even insulin molecules, behave as their de Broglie waves, and are diffracted just like photons. It poses a serious problem in understanding what becomes our familiar geometry when extended to microphysics. My colleague was not trained to suspect this geometry, though we strongly suspect it.

Curious:

- Say! They are not big, your impurities you have accused to punch away or back the conduction electrons! So, to collide these small impurities, are your conduction electrons compelled to shrink to become as small as the stranger atom?

Open-Eyes:

- The premise of your reasoning is almost true: as it differs in size from the substituted atom, or as it is inserted, the impurity atom elastically distorts the crystalline lattice, up to two to three times its diameter around. Moreover, if it has a different valency, like zinc or tin in copper, it locally changes the electronic state, mainly the electronic density and the electron affinity. The impurities also block the run of the dislocations; so the use of alloys to obtain much more hard metals which at pure state would be far too soft (too plastic) to be useful. Examples: copper-tin, or bronze, copper-arsenic when tin is lacking, iridioplatinum, iridiogold, manganese-carbon-iron, etc.

Professor Castle-Holder:

- Would you give some exercises to the reader?

Open-Eyes:

- This time, I break the tradition: not a calculus exercise, but a documentation exercise. It is up to the reader to collect documents on the structural hardening of duralumin, alloy with 4% copper and a few magnesium (AU4G in AFNOR norm). Both aluminum and copper crystallize in face-centered cubic when pure, but their cell parameters slightly differ. The trick is a thermal treatment which efficiently blocks the run of the dislocations. It is up to the reader to find and compare these parameters, and to find which is the thermal treatment producing this optimal hardening. "Hardening" does not imply a change in the elastic constants, but hindering the beginning of plasticity.

In balance to the increased mechanical properties, which opened a wide use in aviation, the duralumin has poor resistance to aqueous corrosion, especially with marine water. The Alclad is a co-rolling: pure aluminum, then duralumin, next pure aluminum.

Professor Castle-Holder:

- I do not agree with your breaching the tradition. I give this elementary exercise: Given that the body-centered cell contains 2 atoms, that the cell-parameter of iron is 2.856 Å, and that the average molar weight of iron is 55.85 g, calculate the theoretical density of the pure iron, supposing it is only good crystal, without lacunae nor dislocations. The Avogadro constant was given earlier. That very pure iron has no technological use: it is far too soft. Excepted maybe for anticathodes for radiocrystallography of the phyllosilicates; it has only some scientific uses, in some laboratories of solid state physics. It costs more or less as dear as gold.

Transactional (Quantum) Microphysics, Principles and Applications

CHAPTER 5

Physical optics; the photon

Recalling the laws of optics: the refraction, the length of coherence which permits the interferences, definition, and properties of the photon, orthochronous causality form the emitter, retrochronous causality from the absorber, nature and laws of the handshake, de Broglie frequency, Planck quantum, Stern and Gerlach experiment. Quantitative laws of the astronomic optics.

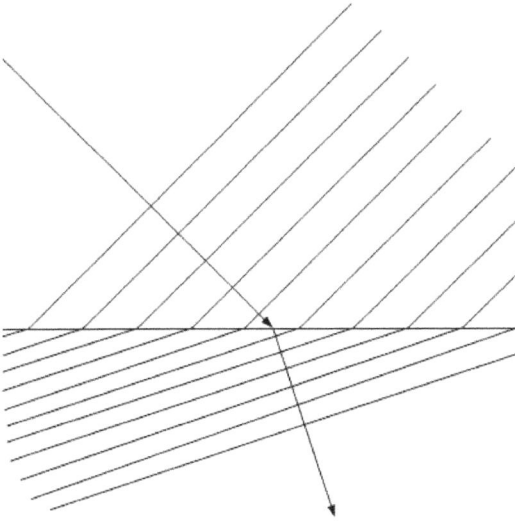

Figure 5.1.

5.1. The law of the refraction, Snellius and Descartes

Mr. Open-Eyes and **Professor Castle-Holder**, together:
- Let us begin with a recall: the refraction of a ray of incident light by a diopter under the law of Snellius and Descartes.
Here is drawn the incident light ray, coming from left and high in the fast medium, and meeting on the diopter plane another transparent medium where the propagation velocity is the half slower. Supposing the light is monochromatic, the distance between two wavefronts is divided by two; it alters the direction of propagation, but an invariant remains: on the diopter plane, the pace of the wavefronts is common to the two media.

Open-Eyes:
- You do not more have to worry how to make the refraction law *compatible with the photons*, as the photons are still a unit of light, still have wavefronts. . . In other words, this image is strictly the same with a collimated laser beam, or with one photon: no physical law of propagation has changed.

Professor Castle-Holder:
- You are satisfied with the drawing, but you should deduct the law of Snellius and Descartes, in general:

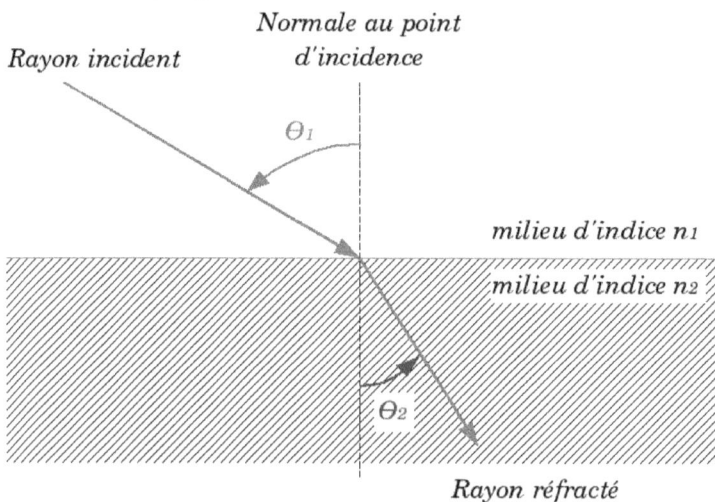

Figure 5.2.

The angle of incidence is from the perpendicular to the diopter. So we have the ϑ_1 angle in the first medium of relative index n_1, and ϑ_2 the angle in the second medium of index n_2. We write that the common property is the trace on the diopter plane: $\lambda / \lambda_d = n_1 . \sin(\vartheta_1) = n_2 . \sin(\vartheta_2)$ where λ is the wavelength in the vacuum, and λ_d is the trace of this wavelength on the diopter plane.

5.2. Airy's disc, and maximal resolution in optics

These discoveries in optics were made by astronomers, do not concern the photon directly, but collectives of waves. While the laws of diffraction, discovered by Fresnel and Arago, were known from about fifteen years, in 1835 the astronomer George Bidell Airy (101-1892) described the physical limit of any optical instrument, in "**On the Diffraction of an Object-glass with Circular Aperture**".

Already in 1828, John Hershell had described the appearance of a bright star, in an article of the Encyclopedia Metropolitana : « *...the star is then seen (in favourable circumstances of tranquil atmosphere, uniform temperature, etc.) as a perfectly round, well-defined planetary disc, surrounded by two, three, or more alternately dark and bright rings, which, if examined attentively, are seen* *to be slightly coloured at their borders. They succeed each other nearly at equal intervals around the central disc... »*

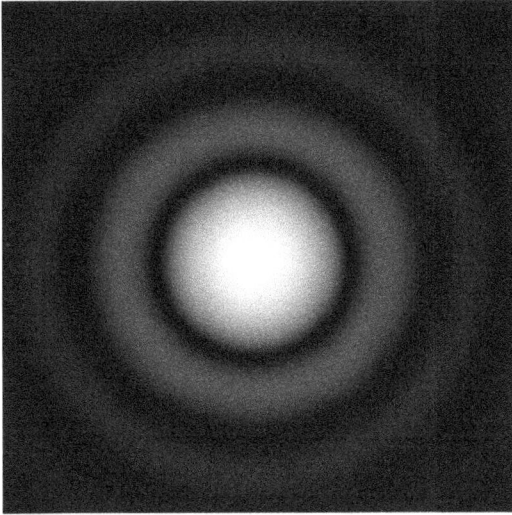

Figure 5.3.

5.2.1. Illuminance on the Airy spot.

Figure 5.4.

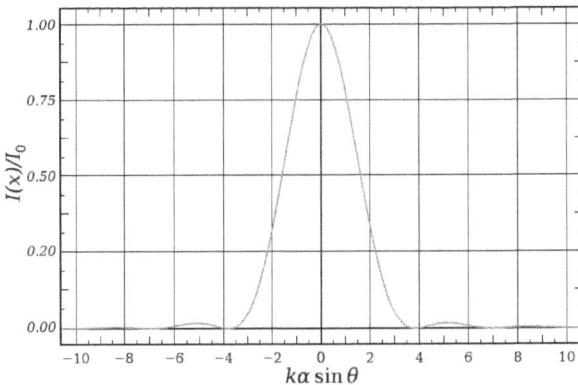

Figure 5.4 shows the illumination as the function of distance to the center of the Airy's disk. In Y-axis, is the intensity of radiation, as the square of wave amplitude. When the figure is observed far from the diffracting hole, the illumination varies in function of angle ϑ between the considered point and the center of the disk :

$E (x) = E_0 \left(\frac{2J_1(\pi x)}{\pi x} \right)^2$ with $x = \frac{d \sin \alpha}{\lambda}$ where E_0 is the illumination at the center; J_1 is the Bessel's function

[1] of the first order; d is the diameter of the opening; ϑ is the considered angle, with its summit at the center of the hole; λ is the wavelength of the light.

This equation is obtained by the theory of diffraction from Fresnel, applied when the object is on infinity. It corresponds to the square of the norm of the Fourier transform of the characteristic function of disc representing the opening. The diffraction figure may also be observed when placing at the center of a lens. So is the case when considering the diffraction caused by the aperture of this lens. In this case, one may use the approximation of the small angles. So one has the same expression for E(x), but with $x = \frac{rd}{f\lambda} = \frac{r}{N\lambda}$ where

• **r** is the distance, on the image, from the center.

• **f** is the focal distance of the lens.

• **N** is its numerical aperture.

The radius of the central spot is then 1.22 **N** λ at the first dark ring, and the diameter at half height is 1.029 **N** λ.

---

[1]In mathematical analysis, the Bessel's functions, first discovered by the Swiss mathematician Daniel Bernoulli, then generalized by the German mathematician Friedrich Wilhelm Bessel. Bessel developed the analysis of these functions in 1816 when studying the movement of planets in a gravitational field. These functions are canonical solutions y(x) of the differential equation of Bessel: $x^2 \frac{d^2 y}{dx^2} + x \frac{dy}{dx} + (x^2 - a^2) = 0$

for all real or complex number α. The most often, α is a natural integer (then said the order of the function), or a half-integer.

Drawing of the three first Bessel's functions of first kind **J**.

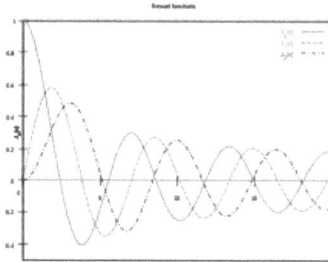

Figure 5.5.

Two kinds of Bessel's functions exist:

First kind of Bessel's functions Jn, solutions of the previous differential equation, defined in 0;

Second kind of Bessel's functions Yn, solutions which are not defined in 0 (but have an infinite limit in 0). The graphical representation of the Bessel's functions looks like sine or cosine functions, but damper as sine or cosine divided by a term similar to $\sqrt{x}$.

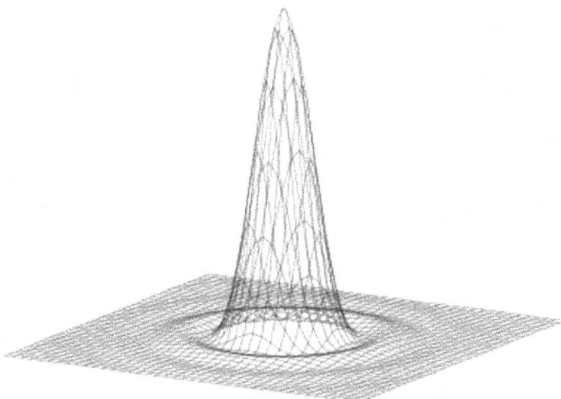

Figure 5.6.
Remarquable values

point	x	$E(x) / E_0$
At half height	0.514497	0.5
first zero	1.219670	0
Local maximum	1.634719	0.017498
second zero	2.233131	0
Local maximum	2.679292	0.004158
Third zero	3.238315	0

5.2.2. Resolving power. A most important consequence of the Airy spot is that it limits the resolving power of any optical apparatus, such as telescopes or cameras. One may calculate the Rayleigh criterion by this profile of the Airy spot, giving a limit of separation between two objects. This spreading of the image may be somehow compensated by numeric techniques of deconvolution.

5.2.3. A limitation on the performances of an optical system. The Airy's spot intervenes too in the reading of an optical storage (CD, DVD, blue-ray...). The storage capacity of these media depends completely on the diameter of the Airy's spot: this spot must cover only one track of the optical disk, and the larger this spot is, the wider must be the spacing between tracks, and the lower is the capacity.

5.2.4. The photographic systems. For photographic systems, the performances of the lenses are limited by two main factors: the optical design, and the diffractions. Though the main limitation comes from the opening, the real result is always compared to the diameter of the Airy's spot, which is the best possible result. We note **f** as the focal length, and **1/n** is the numeric aperture, so the diameter of the front lens is evaluated as **f/n**. Beyond **f/16** any optical system is limited by the diffraction but is limited by optical design in openings wider than **f/8**.

5.3. The telescope resolving, limited by the diffraction

Open-Eyes:
- The Airy's disk is inevitable, but can the astronomer at least reduce it? He does not master wavelength λ. He cannot so much widen the numeric aperture, as the aberrations become impossible to correct. So to widen the front lens, he must lengthen the focal length in the same proportion; so came the race to gigantism for astronomic instruments, which reached its limits in the middle of the 20th century. Two stars of the same brightness, whose angular distance is ε are just separated if $\boldsymbol{\varepsilon} = 12''/\mathbf{D}$, where $\boldsymbol{\varepsilon}$ is in seconds of arc, and $\mathbf{D}$ in centimeters. A front lens of 1 meter discerns two stars whose angular distance is more than 0.12 second of arc. With its entry pupil of 4 mm, the human eye is a well-built device, as it discerns about 30 ".
Going beyond the 5 meters of the telescope on Mount Palomar? But three exceptions, the answer was interferential: using several telescopes or radiotelescopes at some distance, even large distance, and recomposing the angular resolution by aperture synthesis.
Well! Surrounded by all these experimental results, by what trick could we insert a corpuscular theory in it? How to plug out the evidence that the laws of light and optics are undulatory, ruled in the first instance by Maxwell's equations? As seen above, no one of the laws of reflection and refraction, which found the astronomical instrumentation, are obstacles to our strictly undulatory definition of the photon: A photon is a successful transaction between three partners: an emitter, an absorber, and space or optical devices between them. This transaction transfers by electromagnetic means, a quantum of looping $\mathbf{h}$, and an energy-momentum that depends on the respective frames of the emitter and the absorber.
No matter I could wriggle out of it, the image of the Andromeda galaxy from my 150 mm mirror is poor and uncomfortable, compared to the one given to a more passionate and wealthy amateur astronomer by a *Dobson* with a mirror 240 mm wide, rectified at $\lambda/14$.

Curious:
- *Dobson*?

Open-Eyes:
- A *Dobson* telescope does not have any equatorial mounting, but is on smooth bearings, and is oriented by hand. They preferably equip it with wide eye-pieces with long focal and wide field, and use it on far objects from *the deep sky*. No mechanics nor motorization leaves more money to invest in the mirror and the focuser, especially in a larger mirror. So the observations are only with the eye, as photographing requires motorized equatorial mountings, with high precision gears. For astro-photographing, count three times the cost of the optics, for the mechanics.

Curious:

- *Aperture synthesis?*

Professor Castle-holder:
- Aperture synthesis or synthesis imaging is a type of interferometry that mixes signals from a collection of telescopes to produce images having the same angular resolution as an instrument the size of the entire collection. Of course, the quantity of received light is far below a hypothetical telescope of the diameter equal to the basis. At each separation and orientation, the lobe-pattern of the interferometer produces an output which is one component of the Fourier transform of the spatial distribution of the brightness of the observed object. The image (or "map") of the source is produced from these measurements.
To produce a high-quality image, a large number of different separations between different telescopes are required (the projected separation between any two telescopes as seen from the radio source is called a baseline) - as many different baselines as possible are required to get a good quality image. The number of baselines (n_b) for an array of n telescopes is given by $n_b=(n^2-n)/2$. For example, the Very Large Array has 27 radiotelescopes giving 351 independent baselines at once, so can give high-quality images. In visible optics, the largest optical arrays currently have only six telescopes, giving poorer image quality from the fifteen baselines between the telescopes. Aperture synthesis imaging was first developed at radio wavelengths by Martin Ryle and coworkers from the Radio Astronomy Group at Cambridge University. Bibliography: http://www.lchr.org/a/36/il/page3.html from *Astronomical Optical Interferometry*, A Literature Review by Bob Tubbs, Cambridge, 2002. Many military radars use the aperture synthesis, too.

Curious:
- Professor Marmot? What is your point of view?

Professor Marmot:
- Pff! So simple! I square your waves of yours, and I obtain the intensity, then the probability to observe a photon.

Open-Eyes:
- But which is the physical miracle that compels your corpuscles-photons to adhere a statistical law?

Professor Marmot:
- I do not bother with that! Our *Great Ancestors* put it as a postulate. This is enough for us.

Open-Eyes:
- Yeah! This is precisely one of the criteria for diagnosing a pseudo-science: their fanatical isolation from the remaining of the other scientific results. Anyway, the quantification comes back in each toggling of opsins in the rods and cones in the retina. And so on in any toggling of photosensitive receptors, in

any photo-excited chemical reactions, in the photosynthesis by the chlorophyll, etc. In any absorption of radiation, disregarding whether we are conscious or not, attentive or not. *In fine*, the quantification intervenes only at the emitter and absorber reactions. That's all.

Professor Castle-holder:
- But what is your synthesis, about these applications of physical optics to astronomical observation?

Open-Eyes:
- Never the laws of optics can compel the photon to transmute into some other thing than the Fermat's spindle, or a little multiplicity of Fermat's spindles between the absorber *and the emitter. But the absorber* and the phase constraint, nothing can restrain the width a Fermat's spindle. On the contrary, all the laws allow, and all the experiments confirm that the Fermat's spindles expand in width as much as the final phase condition allows. So we do not need any magic postulate, nor any magical and unknowable physics. Our only constraint is that the potential emitters are a heck of many in a star, and the potential absorbers are *a heck of many more times many more in the remaining of the whole Universe.*

The theoretical constraints have changed: the astronomical optics is essentially macrophysics, its material is large collectives of individual waves. So it was founded in considering that any absorbers *have the same value, and that only the emitters and space are causal. Only so!*
In transactional microphysics, we also have to explicit the conditions at the absorbers. *As soon as we have stated that at the center* of the Airy's disk, many are the transactions that may achieve between the star and the sensor, while at the Airy's minima the transactions are rather rare, we have to write down for each point of the sensor, of the photosensitive area, a transfer function, whose result looks like an admittance (the inverse of impedance), we could say **optical admittance**. This admittance rules the setting of the transactions; it depends on the polarizations; the question remains: in what physical unit to express it?

Professor Castle-holder:
- Do you anticipate a violation of the relativistic principle: c is the maximal speed for transfer of energy? As you have supraluminic phase velocities, and in your ground noise, half of the components are at reverse time.

Open-Eyes:
- Not any violation of Relativity is expected: all the retrochronous components under the Dirac equation are also at negative energy. We have not invented new particles; instead, we better describe the already known particles.

Professor Marmot:

- And what about the transmission of information?

Open-Eyes:
- Only the perceiving animals (eventually our artifacts) can define what is for him/her an "*information*". Not any law in microphysics deals with any kind of "*information*". And too bad for the over-mediatized hoaxes, as those from Stephen Hawking.

Curious:
- The popularization in astronomy likes to sum the telescope as a *funnel for photons*, silently thinking that the photons are very small. Let us fantasy the opposite: Do you consider incoming photons as large as the opening of the telescope, say five meters at the Mount Palomar, even ten meters on more recent and even more gigantic mirrors? Photons that would then be concentrated by the optical apparatus down to the suitable smallness?

Open-Eyes:
- Two optical limitations must intervene well before:
1. The defects in stigmatism of the mirror, so the image of the star is spread on the photosensitive area. We also have to proceed the inverse operation: given the size of the grain in photography, or of the photosensitive pixel for digital imaging, what is its angular pre-image? And in this silver bromide grain, or CCD sensor, we have to investigate at the molecular scale, or sub-crystalline to have the real size of the capturing site: the scale of the transaction on the absorber side. Luckily, the bosonic cooperation of the photons comes to help: they bunch, which may remedy a low yield of the sensor.
2. The atmospheric turbulence. It is no more to be demonstrated, that it is a major hindrance in astronomy. However, we should proceed a fine temporal investigation to see whether it is fast enough to perturb the passing of one photon, or perturbs only the crowd of photons, in one sub-cell of atmospheric convection. I do not dare to cut dry on this question. In doubt, I remain pessimistic.

Professor Castle-holder:
- In sum, you answer negatively, but cannot be more precise.
Here I share the hoax:
Santo Domingo - APF - 22 February 2007. An unusual accident happened on the avenue of 27 February in Santo Domingo (one of the main highways of the capital of Dominican Republic). This morning at 8 h 35 (local time) a car was suddenly smashed by a then-unknown object, causing a monster traffic jam in the center of the town. After a thorough inquiry, it seems that the smashing object was a very old photon, which had inordinately fattened during the past milliards years; some specialists evoke a "primal photon", one of these rare photons coming from the very first ages of the Universe one may sometimes observe in the Solar System. It is not yet understood why this "primal photon" could

end in the radiator grill of the car.

Curious:
- Maybe the chromes were tarnished? Not reflecting enough?

5.4. Prisms and gratings, monochromatizing devices

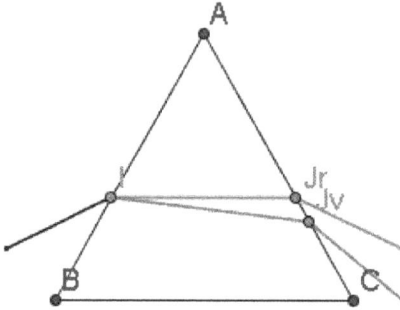

Figure 5.7.

Open-Eyes:
- Between a polychromatic light source, or maybe without a very defined emitted frequency, and absorber *not very selective in frequency, we may insert a prism, which spreads the beams according to their wavelengths, or more modernly and more efficiently, a diffraction grating. The result is that when transmitted to a precise zone of the screen, the photons have a fair resolution in frequency, though nor the emitter nor the absorber may be held responsible* for this fair resolution. It was the crossed-through medium, with its optical features, which has provided the selectivity.

It confirms that each photon is a three-partners transaction, including the crossed-through medium.

Also the neutrons, for fine radiocrystallographic experiments, may be monochromatized and focussed by an elastically bent monocrystal.

CHAPTER 6

Quantification. A real mystery? Or a fictitious mystery?

6.1. "Small grain of light?" Or what?

Curious:
- A photon, is it a small grain of light? No? Why not?

Open-Eyes:
- Ouch! Thanks for so recalling the extent of the disaster. For your release, I read on researchgate.net some researchers, who think to be experienced, who commit the same blunders as you. Please see the annex B and C: we must carry you from the basics, with electric fields, magnetic fields, propagation of the electromagnetic fields in radio-electricity, optics, and X-rays. Again for your release, among the scientists at the Solvay meeting in 1927, only Louis de Broglie had been mobilized in the radio-electric transmissions, only de Broglie had practical notions in wave propagation, antennas, and alternators which were then the generators of the carrier wave. And as in Brussels, he was beaten by the victorious gang, they never learned anything from him.

Curious:
- So a photon is not a small grain of light? Sooner you answered negatively.

Open-Eyes:
- Sure! Negative! To say it shortly: not only, during its propagation, a photon has a beginning, an end, and a duration in between, it stays an electromagnetic wave, but moreover, it has an emitter and a final absorber. These emitter and absorber have properties, and some properties are quantic, depend on the quantum of action-per-cycle **h**.
I have to articulate my answer in two times: the continuity with all the already known macroscopic electromagnetism on the one hand, and on the other hand all the coherent optics; even its historical approach with incoherent light, since the experimental results by Thomas Young in 1801, then a century of radiocrystallography.

Curious:
- Let us begin with the continuity. I presume you think the continuity with the practice of radio-electricity, as you have sketched out?

Open-Eyes:
- Exact. Not to be a monkey business, an electromagnetic theory must be valid all along the spectrum, from the electrostatics up to the gamma rays coming from nuclear reactions, passing through the whole domain of the radio-frequency waves. Now one end is surely quantic, at least with emission from nuclear reactions, while a very large useful domain is surely not quantic. In energies, the transition lies in the domain of the microwaves where some absorptions lines are due to resonance of rotating molecules and are surely quantic, while the remaining of the centimetric and millimetric waves are not. At the other end, in near gammas and hard X-rays, you have in Grenoble a remarkable facility, 268 m in diameter (320 in bulk), the ESRF, where you reserve your time of beam one year in advance, after about one year of preparation of the experiment you plan to achieve. Now in the synchrotron emission that each bending supplies, nothing is quantic at the emission time: the electron when transversally accelerated by the magnetic field had no stationary state before the bending, nor after the bending either. The experimenter selects which frequencies interest him (or her), by a filtering device, then eventually monochromatizes with an elastically bend crystal. However, all these receiving devices are bound by quantic laws and quantification, even when they just inform us about the spectrum of the emitting electrons beam bending.

Professor Castle-Holder:
- Please go up to the facts. We have the habit of saying virtual photons for electrostatics and magnetostatics. Any objections?

Open-Eyes:
- Objections. Experience taught to the radio-amateurs and to the antennas mounters that a Yagi antenna pinches the EM field for its profit, and bereaves the neighborhood. Have several antennas on neighboring roofs, and each one modifies the field available for the others: each one is a resonator on the frequencies band used in radio-broadcasting or television in this region. Pinching the field to its profit, then dissipating a part as losses in the coaxial cable, next on the input stage of the tuner, all the antennas work so, whichever be the designed-for range of frequencies, whichever directivity, whichever polarization. Now pinching the field for the resonant frequency, it is indeed what does a molecule of carbon monoxide, from which started all the investigation. If the phenomenon is the same – but few imperfections, easily understandable in radio frequencies – so the causes must be the same, strictly electromagnetic, without any "quantic" magic.

Professor Marmot:
- Your strange tales about antennas which pinch the field? Can you refer to papers published in peer-reviewed scientific publications?

Open-Eyes:

- Did you ever see antenna mounters who access to writing in peer-reviewed scientific publications? It is up to us to inquire ourselves. Now, the evoked technology has drastically evolved, and the witnesses to interview become old. If you can access to old collections, for instance, "Le haut-parleur" when it existed, look at the advertisements for the field-meters sold to the professionals, or look at the grid-dips (measures the resonance) sold to amateurs; they even build themselves their field-meters. Just pointing at sight the emitting antenna only gave a rough first stage; the surprises were many.

What Galileo learned in the shipyards in Venice, on the strength of materials, did he learned it in peer-reviewed physical reviews? Before Ignác Füllöp Semmelweis published his first paper on aseptic, the women of Wien who preferred to give birth in the street than in the clinic bossed by Dr. Johann Klein, where they were quasi-assured to die of puerperal fever, did they read or write in peer-reviewed scientific publications?

Professor Castle-Holder:
- The bulk of the technical knowledge, necessary to the reader, is in the Appendix B and C. Let us go on in the domain of microphysics, as you are willing to integrate the quantics into a more general microphysics, instead of leaving it a separate kingdom.

Open-Eyes:
- Every photon has a (fuzzy) length of coherence, and has a width of propagation, evolving along the propagation: this width is pinched at the two ends by the reaction of emission and absorption. Our curious is not guilty of his error: it is the everywhere taught blunder. In the real world, never the light ceases to be undulatory and electromagnetic. The physical optics from Fresnel, 1819, and the Maxwell equations, 1873, still apply always and everywhere. The only novelty, due to Max Planck in December 1900 for theoretical reasons, and extended in 1905 by Albert Einstein thanks to the features of the photoelectric effect, is that you cannot buy nor sell some electromagnetic interaction by less than one Planck's quantum of looping. Never less than $\mathbf{h}$, never else than $\mathbf{nh}$, where $\mathbf{n}$ is an integer number, and $\mathbf{h}$, is the Planck's quantum. This quantum of looping $\mathbf{h}$ proved to be a universal unit of account, all along the experimental corpus of the twentieth century.

Curious:
- So Einstein was wrong when he invented the "*small grain*" of light?

Open-Eyes:
- Affirmative. If each researcher wrote only one erroneous half-sentence per paper, it would be Land of milk and honey. I see lots of papers where the whole theoretical part is *strictly for the birds*. With Einstein in 1905, there was not more than a normal rate of errors for an excellent researcher. But the scientific community in its bulk is guilty: unable to correct a blunder contradicting so

much the whole experimental corpus.

Curious:
- How do you explain this error, if an error is?

Open-Eyes:
- In the universities, departments of hard sciences, no one learned the discipline of heuristics – the art of finding, to attain a "*Eureka*" which is not fallacious. It is never taught. What I have learned, it was by "compagnonnage" in a small company of inventors, by many readings, and in part at the CNAM (Conservatoire National des Arts et Métiers), in the unit of Technological Forecasting and Management of Research and Development (in a curriculum of managing engineer, micro-economist). So my scientific colleagues do not know how to explicit nor call into question loads of surreptitious and clandestine postulates. They were never taught to do, in contrast with some clever finders in biology (like Baruch Samuel Blomberg), learned to use a creativity and rigor which remain unknown by the physicists - still clung to the heritage of a Göttingen-København small pack which became hegemonic by the Solvay meeting in 1927. A methodical heuristician would explore many alternatives to the "*small grain*" of Einstein, and should find the solution about 1932. Instead, we independently discovered the basics of the transactional microphysics much later: partly Dirac in 1938, partly Wheeler and Feynman in 1941, finally Cramer in 1986, and myself in 1995-1998. When someone has professional training in heuristics – some renamed it *inventics* – this one far more quickly discerns the problems, and lists many solutions, which are then to sieve, while the specialized by the university remains the nose in its specialized handlebar, and does not see anything. The finder comes up, as he practices horizontal transfers of technology, from craft to craft, while the specialist too soon specialized *knows very much on very little, and at the limit all on nothing at all.*

Curious:
- And what do you prescribe instead?

Open-Eyes:
- Here is our definition of the photon, when a photon is indivualized:

6.2. A photon is a successful transaction between three partners: an emitter, an absorber, *and the space*, and/or the transparent or half-transparent medium in between, which transfers by electromagnetic means, a quantum of looping h, and an energy-momentum whose value depends on the respective frames of the emitter and the absorber

Professor Castle-Holder:
- Stop! Here you have introduced a big postulate: you postulate that each photon has an absorber! *You do not have any experimental fact to found on.*

Open-Eyes:
- Sure! A researcher who would not dare to take chances, he is worth a career officer who does not like the fight: he would be a crook. I have established one postulate, and on balance I have thrown away many others; the economy of postulates is on my side. All the *delayed choice* experiments are on my side. And they are a lot.

The Relativity taught us that a photon travels at a null proper time, so while the duration and the distance are of great importance for us, they do not have in the physics of the photon: the emitter, as well as the absorber, **are equally causal**. But the optical space between them, and its population of other photons, is causal too; if not, no more interferential astronomy with large bases! Too bad for our pride and our egocentrism, which are cruelly disdained by nature: no matter whether a photon was emitted fourteen milliards years of our frame ago, and will meet its absorber in sixty-five milliards years of our frame. We will repeat it in more familiar words:
You do not send a semitrailer of cement to cruise at random on the roads, in search of a customer who by miracle has precisely the right place in his hoppers for your twenty tons of cement, has the use of it, and his ready to pay the price of this delivery. You expedite only when you have a contract, also said an order, sent by telex or another sure mean, an order that you have accepted. You do not send a wagon of phenol in cruise at random on the rails, in search of a customer who, etc. The industrial buyers, as well as the industrial vendors, canvass, phone call, look through professional directories. But the inhabitant of Sirius, whose optical telescope does not discern smaller than lorries and wagons, does not suspect these quasi-immaterial commercial communications.

Professor Marmot:
- Ignorant! Heretic! Nothing such is in the ten volumes from Landau and Lifshitz! Moreover, you are far too dumb in maths to understand the Landau and Lifshitz!

Open-Eyes:
- In quantics, the equations for evolution are strictly determinist and strictly undulatory. We, transactional physicists, do not see as applicable the cruel uncertainty of the prophet Heisenberg to the trajectory, where the emitter and the absorber *hold each one like two telex in communication*. On the contrary, the Broglian ground noise, implied by the discovery by Louis de Broglie in 1923 (but a ground noise that himself never imagined), where sometimes emerges a successful transaction, escapes to all our means of investigation; we cannot predict its results, it seems to us a full random. It corresponds perfectly to the canvassing evoked above, by the industrial buyers and vendors. Moreover, since Dirac in 1928, we know that this ground noise comprises retrochronous components as well as the orthochronous ones. So it escapes to our macro-time.

Another way to say it: we do not more believe it could be possible to isolate a quantic system. While it was a sensible idea in macrophysics, often feasible.

First, we have rejected the postulate that the emitter could be isolated from the absorber: in our minds, no photon exists without a transaction between an absorber *and an emitter*. Though nothing such exists in artillery: the existence or nonexistence of the desired target, it matters on the intentions of the pointing officer, but has no influence on the mechanics of the gun (not even on the servants), it will shoot the same whether its shells fall at sea or on the targeted ship.

So was the rebuttal of the myth of the forward isolation, towards the absorber. Many "*delayed choice*" experiments confirm this rebuttal.

Moreover, I have rejected the postulate that the lateral or omnidirectional isolation exists; I state that no mean exists for shielding the Dirac-de-Broglie ground noise, and nobody has ever exhibited any experimental counter-example. In opposition, in macroscopic mechanics, one could not so badly isolate a gyroscope, including when the gyroscope is in charge of driving a compass, or an inertial navigation. At the quantic scale, all that is over: the lapping from the others is always of the same magnitude as the quantic particle itself. BUT, on the big numbers, this lapping is averaged out, and statistically speaking, one may forget the Broglian ground noise, and do as it did not exist. So are the laws taught in the lecture rooms.

Curious:
- Do you have proofs? Does it work well, your b..., err, business?

Open-Eyes:
- They are many, the experimental facts that the dominating coterie, heirs from Göttingen-København, carefully avoid, as they are incompatible with their corpuscularist ideations.

Curious:
- Corpuscularists? Göttingen-København?

Open-Eyes:
- Supporters for believing in the corpuscles. Niels Bohr acted in København (Copenhagen for the Englishmen), and the Knaben-Physiker – but Dirac, discreet dissident – all came from Göttingen University: Heisenberg, Jordan, Born...

Harasser Marmot:
- But who is this *science-venturer* who should never get out the hold of the cargo ship? If the highest scientists need corpuscles, so they must be! Moreover, if you speak of "*looooooong photons*", then you have never understood QED!

Curious:
- But your experimental facts? Coming?

Open-Eyes:
- All the spectral absorption phenomena. All. I had begun by a beginning: only one spectral absorption, the carbon monoxide in infrared. As it was for my personal instruction, here is the story: in 1995, our pupils in electronics had to study an apparatus for measuring and ring alert on the carbon monoxide in the air, for industries like petrochemical plants, or oil refineries. I was in charge to teach them all the complements in physics (quantic physics), on siderurgy, petrochemical industries, and in biochemistry – the hemoglobin and the fixation of CO on it. Now, the indubitable efficiency of the device with spectral absorption relies on the huge and mysterious capture cross section of this very tiny molecule: 3 Å of short axes, and 4.7 Å of long axis, which captures precisely the infrared photons of 2170 cm^{-1} of *wavenumber*[1], that means millions of times bigger than the molecule. No physical solution as long as I do not admit that the infrared wave converges on the absorber, *by its resonance in frequency. A solution mathematically unavoidable, but rejected with horror by nearly everybody, as it breaches our familiar* notions of irreversibility of our macroscopic time, and of our macroscopic causality, only orthochronous in a static universe where reigns the universal time of Isaac Newton and his god. However, the absorption of ONE photon by ONE molecule is not a statistical phenomenon, but an individual one. There, our familiar notions, though valid in statistical thermodynamics, have never been validated at the scale of the individual photon.

Another similar phenomenon is the selective absorptions of light by some organic molecules, and consequently, their spectral signatures, used in analytic chemistry. In the handbooks of organic chemistry, for instance from the Mir editions, a chapter is devoted to the dyes, and to the hypsochromic or bathochromic variants of each molecule already noticed as a dye, to produce the dyeing precisely asked by an industrial customer. Here again, they are very small molecules, compared to the visible photon they capture. The convergence of the photon on the dyeing molecule, somehow scarce between the micelles in the micellar matrix of high polymer, is unavoidable.

Even smaller are the **F centers** (F as Farbe = color) in the colored or even black crystals: they are lacunae, sizing at most the size of the missing atom, and containing one or few delocalized electrons. Now, these F centers absorb,

---

[1](2169.83 cm^{-1} for the isotopic main population $^{12}C^{16}O$, according to
http://nvlpubs.nist.gov/nistpubs/jres/55/jresv55n4p183_A1b.pdf)
Depending on the craft, two definitions for the **wavenumber**, differing by a 2π factor:
1. The inverse of the wavelength in the vacuum, written in how many wavelengths in the unit of length. To convert it in frequency, multiply by the celerity of light **c**. Here it is 4.608 μm wavelength, convertible in frequency: 65.05 THz. Context: for spectroscopists; for them, the unit of phase is the cycle.
2. For solid state physicists reasoning in Brillouin's zones, whom for the unit of phase is the radian, their wavenumber is the norm of the wave-vector. When multiplied by **c**, it gives the pulsation $\omega = 2\pi\nu$ **rad/s**.

so concentrate on them the resonant photons.

Professor Castle-Holder:
- Do you have proofs that these punctual defects, these F centers, are responsible for these light absorptions?

Open-Eyes:
- That is a very good question! So I verify in my handbook of mineralogy: in the biotite-phlogopite group, the correlation is loose between the global composition and the color. So there are in these sheets some punctual defects whose the global chemical analysis cannot account; some deviations from the stoichiometric[2] are compensated by lacunae. When you study the electrochemistry of the corrosion of metals, you learn that the rust of iron, mainly Fe_2O_3, is not only cracked, but also spread of lacunae, always out of stoichiometry. In the handbooks of solid state physics (Kittel for instance, and Ashcroft & Mermin), that is mainly alkaline halide crystals which were studied for their F centers, thanks to methods to produce these lacunae at will. So it was proved that the colored crystals are always less dense than the normal crystals: lacunae!
Other punctual defects are known, also responsible for the absorption of photons. Coming back to biotites, when weathered by the surface waters during some thousand years, they are hydrolyzed and lose their black color, turning into gold yellow, and next the clayish final result is hardly colored: the lacunae in anions have been replaced with water and OH^- hydroxyls, and some cations (K^+ or Mg^{++}) have been replaced by H_3O^+ hydronium. I have stated the discoloration of biotites into gold yellow in laterites from Madagascar: ferrallitic[3] weathering (or *laterization*) of granites. The biggest feldspars were hydrolyzed into white gibbsite, while the quartz and the magnetite withstood. Otherwise, the two main resulting minerals were kaolinite and an iron oxide (yellow goethite if the water was permanent, hematite where dry seasons were desiccant). Much

---

[2]**Stoichiometry**: a word composed in 1792 by Jeremias Benjamin Richter (1762-1807), from the greek *stoekheion* meaning element, and *metry*, measure. The stoichiometry studies the proportions in which the elements combine and the products come in a chemical reaction. The stoichiometric number of a chemical species in a chemical reaction is the coefficient which preceeds its formula in the considered equation (defect value = 1). Example: $CH_4 + 2\ O_2 \rightarrow CO_2 + 2\ H_2O$, where the stoichiometric number of the methane is 1, of the dioxygen is 2, of the carbon dioxide is 1, and of the water is 2.

[3]**Ferralsols** come from **ferrallitic pedogenesis**: The class of ferralsols has been created by soil taxonomists in order to group the soils which are commonly found at low latitudes and present specific properties related to genesis, geographic location, and management practices. These soils occur mainly under tropical climates, and cover extensive areas on flat, generally well drained land. They are considered as being strongly weathered, and to be associated with old geomorphic surfaces.
The criteria which have been selected to define the ferralsols relate to properties that are characteristic of strong weathering in at least one horizon: an almost complete decomposition of primary weatherable minerals and a clay fraction which is dominated by kaolinite and/or sesquioxides. All soils which have a horizon with such properties which is at least 30 cm thick are grouped in the ferralsols. The diagnostic horizon is called the oxic horizon or the oxic B horizon... (FAO, http://www.fao.org/docrep/x5867e/x5867e03.htm)

more details at this address: http://www.webexhibits.org/causesofcolor/12.html

Curious:
- It should be interesting to verify on other black minerals, such as cupric oxide CuO, magnetite Fe_3O_4, etc.

Open-Eyes:
- And to verify also on gunmetal finishings on steel, one using caustic soda, the other by phosphatizing, both giving a black surface. And a similar question about the green color which gave their name to *chlorites* (magnesian phyllites, 14 Å sheet).
Let us return to the thread.
Finally, we have for us the whole of the spectral absorption phenomena, from Mössbauer spectrography at one end, passing by the dark lines noticed in the spectra by the astronomers, and even by the exochemists, to the whole radio-electricity.

Curious:
- But it does not tell me why you already had grief against the *"grains of light"*.

Open-Eyes:
- On the one hand: plane polarized light exists; on several tens meters at the *Palais de la Découverte*, on several kilometers in the blue of the sky – and the bees exploit it. This fact is incompatible with the myth of photonic corpuscle. A corpuscle-photon could transmit a helicity, not a plane polarization, and the quarter-wave plates, transforming a polarization type into another, could not work. So the corpuscle-photon never exists, at any time of the propagation - and Einstein was wrong in 1905 when imagining the contrary. And what would be the physical laws of the transmutation from wave into corpuscle at the arrival, from corpuscle into wave at the departure?

Professor Marmot:
- But you have not proved anything! One may obtain any plane polarization just by superposing two opposite helicoidal polarizations!

Open-Eyes:
- OK, let us perform the calculation of coherence, that you should have already done several decades ago. You pretend that the erroneous half-sentence by Einstein is correct: *"Therefore, light travels in grains"*. So you make a pair of two corpuscles-photons, one twisting on the right, the other twisting on the left. First, we examine the journey. They must keep their (frequency) tuning in phase all along about 3 km in air, for the blue sky that the bees and the photographer exploits. Yet for a wavelength of 0.5 μm, it is twelve milliards of wavelengths. You will be given a discrepancy in phases of at most 20% of a quarter-wave; that is a frequency tuning of the two Siamese-twins photons better than $4 \cdot 10^{-12}$ in relative precision, or a coherence length about 15 km.

One may imagine that in a crystal, two atoms may be near enough, and accomplice enough to send a pair of tuned twin-photons of opposite polarizations, while it remains an impossible mission in the atmosphere, where the N_2 and O_2 molecules are independent, with incompatible speeds. And when one states plane-polarized lights in astronomy, I let you do the calculations of the frequency tuning and the length of coherence that your mythical Siamese-twins photons need.

Therefore, the corpuscle-photons do not exist, while the absorbers and the individual waves do exist.

Professor Castle-Holder:
- Let us admit that you have made your point, with polarized light. Now do not forget the other point you had let waiting.

Open-Eyes:
- On the other hand, the laws of the interference of light, learned and practiced when we were 18 years old for incoherent light, imply that each photon is undulatory, for each emission, and since Gustav Kirchhoff and Robert Bunsen in 1861, for each absorption too. The incoherent light was the only kind available in 1961: whether heated tungsten, or heated barium sub-oxide or ionized gas, each atom or molecule unexcitizes independently from the others, and each sends its own wave train – whose duration does not much exceed the nanosecond, for instance because of the free path in the gas in the gas tube. The lengths of coherence were known then (1961) by experiments with a long difference of path, and then they knew that each wave train from an atom had a ceiling of length about one meter, in the visible range. Source : D. V. Sivoukhine. *Cours de physique générale, tome 4 : optique*. Ch. 3, § 30 and 31. Éditions Mir, 1984.

Curious:
- I see: 1 m long corpuscles; it does not fit in the picture. But why in 1905 Einstein imagined corpuscles anyway?

Professor Marmot:
- But if they are not grains, why quantic? There are no other solutions!

Open-Eyes:
- Haha? But how do you do with the synchrotron emission, as the ones provided by the ESRF in Grenoble? How do you put and define your *small grains* whose frequency cannot be defined by the emitter? Please answer, professor Marmot!

Professor Marmot:
- So you insinuate that it is the recipient who defines the frequency, then?

Open-Eyes:

- Not on the broad range, but on the detail, yes. It was fully written in the definition of the photon, above.

Curious:
- At the beginning of the discussion, you had postponed the twentieth postulate – that you qualified as *"surreptitious and clandestine"* - which you transaction-ists have rejected:
Göttingen postulate. Only *"states"* exist, forget the transitions.
It is time to return us to it now. Here is the heart of the matter: for a spectro-scopist an emitted or received photon is a transition from one stationary atomic state to another.

Open-Eyes:
- Already in 1927, when returning perplexed from the Solvay meeting, Erwin Schrödinger wrote: *"What an odd physics, who concentrates on the states, and omits the transitions!"*. The durations and the properties of the transitions, such as the lengths of coherence of the photons, as revealed by the interferences described for 1802 by Thomas Young, are incompatible with the corpuscularist postulate.
In a standard handbook, you could find the detail of the hyperfine structure of the state **1s** of the hydrogen atom: the base level **1s** is split in two by the spins of the proton and the electron. They cost slightly more to be parallel than antiparallel. Thanks to the hydrogen maser, we know with high precision the difference of intrinsic frequencies between those two states: 1,420,405,751.768 Hz, in the precision of the last significant digit. Moreover, we know that the transition is officially *"highly improbable"*, as the mean duration of the state S = 1 is about $3.5 \cdot 10^{14}$ s, that is about 100,000 years. But even under torture, they will never say the magnitude of the duration of the photon emitted at that frequency, corresponding to the line at 21 cm, highly scrutinized by the radioastronomers. It is because in the Göttingen-København tribe, it was ad-mitted at the beginning of the 20th century, that the photons are instantaneous grains, and that also the electrons are punctual grains.
One who has a minimum of practice with the lasers, knows that not any laser cavity is built, nor even definable with such level of precision, 10^{-13}, very far from that. So it is indeed the collective of photons which has an enormous intrinsic precision, so each photon is very long: say at one milliard periods, it lasts 0.7 s, on a length about 210,000 km, that is a plausible magnitude. Indeed, the *"very long lifetime"* of the state with parallel spins implies an ex-cellent definition in energy and frequency of the initial state; concerning the final state, it has no threats on its lifetime – virtually infinite – so not either on its excellent definition in frequency. So both have a frequency definition better than the millihertz. The duration of the photon gives us lower bound of the du-ration of the average free path in that kind of interstellar gas, emitting on 21 cm.

Professor Castle-Holder:

- "Lower bound", in french "*minorant*" seems unfamiliar to the laymen. It denotes a number or a quantity which you are sure it is smaller than the not-so-known quantity you wish to determine. About the mass of the neutrinos, we have some *majorants*, upper bounds. Up to now, the stake is to find the smallest possible upper bounds.

Open-Eyes:
- Another indirect proof of the great lengths of these photons in the decimetric range is the success in radio-astronomy of the interferential interferometry with aperture synthesis on large bases. It relies on the gregarious bunching of the photons on astronomical distances, from "neighboring" sources, "neighboring" at the astronomical scale, of course.

Professor Castle-Holder:
- I hope that now, our curious reader is no more confused by this habit of our dissident Open-Eyes: for him, saying mass-energy or frequency in a relativistic frame, is saying the same, as the Planck's constant **h** is universal; only the units differ. He has integrated the fact that in a relativistic frame, a zero of the energies exists, and that the de Broglie's frequency at rest is always $\mathbf{mc^2/h}$. In a non-relativistic and ante-Broglian seeing, this zero of energies did not exist, so you were only allowed to speak differences of energies.

Open-Eyes:
- Only in 1924 with the thesis of Louis de Broglie, we had a beginning of the solution, and the remaining with the paper from Uhlenbeck and Goudsmit (both Dutchmen) in 1925, with the Schrödinger equation in 1926, then updated in 1928 by P.A.M. Dirac. I give the shortcut first, and the demonstrations later: at the low energies that the spectroscopers state, the emitting atom or molecule, as well as the receiving, are bound to quantification rules, that are inescapable. They can only jump from one stationary state to another stationary state, which are almost stables or fully stable. Each of these states is of long duration, compared to the duration of the emission or the reception. In 1905, to postulate grains was a forgivable error. A hundred and thirteen years later, in 2018, this blunder is no more excusable; however, it is still in power. So at the two apexes of its existence (at least at one), the photon is bound to the quantifications by the stationary states of the atoms. Moreover, in QED (Quantum Electrodynamics), it is said that space also demands a quantification for the electromagnetic field, so to the photons.

Curious:
- Please precise what is a stationary state.

Open-Eyes:
- The state of a stationary wave, like a cord of guitar. All the electrons which surround a nucleus in an atom or an ion are **stationary waves**, which last

between two events like a collision.

Curious:
- But *wave* of what? *Stationary wave* of what?

Open-Eyes:
- Of electron. Wave of electron. There are no two distinct *hypostases*, with one being the electron and the other being the wave. They are exactly the same thing. For 1911, they had a planetary model of the atom, where the electrons were small balls which orbited like planets around a nucleus; it was the Rutherford's model. In 1913, Niels Bohr stayed on the planetary ideation, but added the hypothesis that those orbits are quantified, and set three main quantic numbers, all integers, which resolve the first big problems of the spectroscopy. Since, physicists and chemists have replaced those nonexistent "*orbits*" by the fuzzy concept of "*orbitals*", which denotes the habitus of the electron around an atom or a molecule. The Schrödinger equation, next updated by Dirac, has no other bound and stables solutions other than a discrete set of solutions; all are stationary solutions for the whole of electronic waves for the considered atom or molecule. What makes the spectra, whose laws were established along the 19th century, and at the beginning of the 20th for the hyperfine structure of the spectra: in a spectrum, each emitted or absorbed photon is the difference between two stationary states. Electronic states for the atomic physics, nuclear states for the gamma rays emitted or captured by nuclei.

Example of a simple spectrum, this of the cesium in vapor, in the visible domain: Figure 6.1.

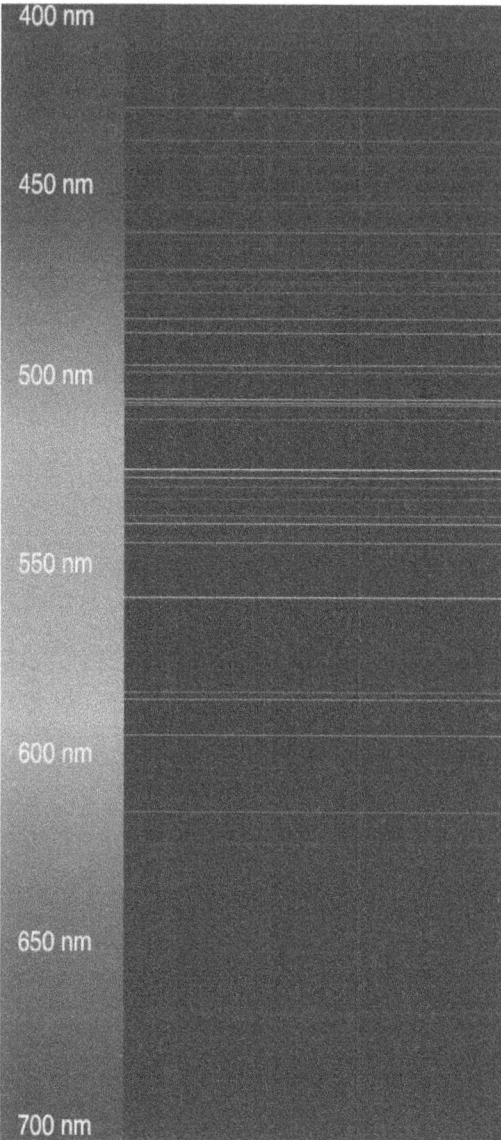

It took a whole century of battle, from the beginning of the 19th to the beginning of the 20th (1913: Jean Perrin) to obtain the admittance in chemistry of the existence and the properties of the atoms. Another challenge, begun soon in the 19th century, was the observation through the spectral decomposition of light with prisms, of spectra-lines with flames, and of dark absorption spectra-lines in the spectrum of the Sun. The first great pioneers: Wollaston, Fraunhofer, Bunsen, Kirchhoff...

Historical reminder: The spectra were already described by Newton. The spectroscope was invented in 1802 by William Wollaston; he discovered that the Sun spectrum was spread with dark lines, but he thought they were just for

delimiting the colors. In 1811 the German optician Joseph von Fraunhofer performed the first spectral analysis of the Sun, he listed 600 spectra-lines. Since
then, the solar spectrum is said the Fraunhofer spectrum. Nowadays, more than
26,000 spectra-lines are listed in the solar spectrum, whose 6,000 are due to iron.
http://www.astrosurf.com/luxorion/spectro-principes.htm http://pagesperso-orange.fr
Read at http://www.astrosurf.com/luxorion/Sciences/kirchhoff-lines-sun.pdf, the
paper from Gustav Kirchhoff, 1866, concerning the link between absorption
spectrum and emission spectrum. It established the equivalence between laws
of absorption and laws of emission and the *complementary symmetry of the
absorber and the emitter of an electromagnetic radiation.*
The spectrum of iron is much richer than that of cesium, so it is used for
the calibration of each film, aside the unknown spectrum to analyze. It is
completely listed in tables. The frequency of each emitted frequency is the
difference, the beat of the initial and final intrinsic (= Broglian) frequencies of
the atom.

6.3. Beginning of the undulatory mechanics in microphysics

In 1923, when preparing his thesis, Louis de Broglie remains in the planetary
ideation, without suspecting he will supply enough to dynamite it. He begins
by joining together the two formulas from Einstein and Planck-Einstein:
$\mathbf{E} = \mathbf{mc^2}$ for anything which has a mass,
$\mathbf{E} = \mathbf{h.\nu}$ for the photons, which are massless, where ν is the frequency.
Hence the de Broglie intrinsic frequency for all particles with mass:
$\nu = \mathbf{mc^2}/\mathbf{h}$. This frequency is intrinsic; it depends only on the mass, and is
evaluated in the frame of the particle. It remains ferociously banned by all the
clergy, heir of the Göttingen-København sect. Never the students are informed
of it. Never: *Strengt verboten* !

All the spectral lines the astronomers, the physicists and the chemists in the
19th century have observed and listed come from transitions between two stationary states apt to last. For emitting a photon, the initial state had the more
energy, so a higher frequency than has the final state. At absorption, it is the
reverse: the final state is the more energetic.

Curious:
- And what about the magnitudes?

Open-Eyes:
- Broglian frequency of the free electron: $123.56 \cdot 10^{18}$ Hz.
To easily generalize this calculation, we store the intermediate quantity:
$c^2/h = 135.639 \cdot 10^{48}$ kg^{-1}.s^{-1}. So we will have only to multiply by the mass of
any object to obtain its Broglian frequency.
Broglian frequency of the neutron: $227.2 \cdot 10^{21}$ Hz.
Broglian frequency of a 238 uranium atom: $54.1 \cdot 10^{24}$ Hz
Broglian frequency of a fullerene (60 carbons): $163.6 \cdot 10^{24}$ Hz.
The experiments of interference with fullerenes have been done, and confirm

the reality of the spatial wavelength according to de Broglie, for such a big molecule (720 units of atomic mass).

Curious:
- Do they? Interferences? Do you mean that this fullerene wave shares on all the planes of the grating?

Open-Eyes:
- At least on five to six of them.

Curious:
- That the helium atom passes through both holes simultaneously? Are you kidding me?

Professor Marmot:
- You see that one must avoid popularizing! This layperson is completely lost.

Open-Eyes:
- The problem is that you want to extrapolate to the atomic scale, even below, our familiar macroscopic geometry, which is reliable at our scale, and without caution, you pretend to apply it below its horizon of competence.

Professor Marmot:
- What? Never heard that! We, we descend down to the Planck length.

Open-Eyes:
- You do, we do not. I return you the positivist argument around: can you experimentally set your mythical problem of a Planck length? We had the same problem of size with the conduction electrons. Our choice, which is not yours, is to stop to blindly trust in this geometry we had learned in class; we set the technical specifications for what is to elaborate instead, for the needs of the microphysics. Instead of accusing the helium atom to have a treacherous double-face nature, *either corpuscular, either wavy*, we move the accusation: it is the macroscopic space-time, as a statistical emergence from **all** the interactions between all the quantic entities, which diverges in its properties; these properties are different if we apply it nearby its lower horizon of competences, and differ from and even contradict what we are familiar in our world of macroscopic animals.

Professor Castle-Holder:
- Maybe we should risk a neologism: the **macro-time**? In 1905, Albert Einstein had already split the *divine* and *universal* Newtonian macro-time into as many local proper times as they were frames, whose only the frames mutually immobile are mutually synchronizable. In 1915, with General Relativity, again a new splitting: the proper times differ along the gravity potential. Further, you

observed that each photon drills a hole or a short-circuit in each macro-time, including the laboratory one.

We could reserve the expression **micro-time** to the time of a particle, say a gas molecule, an atom, or one of the electrons according to the orbital it occupies.

Open-Eyes:
- Accepted! Macro-time, micro-time, we will precise them in the chapter for the relativistic frame. I come back to the historical thread. Then de Broglie[4] worries how, relativistically, an observer sees this intrinsic frequency when an electron passes in front of him. So he reaches his **theorem of Harmony of Phases**, implying that the electron occupies a noticeable extent, which he is not able and not even willing to evaluate. Hence, he deducts the celerity of phase of the electronic wave, so the product of group speed by the phase celerity is always c2. Hence, he deducts the wavelength of this phase, when the electron has a speed. As this last result looks no more relativist, this is the only one that the Göttingen-København clergy has not **thrown into the *Memory Hole*** (according to George Orwell).

Professor Castle-Holder:
- I interrupt here your indictment, as it is time to let Louis de Broglie tell it himself, expose his Theorem of Harmony of Phases. After pleading the necessity to join together the theories of Relativity and the Quanta, as the exchanges of energy occur by quanta in the photo-electric effect, Louis de Broglie writes:

> *Hence it can be conceived, according to an important law of nature, that to each bit of energy of the proper mass m_0 there is connected a periodic phenomenon of the frequency $\nu 0$ stated thus, $h\nu_0 = m_0\ c^2$, ν_0 being measured of course, in a system which is at rest with respect to a certain amount of energy. This hypothesis is the base of our system, and is worth, as are all hypotheses, what the conclusions which can be drawn from it are worth. Further in his work, he explains what induces that this periodic phenomenon is not to be considered as confined: so it should be a wave, propagating in space.*

De Broglie explains his conception of the (electronic) atom or particle: "*What characterizes the electron as a particle of energy is not the small place it occupies in space, and I repeat it occupies it entirely, but the fact it is insecable, indivisible, that it forms one unity.*" Then he presents an apparent contradiction between his hypothesis and the relativity, saying it was "*a difficulty which had long intrigued me*". Implicitly he deals the case of the plane wave. Notation: β = v/c.

---

[4]"Broglie", or "de Broglie"? The rule is: when the name is monosyllabic, keep the preposition. Now the Piemontese « Broglia » is monosyllabic, and is pronounced « Broï » in french. A doublet coming from the gaulish « *broglium* » is « Breuil », meaning a grove of trees.

The Lorentz-Einstein transformation of time teaches us that a periodic phenomenon connected with a body in motion appears to the fixed observer to be slowed down by a factor 1 divided by $\sqrt{1-\beta^2}$ ($\beta = v/c$), which is the famous slowing down of the clocks. Then the frequency observed by the fixed observer will be

$$\nu_1 = \nu_0\sqrt{1-\beta^2} = \frac{m_0c^2}{h}\sqrt{1-\beta^2}$$

On the other hand, since the energy of the moving body is equal to $\frac{m_0c^2}{\sqrt{1-\beta^2}}$ for the same observer, the corresponding frequency according to the relationship of the quantum is $\nu' = \frac{m_0c^2}{h\sqrt{1-\beta^2}}$. The two frequencies ν_1 and ν' are essentially different as the factor $\sqrt{1-\beta^2}$ is not involved in the same way. This is a difficulty which has intrigued me for a long time; I have succeeded in eliminating it by demonstrating the following theorem I shall call the theorem of the harmony of phases: the periodic phenomenon connected with the moving body whose frequency is for the fixed observer equal to $\nu_1 = \frac{m_0c^2}{h}\sqrt{1-\beta^2}$ appears to be constantly in phase with a wave of frequency $\nu' = \frac{m_0c^2}{h\sqrt{1-\beta^2}}$ emitted in the same direction as the moving body, with a velocity $V_\varphi = \frac{c}{\beta} = c^2 / v$. The demonstration is very simple. ...

The phase velocity V_φ is faster than the light. And it is the phase frequency ν', the frequency of the wave, which verifies h ν' = E.

Open-Eyes:
- The group frequency has been proved in the experiment performed in half-clandestinity at the Linear accelerator of Saclay, by the team conducted by Michel Gouanère:
http://aflb.ensmp.fr/AFLB-331/aflb331m625.pdf
In statics, the group frequency is obvious in any atom, and the electromagnetic radiations it emits or intercepts: the most an electron is bound, the lowest is seen its intrinsic frequency, the from the outside. The group frequency requires relativist speeds to be showed off in dynamics; though the usual experiments of diffraction put in evidence the phase wavelength, without the need of relativistic speeds.

Professor Castle-Holder:
- The geometric mean of the two speeds is the velocity of light: $\mathbf{v_g}.\mathbf{v_\varphi} = \mathbf{c^2}$. The theorem de Broglie names « Harmony of phases » points that the phase does not change when changing of frame: $\mathbf{\nu_0}.\mathbf{t_0} = \mathbf{\nu'}\left(t' - \frac{\beta.x}{c}\right)$. This property is demonstrated by two ways, whose one uses the Lorentz transform: it is a phase wave. De Broglie does not set any other hypothesis on the wave, excepted it is **not** the electron itself. He persists in thinking the electron as a

corpuscle, and with spherical symmetry.

Open-Eyes:
- And so, de Broglie seals the failure of his life: he still believed, just as Raymond Poincaré en 1905, that an electron is a small object, with spherical symmetry. Oh sure, he admits it as not confined, but with a dense center of its spherical symmetry. Though he could have the second genius idea of his life, if he caught his mysterious wave-piloting-a-corpuscle as the electron itself.

Professor Castle-Holder:
- Indeed, this second genius idea, Erwin Schrödinger had, the year 1926.

Open-Eyes:
- Remaining in the planetary ideation from E. Rutherford's model, de Broglie laid his corpuscle-electron on a trajectory far larger than a corpuscle, orbiting around the nucleus, an orbit whose diameter was the one calculated by Bohr, and worried how the phase wave behaved. He went to find that the wave lapped exactly one turn when $n = 1$, two turns for $n = 2$, three turns, four turns, etc. Exactly. It was this supraluminic phase wave which remained when Erwin Schrödinger found first the relativist equation now known as Klein-Gordon equation, then its non-relativistic approximation which is still taught, but ferociously disfigured and un-schrödinger-ized.

Professor Castle-Holder:
- We will write it down now, this Schrödinger equation, and retrace its history. The coordinates will be taken as Cartesian, and orthonormal, as here it is not a specialty handbook, but only for initiation purpose.

6.4. Wave equation of the de Broglie waves

We consider the equation of propagation in dimension 1 (d'Alembert[5] equation), of the wave of amplitude $\Psi(\mathbf{x}, \mathbf{t})$ and phase speed v_φ :

$$\frac{\partial^2 \psi}{\partial x^2} + \frac{1}{v_\varphi{}^2} \frac{\partial^2 \psi}{\partial t^2} = 0$$

Curious:
- For the laymen who have not yet handled differential equations, nor partial derivative equations, let us explain more. The symbol ∂ (rounded **d**) means a partial derivative of the function, in respect to only one variable, opposite to **d**, which means a total derivative. So for a small (infinitesimal) variation of

---

[5] Jean le Rond d'Alembert: 1717-1783. Co-directed the *Encyclopédie* with Diderot up to 1757. The equation of the vibrating strings in 1747. The second-order operator Dalembertian is explicit above in the text; in a quadri-dimensional space, as the Minkowski's style, it may be considered as the extension of the *Laplacian* operator. In Cartesian and orthonormal coordinates le Laplacian of the function Ψ is $\frac{\partial^2 \psi}{\partial x^2} + \frac{\partial^2 \psi}{\partial y^2} + \frac{\partial^2 \psi}{\partial z^2}$
and for a celerity **v**, its Dalembertian is $\frac{\partial^2 \psi}{\partial x^2} + \frac{\partial^2 \psi}{\partial y^2} + \frac{\partial^2 \psi}{\partial z^2} - \frac{1}{v_2} \frac{\partial^2 \psi}{\partial t^2}$.
On other systems of coordinates, it is a good exercise: you need the metric tensor and the Christoffel connectors for this other system of coordinates.

the considered variable, $\partial\,\Psi/\partial x$ is the quotient of the variation of the function Ψ, by the variation of the variable x. So we pronounce the above equation as: Second partial derivative of Ψ in respect to x, minus the square of the inverse of $v\varphi$ times the second partial derivative of Ψ in respect to t, equals null.

Open-Eyes:
- The situation is even worse than you think, as for the bulk of the laymen even « *equation* » is an alien word, generally misunderstood. They confuse with any mathematical writing. Now, $v_1 = v_0\sqrt{1-\beta^2}$ is simply a prescription of operations, to obtain the desired result at the end. It is not an equation. Though writing an equation uses the same symbol « = », it has not the same meaning. It does not express an unconditional equality: it just expresses a constraint on the variables. Next, the calculator may operate some regular transforms on an equation or a set of equations to make it more practical, and even to solve it completely.

Professor Castle-Holder:
- Shorter for us: Dalembertian of Ψ is null.
This equation is not relativistic, as $v_\varphi \neq c$ (the speed of light); it is not conserved nor in the Lorentz transform, nor in Galileo transform. We have seen that c the speed of light is the geometrical mean of phase velocity v_φ of de Broglie waves, and particle speed v_g, or group velocity according to $v_g \cdot v_\varphi = c^2$. So one may write the equation of waves of matter with the group velocity:
$$\frac{\partial^2\psi}{\partial x^2} + \frac{v_g^2}{c^4}\frac{\partial^2\psi}{\partial t^2} = 0$$

6.5. Klein-Gordon equation

Again we write the equation of the de Broglie waves, using the speed of the particle:
$$\frac{\partial^2\psi}{\partial x^2} - \frac{1}{c^2}\frac{\partial^2\psi}{\partial t^2} = -\frac{1}{c^2}\left(1 - \frac{v_g^2}{c^2}\right)\frac{\partial^2\psi}{\partial t^2}$$
We use the Planck-Einstein relation: $\dfrac{m_0c^2}{\sqrt{1-\frac{v_g^2}{c^2}}} = h\nu$, where ν is the phase frequency, as seen from the laboratory.

Curious:
- Breakneck! For already three pages, in your calculations you invoke relativistic laws, when you will set the relativistic frame only two chapters later, in chapter 8.

Professor Castle-Holder:
- Well, it is all his guilt! He fanatically wants to show that the grounding of the equations of de Broglie, Schrödinger, and Klein-Gordon, is relativistic. We use the Planck-Einstein relation: $\dfrac{m_0c^2}{\sqrt{1-\frac{v_g^2}{c^2}}} = h\nu$, so the group velocity disappears,

and the unidimensional equation of the de Broglie waves becomes:

$$\frac{\partial^2 \psi}{\partial x^2} - \frac{1}{c^2}\frac{\partial^2 \psi}{\partial t^2} = \left(\frac{m_0 c^2}{h\nu}\right)^2 \frac{\partial^2 \psi}{\partial t^2}$$

For quasi-monochromatic and quasi-stationary waves (as they are around an atomic nucleus, for instance), the wave function of time is sinusoidal:

$\Psi(\mathbf{x}, \mathbf{t}) = \varphi(\mathbf{x}).\sin(\boldsymbol{\omega t})$ where the phase pulsation $\boldsymbol{\omega} = \mathbf{2\pi\nu}$ varies with the speed $\mathbf{v}$ according to $\boldsymbol{\omega} = \dfrac{\omega}{\sqrt{1-\frac{v_g^2}{c^2}}} = \boldsymbol{\gamma}.\boldsymbol{\omega_0}$

Its second partial derivative with respect to $\mathbf{t}$ is: $\frac{\partial^2 \psi}{\partial t^2} = - \boldsymbol{\omega}^2 \Psi$

Warning, from here, we will trick you, in confusing $\boldsymbol{\omega}$ with $\boldsymbol{\omega_0}$, and $\boldsymbol{\gamma}$ with 1, in other words, we will do as if the speeds were far to be relativistic. We use this expression in the second part, and the pulsation $\boldsymbol{\omega}$ disappears, but the intrinsic pulsation (of the particle at rest) $\boldsymbol{\omega_0}$ remains:

$$\frac{\partial^2 \psi}{\partial x^2} - \frac{1}{c^2}\frac{\partial^2 \psi}{\partial t^2} = - \left(\frac{2\pi m_0 c^2}{h}\right)^2 \Psi$$

As for a photon the rest mass is null, the above equation reduces to the classical equation of the electromagnetic waves, as the second part is null. The constant $R_C = \frac{\lambda_C}{2\pi} = \frac{\hbar}{m_0 c^2}$ is the Compton radius, and λ_C is the Compton wavelength of the particle. When m_0 is the mass m_e of the electron, you find the Compton radius of the electron is 386.159 fm, and the Compton wavelength is 2.42631 pm.

Now we write the Dalembertian in four coordinates, with the time as a false spatial coordinate, using $\mathbf{w} = \mathbf{ict}$:

$$\frac{\partial^2 \psi}{\partial x^2} + \frac{\partial^2 \psi}{\partial y^2} + \frac{\partial^2 \psi}{\partial z^2} + \frac{\partial^2 \psi}{\partial w^2} = \left(\frac{2\pi m_0 c^2}{h}\right)^2 \Psi = \left(\frac{1}{R_C}\right)^2 \Psi$$

It may also be written by using the Laplacian Δ :

$$\Delta\psi - \frac{1}{c^2}\frac{\partial^2 \psi}{\partial t^2} = \frac{\psi}{R_C^2}$$

The Dalembertian is invariant in the Lorentz transform, and the second part transforms as a scalar, so the Klein-Gordon equation is invariant under Lorentz transform, so it is fully relativistic. It was discovered by de Broglie in 1925 (*Sur la fréquence propre de l'électron*, C. R. Acad. Sci., 180, 1925, p. 498-500), but with the wrong sign for the second part. In 1927, de Broglie had corrected the erroneous sign: *La mécanique ondulatoire et la structure atomique de la matière et du rayonnement*. In **Le Journal de Physique et le Radium**. May 1927. From it, Schrödinger made the stationary and non-relativistic approximation, easier to solve, and which bears his name.

6.6. Static Schrödinger equation, time-independent. (independent of the macro-time)

Let us take the equation of Klein-Gordon:

$$\Delta\psi - \frac{1}{c^2}\frac{\partial^2 \psi}{\partial t^2} = \left(\frac{\omega_0}{c}\right)^2 \psi$$

When the waves are really stationary and monochromatic, a basis of the wave function is the product of two functions, one of the coordinates, one of the time, and the function is a sum of sinusoidals, each of the form $\Psi(\mathbf{x}, \mathbf{t}){=}\varphi(\mathbf{x})$

. $\sin(\omega t)$.

Hence: $\frac{1}{c^2}\frac{\partial^2\psi}{\partial t^2} = -\omega^2\Psi$

So the Klein-Gordon equation becomes $\Delta\psi - \frac{1}{c^2}(\omega^2 - \omega_0^2) = 0$

Instead of the pulsations, let us use the rest mass, and the dynamic mass. With the relation of Planck-Einstein: $\mathbf{E} = \hbar\omega = \mathbf{mc^2}$ we obtain:

$\Delta\psi - \frac{1}{c^2}\left[\left(\frac{2\pi mc^2}{h}\right) - \left(\frac{2\pi m_0 c^2}{h}\right)\right] = 0$

or $\Delta\psi - \left(\frac{c}{\hbar}\right)^2 (m - m_0)(m + m_0) = 0$

Now the relativistic kinetic energy is $\mathbf{T} = \mathbf{E}\text{-}\mathbf{V} = (\mathbf{m}\text{-}\mathbf{m_0})\,\mathbf{c^2}$ where $\mathbf{E}$ is the total mechanical energy and $\mathbf{V}$ is the potential energy. If we add the hypothesis of speeds small compared to the light one, we obtain: $\mathbf{m} + \mathbf{m_0} \approx \mathbf{2m}$ and $\mathbf{m}$ - $\mathbf{m_0} \approx \frac{E-V}{c^2}$

So the equation of the stationary waves becomes the independent of time Schrödinger equation $\frac{\hbar^2}{2m}\Delta\psi + (E - V)\psi = 0$

So we have proved (demonstration by Bernard Schaeffer) that the Schrödinger equation is not a postulate, but the stationary and non-relativistic approximation of the d'Alembert equation, applied to the de Broglie waves, defined by $\lambda = \mathbf{h/p}$. Even not relativistic, it has its origin in the Lorentz transform.

6.7. Dynamic Schrödinger equation, time-dependent

Professor Castle-Holder:

- Here is the standard teaching – I devote myself to introduce it, knowing that my colleague will shoot at it. In the case of a conservative field, when the potential function does not explicitly depend on time, one may admit that the dependency on macro-time of Ψ is expressed by a monochromatic term: $e^{i\omega t} = e^{-\frac{Et}{\hbar}}$. Considering that in this case $i\hbar\frac{\partial\psi}{\partial t} = E\Psi$, then $\frac{\hbar^2}{2m}\Delta\psi - V\psi = -i\hbar\frac{\partial\psi}{\partial t}$

Open-Eyes:

- Handsome! Though you used an erroneous pulsation, you nevertheless came to a formally irreproachable equation! The only fault, written in all the handbooks is to forget to take the total relativistic energy, which gives the correct pulsation, this one from Louis de Broglie, confirmed in 1928 by Dirac (but an essential refinement): $\omega = \mathbf{mc^2}/\hbar$. We will precise this later.

The problem is that any evolution in microphysics depends simultaneously on the final and the initial states, and that between them the Fermat's principle is ruling: the phase wave is spatially concentrated at the beginning and the end. As a transitional exercise, we could write **two** Schrödinger evolution equations, one which downstreams our familiar macro-time and carries positive energy, the other which upstreams from the absorber to the emitter, with negative energy micro-time in reverse of our macro-time. So did Yakir Aharonov, Peter Bergmann and Joel Lebowitz in 1964, after Satosi Watanabe in 1955. Later, I will introduce the innovation: the quotient of a function of transmittance (the inverse of resistance) by a function of concurrency.

Curious:

- It reminds me something learned when I was 18 years old: a stationary wave along a guitar string or a piano wire, may be decomposed into a progressive wave which goes, and another progressive wave of the same frequency, but going in the reverse direction.

Open-Eyes:
- On your piano wire, each progressive wave is with positive energy, and they are spatial mirrors to each other. If there were no dissipation in the metal, nor acoustic emission through the supports and the air, all the cord would remain with constant and positive mechanical energy. As long as an atom does not receive nor emit a photon, as long as it does not endure any collision, nor any vibration by the condensed matter, the electronic waves around a nucleus remain at constant energy in a stationary state. Though when the set of two progressive equations describes an evolution, transferring an electron from one place to another, it is a global change. Only the global charge of the electron is conserved.

Another fool trap: due to the first-order derivative, the solution naturally comes in complex numbers. One cannot simply project it in sine or cosine, as we do in photonic optics to obtain the electric field of the electromagnetic wave. Traditionally, they conclude that every solution of the Schrödinger equation *has not more any reality*, is only good to be Hermitian-squared to give a *probability of presence of the corpuscle*. It is again a confusion inherited from the macrophysics, that they thoughtlessly plated on microphysics. They dream that the wave should be quite another thing than the electron, and should be already familiar in macrophysics.

It does not occur to them to realize that the electron is already an intrinsically periodic and undulatory thing, and that this spinorial entity does not have a valid projection onto our familiar space-time.

6.8. Emission of a photon: a beat of frequencies

Open-Eyes:
- Starting from the relativist reasoning inaugurated by de Broglie, in 1926 Schrödinger drew the logical conclusion that the frequency of the emitted photon is exactly the beat between the intrinsic frequencies of the electron in the initial state (the less bound in the case of an emission) and in the final state (the most bound in the case of an emission). The revolutionary idea, the unthinkable by the corpuscularists[6], is that during all the time of the emission (respectively: absorption) of a photon, the atom and its electrons are simultaneously in the final (the more and more) and in the initial state (the less and less). Application to the Compton scattering: during all the interaction, the electron is simultaneously in its run forward and back, forming a stationary wave to whom the Bragg law is applicable.

---

[6]Inclusively Louis de Broglie.

Rotten luck for Schrödinger in his english translation for the Physical Review in 1926: he forgot to come back under the conceptual relativistic frame, where the origin of the energies is absolute, and so the frequencies are perfectly defined, though discouraging for the experimenter; he does not see how he will succeed to measure such enormous frequencies.

Though it is energetically occulted since, Schrödinger did not more see any reality to the corpuscles. For him, each electron occupied nearly all the volume of the atom, so it had not any property of the solid bodies of our familiar macroscopic world. So the stable or metastables states of the atoms, giving the spectra lines the spectroscopists observe, are all stationary states of the electronic wave (if only one, as in hydrogen atom, or ionized molecule H_2^+) or of the electronic cortege in all the more complicated atoms and molecules. No more orbiting screwy corpuscle, just stationary waves. Several electrons share all the volume of an iron atom simultaneously, yes, and so? Moreover, the conduction electrons in a metal are each as extensive as several tens of interatomic distances; it is so, and not else, even it does not looks like anything familiar in our world of big animals, largely macroscopic.

Curious:
- But all these images and videos in the popularizing literature, showing small green balls orbiting around pink nuclei? No?

Open-Eyes:
- It is an elaborate fib. They exploit your gullibility. We have nothing that could be "*smaller*" than an electron to inform us about the smallness, the "*shape*" or the "*color*" of an electron. The electron is already the lightest thing, and in some meaning, the "smallest". In only one meaning, as the most a particle is light, the least it can be concentrated in a small volume. The only way to concentrate the electron on a small length, and in some cases, on a small transverse diameter, is to accelerate it and mainly to weigh it, in a particle accelerator; it is very far from the ordinary conditions treated by the atomic physics. In atomic physics, we treat on detailed properties of the electronic cloud, or the shared electronic cloud in molecules, or crystals, or other solids and liquids.

6.9. Spin?

Curious:
- For us, a *spin doctor* is a facts twister, an image gilder paid by a rich politician to boast him for winning an election. But for you physicists, what is a spin?

Open-Eyes:
- On the one hand, it is still a mystery. Nothing such exists in macrophysics. On the other hand, we know its mathematical properties, and today, it seems enough to many. Before becoming international, this word names a spinning top. However, the contradictions are many, when we try to apply the origin of the word, to the reality of the electrons, and other particles whose spin is

$\frac{1}{2}$ (also said Fermions, as they are relevant of the Fermi-Dirac statistics) such as protons, neutrons, neutrinos, silver atoms, etc. A spin is either half-integer, either integer multiple of the looping Planck's quantum $\mathbf{h}$, and it has several properties of an angular moment.

The spin is a *precursor* of the macroscopic angular moment we already know, and some of its properties surprise us, we macroscopic beings.

Why this fuzzy word *"precursor"*, which has no precise definition? Precisely because in heuristics, we need temporary words, with the fuzziness and precision dose corresponding to our temporary state of knowledge and ignorance. Why this word, which is simultaneously clear to the bulk of the mortals, and fuzzy for the bulk of the physicists? Because it is false that the quantum should be only spin or angular moment, if not, all light should be circularly polarized, and no plane polarized light could exist, and the only allowed transitions in an atom should those which change the angular moment. Now those three statements are false: the plane polarized light exists, the plane polarized photons exist, more than half of the allowed transitions in an atom are of the electric dipolar type, and a plane polarization is convertible into circular polarization, and vice-versa, through *quarter-wave plates*. So exists some precursor, quantified by the Planck's constant, which is not quite like a macroscopic angular moment.

An alternative conjecture could be a virtuality: virtually, a photon in plane polarization could be converted into a circular photon, thanks to a birefringent medium, especially a *quarter-wave plate*, and be absorbed by an atomic or molecular transition; but here, the transfer of angular moment is from the birefringent plate to the final absorber. Anyway, the initial and the final state both are ruled by the quantification of the stationary states of the electronic wave.

Should I have to say more, it would take a long detour by the rational mechanics, which is a strongly mathematized branch of the physics; in the beginning, it dealt with the movements of the rigid bodies, and it may specialize in fluids mechanics, vibrations mechanics, elasticity and strength of materials, etc.; it may disgust the reader. There are very few principles of rational mechanics, two:

Conservation of momentum-energy; to change it, you must do something, act.

Conservation of angular moment. When the basis of the principles was the material point, the angular moment was identified as the moment of the momentum from a sliding axis. To change it, you must do something, you must act.

If there is no momentum of the body, or if we are in the frame where the speed of the center of mass is null, then the angular moment corresponds to a rotation of the body, with the same angular speed everywhere, and it is the product of a moment of inertia by the angular speed.

And for extrapolating to the microphysical scale, to the spin of a particle? The differences are alas many. The rational mechanics deals with macroscopic bodies, that is: not too small compared to our hands. There we can contemplate the position, the speed or the spinning of a body without screwing things up. For instance, the optical methods are handy for observing bodies well bigger than the wavelength of the light (about 0.5 μm). But all these facilities are over when working at the scale of the individual particles. One cannot more contemplate, but only act to change something, and state the result of the change; there lye new surprises. We have to visit the experiment that Otto Stern and Walther Gerlach mounted in the twenties years, and whose results were published in 1922. The experiment uses the fact that electrons, and atoms whose spin is 1/2 like an electron, have a magnetic moment, and this magnetic moment reacts to an external magnetic field, here an inhomogeneous field. See at:

http://galileo.phys.virginia.edu/classes/252/Angular_Momentum/
Angular_Momentum.html

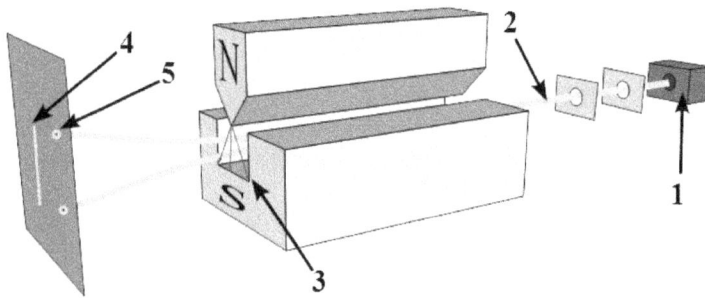

Figure 6.2.
The experiment is in a pumped vacuum. A small oven (1) emit a vapor of silver in atoms. In the vacuum, they propagate in a straight line. The diaphragms (2) collimate the thin silver beam, which then passes between the pole pieces (3) of a strong electromagnet. These pole pieces differ: one is wide, the other is thin. So the magnetic field is stronger near the thin pole piece. We rely on this gradient of field.

Curious:
- The magnetic field, this is the blue arrows. So the field goes down from the north pole to the south?

Open-Eyes:
- Oh dear! The fool trap has won again! Alas, I have borrowed the draw to others. And they do believe that a magnetic field has the symmetries of an arrowed stick, so they draw it as an arrowed stick. With this choice of the south pole above, if you look this field from the top, it turns clockwise. But if you reverse the current, and so the poles, the experiment will remain the same. This Stern-Gerlach experiment has a top-down dissymmetry, but this does not come from the direction of the current in the coil (in a horizontal equiplane), it depends only on the vertical gradient of field, ruled by the difference of thickness of the pole pieces. In other words, the mention of north and south

poles is superfluous: what matters is the presence of magnetic field, and its gradient where the silver beam passes through.

Now, the macroscopic physics should predict any precession orientations of the magnetic moments of the silver atoms, as they issued from the oven with random orientations, and they should give a short straight line, drawn in (4). Instead, we observe quite another pattern: two separate impact spots. Therefore, the magnetic moment of the atom chose, when passing in the strong magnetic field, to be either **all-against**, either **all-supporting** this field. So the all-supporting neared the strong field, while the all-against pulled away.

Professor Marmot:

- What are all those monkey spins? We all know that a spin is **up** or **down** along the z-axis! Stop that hoax!

Open-Eyes:

- And how do you "*know*"? Please describe your experiments, which enable you to get cleared from the rumor and the tribal mess, which enable you to access the true symmetries of the spin. Mmh? We retain that "*up*" means "all-supporting the ambient magnetic field", and "*down*" means "all-opposing the ambient magnetic field". In the swedish tale "*Kjerringa mot strömmen*", the shrew or witch is an all-opposing person. In the stream, she swims upstream, instead of downstream. "*Kjerringa mot strömmen*" is the swedish expression to name the contradictory spirit. So the spin is not an isolated property but a relational one. "*Intrinsically relational*" if we are eager to impress the *pale faces*.

Professor Marmot:

- Grmbl, grmbl, grmbl! Heretic! Complotist! Burn him at pyre!

Professor Castle-Holder:

- Please let us return to the thread. We were on the quantification, and were just dealing with the spin, with the historic experiment by Stern and Gerlach. As I know you for fifteen years, I am surprised that you have not yet quoted the experiment by Shahriar S. Afshar. In its time, it has raised passionate brawls.

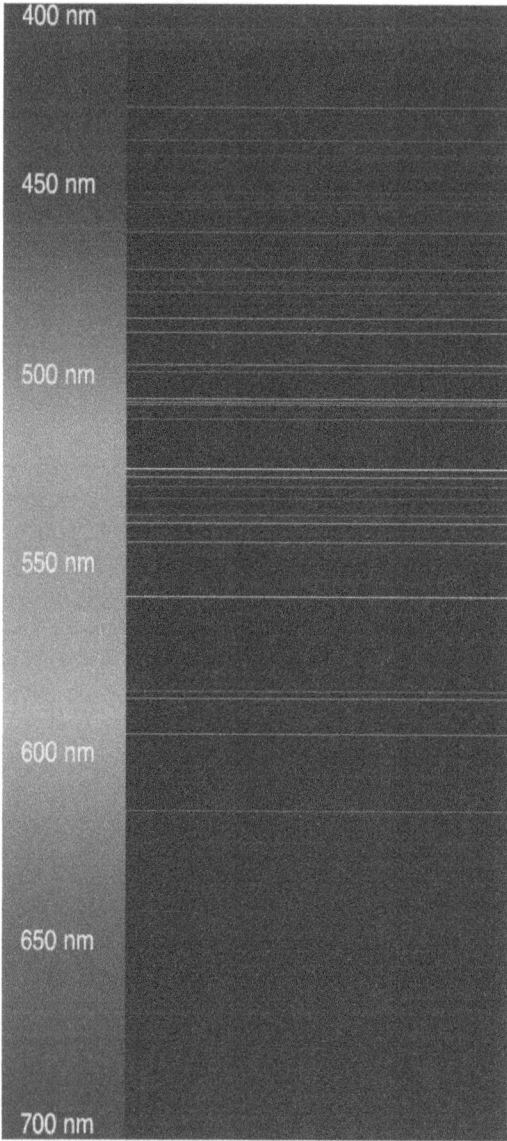

CHAPTER 7

The experiments by Shahriar S. Afshar

Open-Eyes:
- This experiment was done in 2001 and repeated in 2003. It refutes the copen-haguist myth of *duality*, which forbids to measure simultaneously where the photon impacts, and what was its momentum. The commentators persist reasoning as corpuscularists, in the *"which way"* style; it confers them an unintended humor.

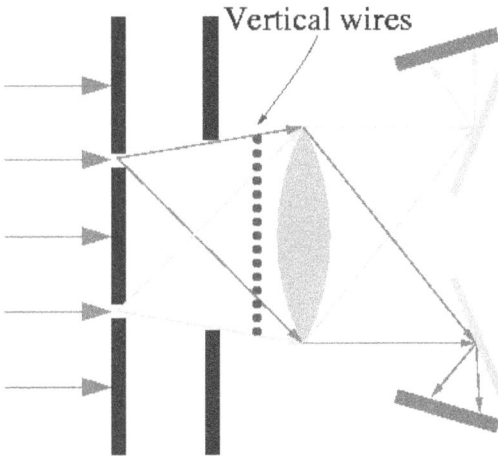

Figure 7.1. The principle is to redo the experiment of the two holes as from Young, but to add a grid of six thin wires, each in a dark zone of the interference. Result: practically no obstruction, more than 99% transmission. So never the photons transmuted into corpuscles, they remained individual waves all the time. Each one of these individual waves passed simultaneously by the two Young holes. *"Corpuscles"* could not.

Afshar S. Violation of Bohr's complementarity: one slit or both? AIP Conference Proceedings, 2006 v. 810, 294--299. The paper is not free: $ 18.

Afshar S., Flores E., McDonalds K. F., Knoesel E. Paradox in wave-particle duality. Foundations of Physics, 2007, v. 37, 295--305.

https://arxiv.org/abs/quant-ph/0702188

Free access at https://arxiv.org/pdf/quant-ph/0702188v1

www.ptep-online.com/index_files/2011/PP-24-07.PDF

web.mit.edu/redingtn/www/netadv/Xafshar.html

https://en.wikipedia.org/wiki/Afshar_experiment or in french at https://fr.wikipedia.org/wiki/Exp%C3%A9rience_d%27Afshar ...

Exercise.
You will take the hypothesis, used by Elitzur and Vaidman in one of their most famous hoaxes, that the photon is a corpuscular particle, so in daring to stop or scatter it, the thin wire will receive a *bloody slamming* blow which makes it shake in Bzoiiiing!
Your mission is to calculate this *bloody slamming* blow, and to compare it to some real shakes in your laboratory, such as the pace of the technician, or a lorry passing on the road at 100 yards, or worse, the strikes by cosmic rays.

Chosen frequency of the photon: 565 THz (in the green range).
Transmitted momentum:
The solution in Appendix H.

CHAPTER 8

Ramsauer and Townsend ruined the corpuscularist postulate

Open-Eyes:
- In the years 1921-1923, and independently, Carl Ramsauer and John Sealy Townsend, by studying the diffusion of the electrons at low energy through a rarefied gas, discovered the equivalent of the anti-reflect coatings on good lenses: when the de Broglie wavelength of the incoming electrons was double the diameter of the met atom or molecule, those obstacles - atom or molecule - became transparent to the electrons, did not scatter them, but only little on the side.

This fact is incompatible with any corpuscularist ideation; it proves that the electron wave, invented later by Louis de Broglie, is precisely the electron itself. In the same way, the excellent success of the anti-reflect layers in optics, and also the quarter-wave plates, converting plane polarized light into circularly polarized, and vice-versa, reject the corpuscular ideation for the light (alas resurrected by Albert Einstein): they work only in undulatory, when each photon is indeed undulatory, and ruled by the Maxwell equations (perfected as bosonic) and has a width and a length very far from any corpuscularism.

When you know that the Ramsauer-Townsend effect exists, then you can investigate the documents, and you find dozens, whose the major part are experimental. Plus some prescriptions for practical work, plus some reports from students who achieved a research work, and even the commercial description for a specialized apparatus or this practical work. When these docs are printed, they weigh 1526 g net, 1833 g with the binder.
What was very difficult, was to know that this knowledge exists. It is unknown and censured in every handbook of quantics, excepted the D. Sivukhin, Ed. Mir, volume 5.

Curious:
- It would be useful that you explain first the anti-reflect coatings. We are so glad to profit from them, but we cannot make them, nor theorize them. Please explain!

8.1. The anti-reflect coatings on the optical surfaces

Professor Castle-Holder:

- Deposing this layer is for avoiding the losses of light by reflection, and so to transmit the maximum of flux up to the sensor. On the intermediate surfaces of an elaborate frontal lens, or of an eyepiece, the anti-reflect layer minimizes the parasitic reflections, very unpleasant. No universal anti-reflect coating exist: the effect depends on the wavelength of the incident light, and it depends on the angle of incidence. The laws of the diopter give a reflection rate for each change of optical index when changing of a milieu. In astronomical or photographic optics, one is careful to keep the incidences near the normal[1], so the cosine remains near 1.
Figure 8.1.

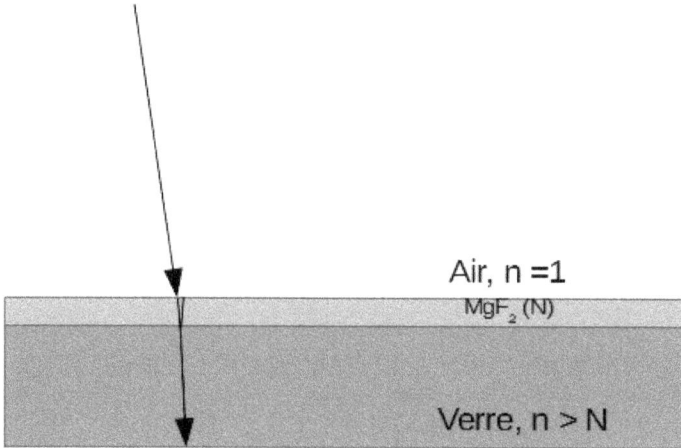

The trick is to obtain that the wave reflected by the first interface air-coating be annulled by the wave reflected by the second interface coating-crown, coming in opposition of phase.

---

[1]Normal incidence: the ray of light is perpendicular to the diopter. In other words, each wave front comes in one blow on the surface, with an infinite speed of phase on the surface.

Figure 8.2

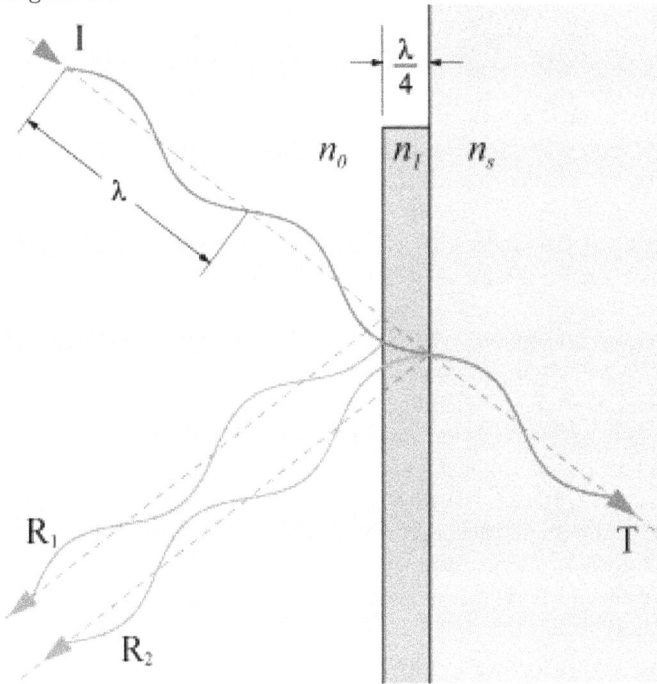

This imposes the optical thickness of the coating: the quarter of the wavelength in this refringent medium for the frequency for which the device is optimized, so the phase delay after two crossings (to and from) of the anti-reflect layer is π radians. Optical thickness = the product of the physical thickness times the index of refraction.

Open-Eyes:
- Excuse me for objecting to a detail. You said " *the wave reflected* ", as it was the habit to say before quantum physics. But now in **transactional quantum physics**, we do not more consider a generic wave flowing from the source towards everywhere, but only **individual** waves, each one converging on its absorber. However, the groping of the ground noise continuously explores the **optical admittances**. Thus the anti-reflect coating enhances the admittance for the refracted path, and lowers the admittance for a reflected path. Thus the reflected path is less attended. End of the objection.
If the incident ray is not perpendicular to the diopter, then the emergent path, coming from reflection, and through twice the layer, emerges at some distance from the incident, according to a law in sine of the internal angle of incidence. If each path were of infinitesimal width, then the two reflected rays would be unconnected and the anti-reflect could no more work. Therefore, each photonic path has a width. Later, we will predict this width with the quantitative law of the Fermat spindles.
Of course, if you remain under the influence of the tribal disinformation telling that you should forget all the undulatory optics *because modern times mean*

corpuscles, then you are finite. We will see later the appalling example given by Linus Pauling and E. Bright Wilson, Jr.

Professor Castle-Holder:
- **Practical realization**: one lays on the first glass (index n_s) of the lens of a photographic optical objective, a coating of transparent material, index N, which we will calculate.

In order to obtain a totally *destructive interference* (no reflected light), two conditions must be fulfilled (the calculation is done at normal incidence): (1) The same reflection rate on the two successive diopters r_1 and r_2. At the end of calculation and simplifying, it comes: $N \simeq \sqrt{n_e.n_s}$. So, with $n_e = 1$ (air) and $n_s = 1.52$, as in crown, then $N = 1.23$. Now, such material does not exist in the minerals. The least bad approximations are the magnesium fluoride MgF_2, of optical index 1.378^2 in the visible range, or some other fluorides, less often used. They are laid by evaporation under vacuum. The best anti-reflect is obtained with the glass of the highest index, say a flint of index 1.62, up to 1.80 with a lanthanum flint, dense. However, almost all achromatic doublets or apochromatic triplets begin with the converging lens in "*crown*", the less dispersive glass. Therefore, the reflection on the first diopter (air - MgF_2) is always higher than the second reflection on the second diopter (MgF_2 - crown), and cannot be completely annihilated.

Figure 8.3.

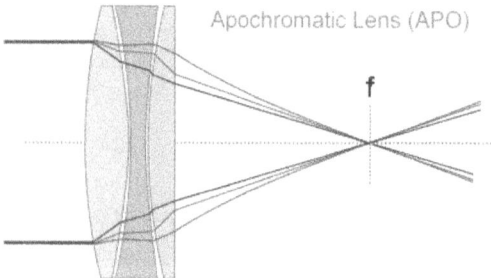

Apochromatic Lens (APO)

f

(2) The phase difference must be π (+ **2kπ**). The corresponding thickness is $\lambda_0 / (4N)$. When the incident light is white, the thickness of the laid coating is fully effective only for the wavelength of the maximal sensitivity of the human eye, or for a sensor mimicking the human eye, that is yellow-green; so it is **e** $\approx$ 0.116 μm. For the other radiations, the interference is not so well destructive of the reflection, so a complementary color, mainly blue and violet is slightly visible by reflection on the lenses of binoculars or for photography. By this technique, one can reduce the reflected energy to less than 1%.

Curious:
- You just said: "*the reflection on the first diopter (air- MgF_2) is always higher than the second reflection on the second diopter (MgF_2 -crown), and cannot*

---

[2]http://refractiveindex.info/?shelf=main&book=MgF2&page=Li-o

be completely annihilated.". But so you have retaken the mode of the "*classical physics*". Now Mr. Open-Eyes should ask you to reason photon by photon.

Open-Eyes:
- The articulation between the macroscopic mode and the individual mode of analysis is a new quantity that I provisionally name "*impedance*" (or its inverse, the admittance). As we make the reflection rarer, we increase the impedance through which a potential emitter and a potential absorber on the reflection side try to establish a transaction. Therefore, the successful transactions on the reflection side rarefy, in comparison to those successful through a complete optical transmission.

Open-Eyes:
- The optical length varies as the inverse of the cosine (the secant) of the angle of incidence in the anti-reflect coating, so its domain of efficiency shifts to the red when the incidence angle increases. Therefore, the reflected light contains more blue and purple and less red.
However, for each precise radiation, one must observe another feature: a loss of efficiency as the wave reflected by the second diopter is more laterally shifted. It provides an experimental cross-checking for the quantitative law of the Fermat spindles: If you have a precise law of dependence of the optical efficiency of anti-reflect upon incidence, and if you know the dimensions of the experimental device then you can assess at least a minorant of the quantitative law of width of the Fermat spindle. If this width was null or negligible, then no anti-reflect coating could work beyond the normal incidence. It would be a very severe hindrance for photographic or observational optics, especially with wide fields, and even with the eyepieces in astronomy.

Example of calculation: We choose $10°$ as the angle of incidence in the coating. Its sine is 0.17365. To and from on a thickness of 0.109 µm ($\lambda = 0.60$ µm) make a shift of 38 nm; about the double for an internal angle of $20°$. Now, this shift should be far lesser than the radius of a Fermat spindle, which is at least about 0.1 to 0.2 µm. We estimate it is so in the majority of the optical devices, as long as the source is far, but is to scrutinize more about the distance to the absorber.
On a compact camera, the focal is not more than about 10 to 15 mm. But what is the final absorber to consider? The CCD sensor? It receives the transmitted photons through the refractor, without any reflection. But if we consider the loss of efficiency of the anti-reflect, the optical absorber to consider is outside, for the reflections. We will proceed further the complete calculation when we do have given the geometric law of the Fermat spindles.

Professor Castle-Holder:
- The paradox is it was enough that our Curious posed an innocent question to prompt our dissenter with Open Eyes to grasp the opportunity to show us his *Fermat spindles*. I foresee I will not issue alive from this venture: as soon I

will go out, our corpuscularist colleagues will kill me, for having you let speak; which is forbidden under the laws of the jungle.

Curious:
- I object. I wish a drawing, with the wavefronts, at the scale, to have an opportunity to compare the certain wavelength with the probable width of the photon.

Open-Eyes:
- Owing to our reduced means of drawing, we need to proceed with some simplifying calculations first. We will take 4/3 as the refraction index of the coating material like MgF_2 (about the index of water). The incident wavelength is chosen as 580 nm, giving 435 nm in the fluoride. Hence the thickness of the coating: 435 nm / 4 = 109 nm. Now choosing the angle for the external incidence, probably exaggerated. Let us take 30° as the external angle. Its sine is $\frac{1}{2}$. Hence the equidistance of the wavefronts on the plane of the diopter: 580 nm x 2 = 1160 nm. The half-equidistance: 580 nm. A constraint is that this quantity is quite larger than the distance of re-emergence of the ray refracted in D_1, reflected in D_2; we will see that this constraint is fulfilled for this angle of incidence. Inside, the incidence is Arcsin (3/8) = 22°, hence the re-emergence of the geometrical ray is at 88.2 nm, after reflection at 44.1 nm (88.2 nm = 218 nm x tan(22°)).
The resulting constraint is that that 88 nm should be very small compared to the optical width of the incident photon, as set by the remaining of the observational device; they so are a minorant. One could dream a not so small minorant. For drawing with a primitive tool, we also need the tangents of the incidence angles: $\tan(30°) = \frac{\sqrt{3}}{3} = 0.577$
$\tan(\text{Arcsin } (3/8)) = 0.4045$.
This approximation is better than this of the graphic tool.
Shift towards the red of the anti-reflect optimum: $\cos(\text{Arcsin } (3/8)) = 0.927$
Hence the new wavelength (in vacuum) of extinction of the reflection: 580 nm / 0.927 = 626 nm, which is in the yellow-orange range.
Figure 8.4.

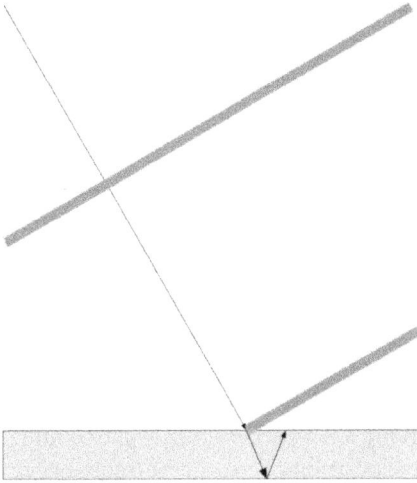

Conclusion: The more we study the boundary conditions of these undulatory phenomena, already thoroughly established and well exploited, the less it remains gaps where could lie the loophole for the corpuscularists and the copenhaguists.

Professor Castle-Holder:
- As well as the more we study biology, the less it remains interstices where could lie the god of the gaps, so cherished by the creationists.

8.2. The surprises of the interferential colors of the ducks.

Open-Eyes:
- Another noticeable experimental fact is the interferential colors on some birds feathers (and fish scales, even reptile scales such as on the green lizard, and also on beetles and butterfly wings): on incidence near the normal, the wing-mirrors of the green-winged teals shine green. But when you observe in the west such teals, when the Sun is lowering in the south-west, then these wing-mirrors shine *magenta*. Is it a mutation, toward a new variety of teals? Nay and thrice nay!: It is only that under grazing incidence, the optical paths of the light from the Sun through the two interferential layers of the teal feathers are seriously increased; so the selected and reflected frequencies are shifted, are not more the same. Here it is a group of at least two main reflected frequencies, as the color *magenta* does not exist in the spectrum; it only exists in our visual cortex (resulting from the wiring of our retina, giving a colored vision specific to human and the big apes of the old world) by superposition of two main frequencies; it results from a red component superposed to a blue-purple component.

Here is a photographer who produces masterpieces from his floating blind; he confirms the changing irisations of the birds, and gives further examples: **Hervé Stievenart.** *Au ras de l'eau ; la vie secrète des marais.* **Éditions du Perron.** Beside the teals, Stievenart mentions the lapwings, the mallards, the shovelers, the kingfishers, the common pheasants... I quote him on the teals, Anas crecca: « *The incidence of the light plays tricks to my sight. The reflects of their eyebrows vary from green to blue. Their cheeks change form brown to crude red. Their rump is sometimes nearly white, sometimes canary yellow.* » On the lapwing, Vanellus vanellus: « *Again a bird whose plumage plays with light. Its reflect can change from night blue to emerald green, softened by exquisite wine-reds.* » On the blue mirror of the mallards, Anas platyrhynchos: « *Here too, the metallic blue mirrors of the female mallard change colors with the incidence of the light. Sometimes they seem nearly black, and sometimes they rival to saphir.* »

As for snorkeling at sea, I mention as evident the interferential colors of many fishes, for instance, the peacock wrass (or *roucaous* in the french of Marseille), *Crenilabrus tinca*, or *Symphodus tinca*; when taken out of the water, the desiccation quickly alters the interferences. Evident too on the green lizards, and many beetles. I suspect several odonates, such as the blue damselflies, or more spectacular, the banded demoiselle, *Calopteryx splendens*.

Curious:
- But I do not see the link with the subject of the chapter. We began with anti-reflect coatings, and we intended to see some anti-reflect for the electrons. Please explain!

Open-Eyes:
- If the photons were "*small grains*", these interferential colors and their changes according to the more or less grazing incidence would be impossible. They are strictly undulatory phenomena. The interference is possible only if the width of

each photon is several times the track of the wavelength on the entry diopter, and its length is several times, at least several times the wavelength. Above, we dreamt a larger minorant, but it was supplied by nature for at least the Cretaceous, even much older (for the Devonian for the fishes? But beetles such as the green rose chafers, *Cetonia aurata* also have structural colors, and it was an independent evolution; the origin of the beetles seems as old as the late Carboniferous). The photon is not any kind of "*small grain*", it is just the minimal quantity of electromagnetic radiation which can be emitted or absorbed by a system when jumping from a stationary state to another stationary state. Those stationary states are constraint by the Planck's quantum.

Curious:
- Objection! You have invoked pieces of knowledge on the colors and the perception of colors which the general public do not share. For instance, my wife will become mad at us and bark her contempt if she reads you and has the unpleasant surprise of not understanding what she was convinced to know for always, and better than anybody, giving her an unquestionable authority on anybody.

Professor Castle-Holder:
- There are three main ways to generate colors:
1. The pigments and dyes absorb one or several bands of frequencies in the visible domain. They do so by absorbing at resonating frequencies, and jumping so onto an excited state. Then they de-energize by non-optical means, preferably thermic ones. For instance, the chlorophylls A and B absorb in the bands: in the blue, from 430 to 490 nm[3], and in the red about 650 to 695 nm (in colloidal clustering or in vivo). Our opsins[4] in the cones of our retina also have frequential absorptions, moderately selective.
2. Some colors may result from fluorescence: a molecule absorbs some UV, then unexcites by emitting one or two photons in the visible range, or in near-infrared. In vivo or in solution, the chlorophylls present a red fluorescence.
3. What interests us now are the interferential colors, like those that make iridescent the thin layers of oil on water, or which make iridescent our credits cards, our banknotes, or stickers on our computers. There, it is the local wavelength that matters, in the thin layer crossed to and from the reflection. What we perceive is reflected by this thin combination. The thickness of these thin layers is about the double of the anti-reflect because the effect (or the purpose? In sexual selection) is the reinforcement of the reflectance of the first diopter for one wavelength. The surprising magenta color from the wing-mirror of teals in grazing incidence came from two reflectances instead of only one, of which one is a second-order one. An online encyclopedia gives much further information

---

[3]It is a usual abuse, to name a light frequency by the wavelength of the same radiation in the vacuum. But to access to the physics of an absorption, one has to translate into frequency, and hence into difference of energy. Dividing the celerity of light by the wavelength gives the frequency of the photon. Multiply the frequency by Planck's quantum of looping **h** gives the difference of energy, at absorption as well as at emission.
[4]**Opsin**: a photosensitive protein in the cones or the sticks, photosensitive cells in the retina.

on the details of the realization by lots of insects, including the butterflies, of structural and interferential tricks, whose results bring forth admiration.

Next, we have to understand how our visual system is organized and wired. When you look at the spectral responses of the opsins that make our colored daylight vision, you are disappointed: the resonance whose peaks are at 419 nm, 531 nm and 559 nm are not sharp at all.

A trap in the habits: 419 nm is the wavelength in the vacuum. Instead, the handbooks should give the frequency, as the absorptions by the opsins and their retinal are frequential. Whereas the opsin is in an intracellular aqueous solution, where the wavelength is shorter by $\frac{3}{4}$. So we translate into frequencies:
419 nm ==> 720 THz (terahertz),
531 nm ==> 565 THz,
559 nm ==> 536 THz,
The following figure is extracted from **Neurosciences**, by Purves, Augustine, Fitzpatrick, Katz, La Mantia, McNamara. DeBoeck editor.
Figure 8.5.

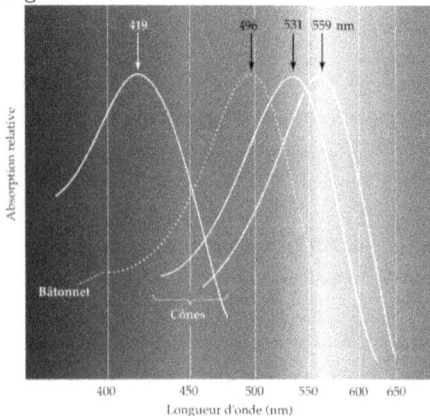

Open-Eyes:
- How in the world have we happened to get such a fine perception of colors, so finely discriminating? Already in the retina, the neuronal wiring multiplies the tricks to enhance the contrasts. So is our genetic heritage. The price to pay for these boosts to the contrast is many visual illusions, which luckily are rare in hunters-gatherers life, but may be proved by clever artifacts. What is transmitted by the optical nerve to the thalamus is already the subtraction between the answers of different cones, not only the response of the cone sensitive to green, but also the differences green minus orange, and green minus blue, etc. Then an ancestral proceeding is done in the thalami, next to the visual cortex, at the end of the occiput, where more elaborate treatments are performed.
So the *circle of the colors*, where the red is neighbor to the purple, which pleases so much to the painting teachers at school, has nothing to do with the spectrum, but results from the organization of our neuronal visual system.

A still more ancestral (at least for the Devonian) proceeding is done in the colliculi, located on the brain stem, which computes the speeds, and coordinates with the audition, to locate something mobile – say an insect when our ancestors where insectivore, or an eventual predator for the totality of our ancestors. The birds do not have our cortex but have hyper-developed basal ganglia, and especially the two thalamic bodies. This architecture privileges the speed of treatment. The diurnal birds have four different opsins for the daylight vision, and it contributes to a far better discriminating power than has any mammal, including ourselves. An eagle spots a green snake from two kilometers.

Open-Eyes:
- My thanks to the professor Castle-Holder. Acting as a duo, we have exposed here that the existence of the interferential colors, which anybody can observe, provides minorants less small than the previous, for the width and length of a photon.

Curious:
- If I do not misunderstand, you have proved that never a photon transmutes into a corpuscle: it remains undulatory, an electromagnetic wave from the beginning to its end. Only the properties of many emitters and many absorbers – those who toggle from a stable or quasi-stable state to another stable or quasi-stable state – oblige the photon to be a transfer of exactly a Planck's quantum.

Open-Eyes:
- And precisely our retinal sensors, and also all the sensors we use in the laboratory to obtain fine details belong to the category of the quantifying absorbers. A bolometer, which only absorbs radiation energy and heats, does not give any piece of fine information, neither quantic nor anti-quantic. On our roofs, we have solar absorbers which at the good season heat our domestic water and do not give any spectral information.

8.3. The resonating transparency in gas: Ramsauer-Townsend.

Open-Eyes:
- A first decision seems to be made, but later we will see it is less serious than feared: when we model in unidimensional the crossing through an atom by an electron, shall we represent the atom by a well of potential, or a wall of potential? In other words, shall we give priority to the Coulombian attraction by the nucleus (charge +), or to the repulsion by charge - of the electrons of the electronic cloud? However, the problem remains concerning the dialectic bind with the modeling of the main mechanism: the elastic scattering of the electrons by the atoms or molecules in a gas. Now we should study the main mechanism, before studying its exceptions. Max Born did this mathematization. Moreover, we have to modify the previous modeling to make it compatible with the role of the absorbers; it is a big change in perspective and method. Even when placing in the center of inertia of the electron-atom system, it is a calculation beyond

the public of this publication.
Figure 8.6.

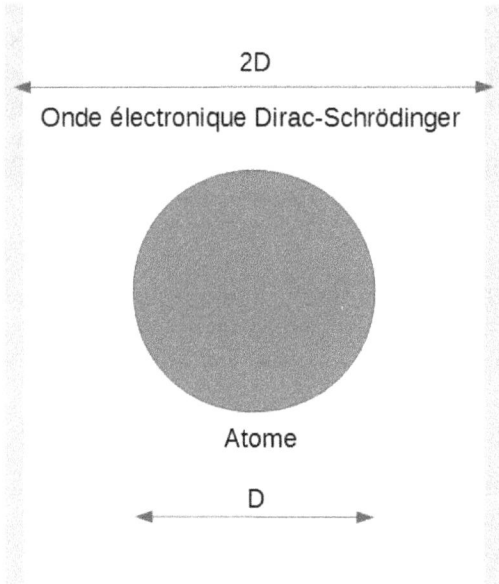

The basic summing up is we again find the same metric relation as in photonic optics for the anti-reflect coatings: the Broglian wavelength of the electronic wave must be four times the diameter of the obstacle-atom. Then the retro-scattering is nullified, and therefore the probability of lateral scattering is minimized. However, beyond this geometrical description, the complete physical explanation remains a delicate affair. Particularly, the diameter of what sees the electron (the electronic wave, it is the same thing) during its propagation, is not evident to define.

Trap! But this trap will be explained only in the chapter on relativity: if the Broglian wavelength is quadruple of the diameter of the atom, then the Dirac-Schrödinger electromagnetic wavelength is only the double of this diameter.

Orders of magnitude for 1 electron-volt:

Group velocity: 593 m/s

Phase velocity: $151.5 \cdot 10^9$ m/s.

Broglian wavelength: 12.25 Å.

Hence we deduct the apparent diameter of the atom as seen by the electron: 3.06 Å, or a little bigger if the resonance occurs at a lesser energy. Such a magnitude is correct. The crossing of this atom by the electron lasts about 2 fs (two femtoseconds). Implicitly, the drawing 8.7 supposes the electron being much smaller than the atom. No experimental fact supports this ideation.

Figure 8.7.

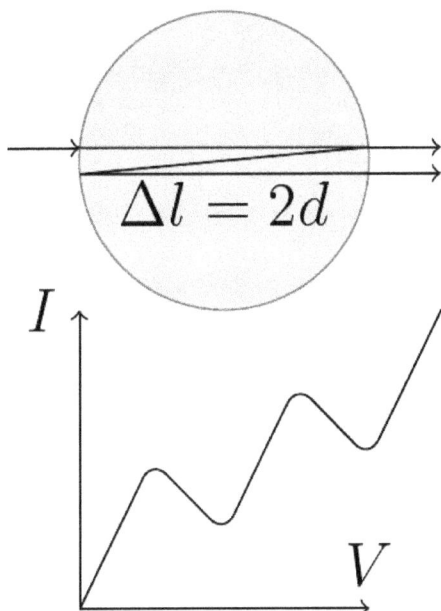

$$\Delta l = 2d$$

8.3.1. The experimental device used by the students of the M.I.T., next by those of the University of Wisconsin. Initial description by Stephen G. Kukolich, in the American Journal of Physics, August 1968. The following images are due to Martha Buckley, also in the M.I.T.

Apparatus: A thyratron RCA 2D21 with a heated cathode[5], contains xenon gas, whose pressure initially at room temperature is about 0.005 torr.
This kind of gas tube, a diode with heated cathode and whose ionization is controlled by a grid were used by the industry before they were replaced by the silicon thyristors, for power controlling, necessarily in alternating current, as once conductive by ionization of the gas, they could not be switched off, but by annihilation or inversion of the anodic tension. This RCA 2D21 can control 100 mA under 117 to 400 V. The grid may hinder or trigger the ionization of the gas in this controlled diode. Why diode? Because only the cathode was heated, and covered with thermo-emissive material (barium oxide, partly reduced). The delay of de-ionization of the gas limited the frequency for the correct working of a diode or a thyratron.
Figure 8.8.

---

[5]See for instance the handbook **Electronique industrielle** by I. Kaganov. Éditions Mir, 1972.

Apparatus for Measuring Scattering Cross-Section

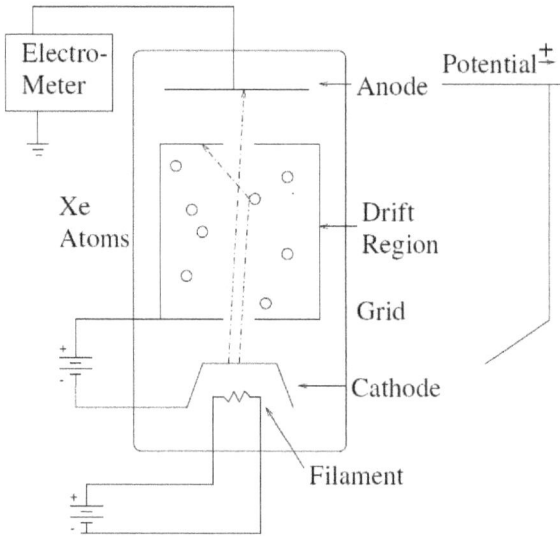

Electro-Meter

Xe Atoms

Potential $\pm$

Anode

Drift Region

Grid

Cathode

Filament

The use of this thyratron is here very different from the primary electrotechnical use. The heating supply of the cathode has been lowered from 6.3 V rms to 4 V, to reduce the spreading in energy and speed of the emitted electrons, at the cost of a lowered emissivity. The anodic tension has been much lowered, from 0 to 12 V. At 13 V, the xenon begins to ionize, which is contrary to the experiment, where we want elastic collisions.

So the apparatus comprises a lowering transformer to 4 V output, and a regulated and adjustable power supply, from 0 to 15 V, and potentiometric tension divider – necessary to master the very low tensions with precision. Here this thyratron is mounted with the head down, with its pinout above, and via an adjustable stand, its body may be dipped into a liquid nitrogen Dewar – after the extinction of the heating of the cathode, and enough time for cooling, softly enough to limit the thermal shock.

A Dewar flask filled with liquid nitrogen. Its purpose is to condense the xenon in the head of the tube, so eliminating it from the run of the electrons. So one can study the effects of the geometry of the tube separately from the effect of the xenon gas.

Three digital multimeters. These are high impedance meters, 3 $\frac{1}{2}$ digits, to measure the anode current Ia through a 10 kΩ resistor, the shield current I_g through a 33 Ω, and the cathode to shield tension. The box grid-shield is an equipotential, placed between cathode and anode. The variable measuring the transparency is the ratio I_a / I_g.

I love the simplicity and pragmatism of this apparatus for the practical work for the students: a thyratron from the shelf, and usual devices in a laboratory: a Dewar flask, liquid nitrogen, a regulated power supply, a 4 V transformer, three multimeters, an adjustable stand. The proverb says, about a prototype: *Make it as simple and stupid as possible.*

In the Kaganov handbook (Mir ed.) you find the scheme of a TG 1-0,1 and following:
Figure 8.9.

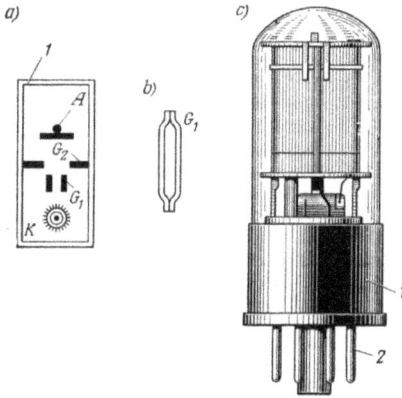

Fig. 6.5. Thyratron à cathode chaude, type TГ 1-0,1/1,3:
-*a*—schéma de disposition des électrodes en plan; *b*—construction de la grille; *c*—aspect extérieur

8.3.2. Results of the measurements. Figure 8.10.

Determination of Scattering Cross Section

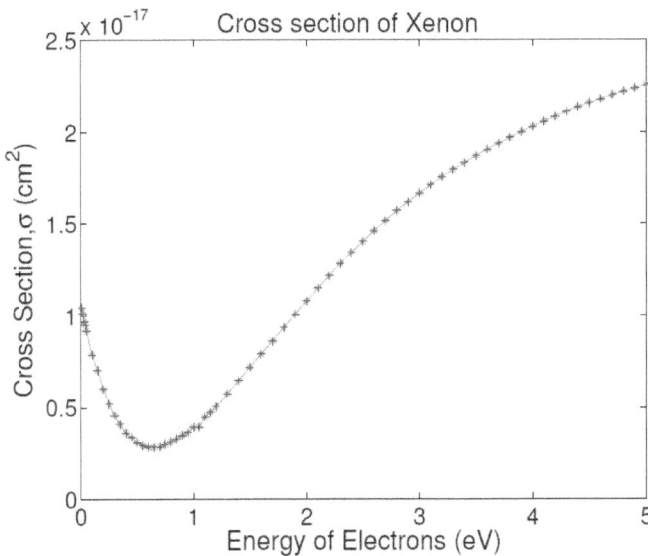

Indeed we state a vanishing of the obstacles that be the atoms of xenon when the electrons have an energy of about 0.7 eV.

Comparing with no gas when the xenon is condensed in the head of the thyratron:

Figure 8.11.

Measured Anode Currents for
Xenon Free and Xenon Frozen

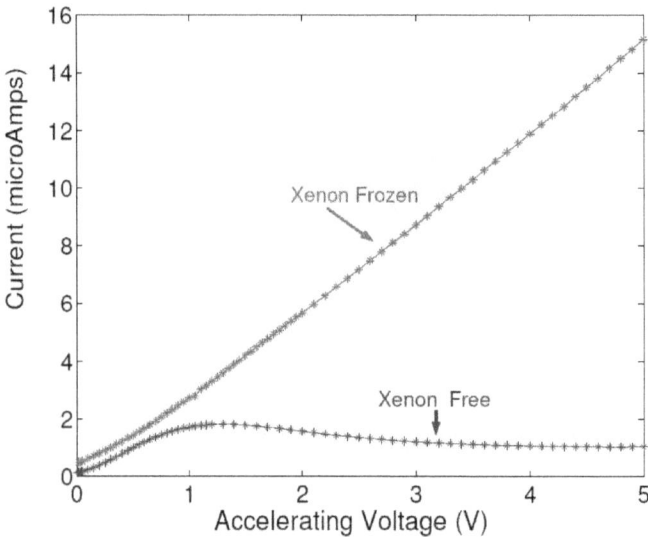

So it is obvious that as long it is not ionized, the gas is an obstacle which efficiently hinders the current in the tube, despite the heating of the cathode. In the industry, a gas diode with a heated cathode and cold anode, withstand far better than a vacuum diode, provided the frequency is low enough to leave time to the deionization.

Figure 8.12.

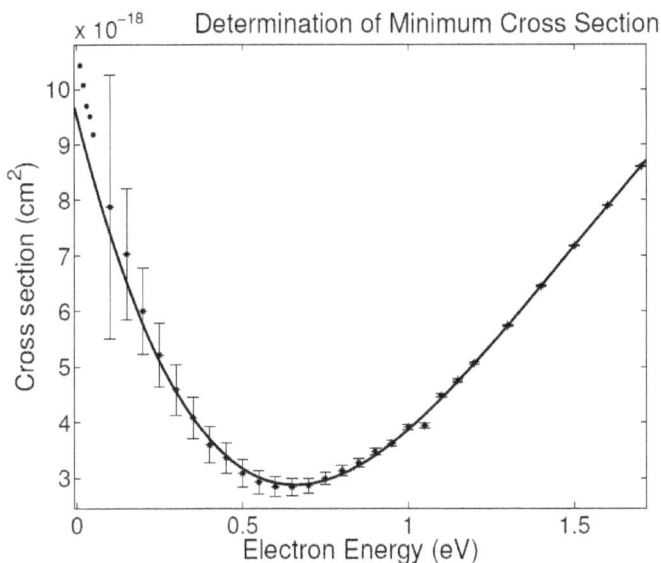

So it is confirmed that the minimum of the cross-section for the xenon is for the energy of 0.7 eV, that is a wavelength of 14.64 Å and an apparent diameter of the xenon atom of 3.66 Å. Supposing however that the celerities of the phase wave would be the same in the atom as outside it; it might be a strong assumption, never yet confirmed.

Curious:
- And what about verifying the coherence?

Open-Eyes:
- It is a devil of a job to find the diameter of the xenon in the literature. Exception in the (Bénard, Michel, Philibert, and Talbot. Masson Ed.) handbook of metallurgy: 2.39 Å mean radius, 4.78 Å diameter (approximative). A value that depends on the experimental method and the question it poses. Another ceiling could be by the density of the liquid at low temperature: 3,520 kg/m^3. The atomic mass is 131.3 g/mol, so we have the molar volume: 3.73 . 10^{-5} m^3. Dividing by the Avogadro constant gives the mean volume occupied by a xenon atom: 6.19 Å^3. If it were a cube, its edge would be 3.956 Å. If it is a sphere, we first have to estimate a coefficient of compacity in the liquid. Arbitrary choose: 0.71, as an intermediate between 0.74 of fcc and 0.68 of the bcc. Hence comes a radius of 2,19 Å and a diameter of 4.38 Å. Not so incoherent with the result obtained in interpreting the wavelength in the Ramsauer-Townsend transparency effect.

The defective approximation by hard spheres does not allow a more precise experimental cross-check: in the real world, the diameters of the atoms are very fuzzy, as we had already proved in chapter 3 when resolving the Schrödinger equation. Each experimental method returns a different answer, according to the question experimentally posed.

8.3.3. The other cases of proved Ramsauer-Townsend transparency.
In history, John S. Townsend and V. A. Bailey had experimented on argon, dihydrogen, next helium. Carl Ramsauer investigated with air, dihydrogen, dinitrogen, helium, argon. Later with krypton, xenon, neon, carbon dioxide CO_2. Brode experimented on methane CH_4, dioxygen, carbon monoxide CO, nitrogen oxide N_2O, next with monoatomic metallic vapors: Cd, Zn, Hg, next with metallic vapors of alkalies Na, K, Rb, Cs. Soon in the twenties and thirties years came new results on tens of organic molecules, and other monoatomic vapors, such as phosphor, such as cations Cs^+ and Ba^{++}. More recently with the vapor of beryllium, and as well electrons or positrons as the projectile.
The projectile also has been varied: were investigated the scattering of helium in helium, of atomic hydrogen on heavy atoms such as krypton. In all cases, the undulatory description from Louis de Broglie was confirmed by the experiments.

8.3.4. Conclusion: The Ramsauer-Townsend transparency effect is universal.
Everywhere, the qualitative conclusions are the same: for every gaseous obstacle, exists a Broglian wavelength which irons out the obstacles and makes the transparent.
The conclusion is unavoidable: the Broglian wave is not distinct from the moving electron, or the moving hydrogen or helium atom. Each of those projectiles **is** its Broglian wave. The dualist ideation is always contradicted by experience. The dualism is a blind alley they should eliminate as soon as the twenties, but they still teach a century later.

Curious:
- So, according to you, electron and electronic wave are the same thing. In your opinion, because it proved so, the Ramsauer-Townsend is struck by a censure.

8.4. Bibliography of the Ramsauer-Townsend effect.

J. S. Townsend and V. A. Bailey, « *The motion of electrons in gases* », Philosophical Magazine, vol. S.6, no 42, 1921, p. 873–891
J. S. Townsend and V. A. Bailey, « *The motion of electrons in argon* », Philosophical Magazine, vol. S.6, no 43, 1922, p. 593-600
J. S. Townsend and V. A. Bailey, « *The abnormally long free paths of electrons in argon* », Philosophical Magazine, vol. S.6, no 43, 1922, p. 1127-1128
J. S. Townsend and V. A. Bailey, « *The motion of electrons in argon and in hydrogen* », Philosophical Magazine, vol. S.6, no 44, 1922, p. 1033-1052
J. S. Townsend and V. A. Bailey, « *Motion of electrons in helium* », Philosophical Magazine, vol. S.6, no 46, 1923, p. 657-664
C. Ramsauer, « *Über den Wirkungsquerschnitt der Gasmoleküle gegenüber langsamen Elektronen* », Annalen der Physik, vol. 369, no 6, 1921, p. 513-540 (DOI 10.1002/andp.1
C. Ramsauer, « *Über den Wirkungsquerschnitt der Gasmoleküle gegenüber langsamen Elektronen. II. Fortsetzung und Schluß* », Annalen der Physik, vol. 377, no 21, 1923, p. 345-352 (DOI 10.1002/andp.19233772103)

MIT Department of Physics, August 28, 2013. *The Franck-Hertz Experiment and the Ramsauer-Townsend Effect: Elastic and Inelastic Scattering of Electrons by Atoms.*

David Bohm, *Quantum Theory*, Englewood Cliffs, New Jersey, Prentice-Hall, 1951

R. B. Brode, « *The Quantitative Study of the Collisions of Electrons with Atoms* », Rev. Mod. Phys., vol. 5, 1933, p. 257

W. R. Johnson and C. Guet, « *Elastic scattering of electrons from Xe, Cs^+, and Ba^{2+}* », Phys. Rev. A, vol. 49, 1994, p. 1041

Nevill Francis Mott, *The Theory of Atomic Collisions*, Oxford, Clarendon Press, 1965, chap. 18

David Whyte, *The Ramsauer–Townsend Effect*, Dublin, Trinity College Dublin, 18 mars 2010

International Journal of Modern Physics A, January 1997, Vol. 12, No. 02: pp. 305-378. *Quasifree Electron Scattering in Atomic Collisions: The Ramsauer–Townsend Effect Revisited.* M. W. Lucas, D. H. Jakubaßa-Amundsen, M. Kuzel, and K. O. Groeneveld (doi : 10.1142/S0217751X97000463)

Ramsauer-Townsend minima in the electron-scattering cross sections of polyatomic gases: methane, ethane, propane, butane, and neopentane D L McCorkle, L G Christophorou, D V Maxey and J G Carter. Journal of Physics B: Atomic and Molecular Physics, Volume 11, Number 17

L.G. Christophorou and D.L. McCorkle. *Experimental evidence of the existence of a Ramsauer-Townsend minimum in liquid CH_4 and Ar (Kr and Xe) and in gaseous C_2H_6 and C_3H_8.* Can. J. Chem. Vol 55. 1977.

F.A. Gianturco, D.G. Thompson. *The Ramsauer-Townsend effect in methane.* Journal of Physics B, At ; Mol. Phys. 9, L383.

W. Aufm Kampe, D.E. Oates 1, W. Schrader, H.G. Bennewitz. *Observation of the atomic Ramsauer-Townsend effect in ^4He-^4He scattering.* Chemical Physics Letters, Volume 18, Issue 3, 1 February 1973, Pages 323-324

W.H. Miller. *Molecular Ramsauer-Townsend effect in very low energy ^4He-^4He scattering.* Chemical Physics Letters. Vol. 10, Issue 1, 1 July 1971, pp. 7-9.

K. Jahankohan, H. Hassanabadi, S. Zarrinkamar. *Relativistic RamsauerTownsend effect in minimal length framework.* Modern Physics Letters A Vol. 30, No. 32, 1550173 (2015)1550173

J. Vahedi, K. Nozari. *The Ramsauer-Townsend Effect in the Presence of a Minimal Length and Maximal Momentum.* Acta Physica Polonica A. Vol 122 (2012) n° 1.

J. Vahedi, K. Nozari, P. Pedram. *Generalized Uncertainty Principle and the Ramsauer-Townsend Effect.* 9 August 2012.

David D. Reid, J.M. Wadehra. *Scattering of low-energy electrons and positrons by atomic beryllium: Ramsauer-Townsend effect.* Aug, 2014. J. Phys. B: At. Mol. Phys.

Stephen G. Kukolich (1968). *Demonstration of the Ramsauer-Townsend Effect in a Xenon Thyratron.* American Journal of Physics, 36(8), 701-703.

David-Alexander Robinson; Jack Denning; 08332461. *The Ramsauer-Townsend Effect.* 25 March 2010.

Martha Buckley, MIT Department of Mathematics. *The Ramsauer-Townsend Effect.* December 10, 2002.

M. Kuzel, R. Maier, O. Heil, D.H. Jakubassa-Amundsen, M.W. Lucas, K.O. Groeneveld. *Ramsauer-Townsend Effect in the Electron Loss from H^0 colliding with Heavy Atoms.* Physical Review Letters, volume 71, number 18; 1 November 1993.

8.5. Corpuscularist dumbness...

Open-Eyes:

- In general terms, each time a believer in the corpuscularism ventures into treating on optical phenomena which are perfectly undulatory, he or she is exposed to ridicule oneself heavily. Linus Pauling was Nobel laureate in 1954. So it would be a terrible *sin of pride* to dare to show up one of his blunders – and he committed some. At the end of his life, Linus Pauling engaged in a fanatic campaign in favor of vitamin C at extreme doses.

There, it was in 1935, when he rode the triumph of the **Knabenphysik**, with his book "*Introduction to Quantum Mechanics, with Applications to Chemistry*", he and E. Bright Wilson Jr re-invented the Bragg's law in their fashion, without any more a word of undulatory, but inventing a quantification of the (linear) momenta; this trick was necessary to reduce the photons to neo-Newtonian corpuscles. I let you enjoy the extracts from pages 34 to 36 of this book.

Figure 8.13.

6e. Diffraction by a Crystal Lattice.—Let us consider an infinite crystal lattice, involving a sequence of identical planes spaced with the regular interval d. The allowed states of motion of this crystal along the z axis we assume, in accordance with the rules of the old quantum theory, to be those for which

$$\oint p_z dz = n_z h.$$

For this crystal it is seen that a cycle for the coordinate z is the identity distance d, so that (p_z being constant in the absence of forces acting on the crystal) the quantum rule becomes

$$\int_0^d p_z dz = n_z h, \quad \text{or} \quad p_z = \frac{n_z h}{d}. \quad (6\text{–}15)$$

Any interaction with another system must be such as to leave p_z quantized; that is, to change it by the amount $\Delta p_z = \Delta n_z h/d$ or nh/d, in which $n = \Delta n_z$ is an integer. One such type of interaction is collision with a photon of frequency ν, represented in Figure 6–4 as impinging at the angle ϑ and being specularly reflected. Since the momentum of a photon is $h\nu/c$, and its component along the z axis $\dfrac{h\nu}{c} \sin \vartheta$, the momentum transferred to the crystal is $\dfrac{2h\nu}{c} \sin \vartheta = \dfrac{2h}{\lambda} \sin \vartheta$. Equating this with the,

Figure 8.14.

allowed momentum change of the crystal nh/d, we obtain the expression

$$n\lambda = 2d \sin \vartheta. \qquad (6\text{-}16)$$

This is, however, just the Bragg equation for the diffraction of x-rays by a crystal. This derivation from the corpuscular view of the nature of light was given by Duane and Compton[1] in 1923.

Let us now consider a particle, say an electron, of mass m similarly reflected by the crystal. The momentum transferred to the crystal will be $2mv \sin \vartheta$, which is equal to a quantum for the crystal when

$$n\frac{h}{mv} = 2d \sin \vartheta. \qquad (6\text{-}17)$$

FIG. 6-4.—The reflection of a photon by a crystal.

Thus we see that a particle would be scattered by a crystal only when a diffraction equation similar to the Bragg equation for x-rays is satisfied. The wave length of light is replaced by the expression

$$\lambda = \frac{h}{mv}, \qquad (6\text{-}18)$$

which is indeed the de Broglie expression for the wave length associated with an electron moving with the speed v. This simple consideration, which might have led to the discovery of the wave character of material particles in the days when the old quantum theory had not yet been discarded, was overlooked at that time.

In the above treatment, which is analogous to the Bragg treatment of x-ray diffraction, the assumption of specular reflection is made. This can be avoided by a treatment similar to Laue's derivation of his diffraction equations.

The foregoing considerations provide a simple though perhaps somewhat extreme illustration of the power of the old quantum theory as well as of its indefinite character. That a formal argument of this type leading to diffraction equations usually derived

[1] W. DUANE, Proc. Nat. Acad. Sci. 9, 158 (1923); A. H. COMPTON, ibid. 9, 359 (1923).

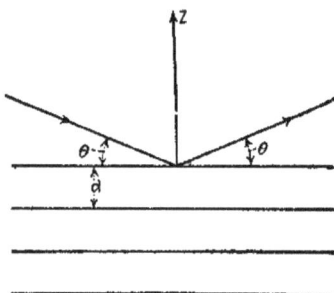

by the discussion of interference and reinforcement of waves could be carried through from the corpuscular viewpoint with the old quantum theory, and that a similar treatment could be given the scattering of electrons by a crystal, with the introduction of the de Broglie wave length for the electron, indicates that the gap between the old quantum theory and the new wave mechanics is not so wide as has been customarily assumed. The indefiniteness of the old quantum theory arose from its incompleteness— its inability to deal with any systems except multiply-periodic ones. Thus in this diffraction problem we are able to derive only the simple diffraction equation for an infinite crystal, the interesting questions of the width of the diffracted beam, the distribution of intensity in different diffraction maxima, the effect of finite size of the crystal, etc., being left unanswered.[1]

Figure 8.15.

However, their postulate of quantification of the (linear) momenta was never confirmed: it is blown up by any change of frame, and is even worse with relativistic changes of frame. An arbitrary postulate, without any basis, even theoretic.

You have read that at the end a shadow of doubt crosses their minds: they cannot more give an account of the effects of the size of the crystallites on the width of the reflections (Scherrer's law), of the effects of he not-so-monochromaticity of the incident radiation, of the law of extinct reflexes according to the atomic pattern in the cell. Indeed?

But then? This crushing triumph of *"the new physics"* (a corpuscularist one), after which *there will be no more prophets after me*, over the *"classical physics"* (physical optics, Augustin Fresnel, 1819), it was not so total and crushing as trumpeted?

In reality, the radiocrystallographers only use the physical and undulatory optics, this one elaborated by Fresnel, and never the phantasmagoric quantification of the linear momenta fantasized by Wilson and Pauling, for the glory of the Copenhagen communautarism. Now, it was by the size of the crystallites regarding the width of the peaks of the diffractogram I could confound an international crook:

http://impostures.deontologic.org/index.php?topic=133.0
http://deonto-ethics.org/resources/Corrige_expertise.html

The engineer Gleizes missioned to start the plant stumbled on the inadequate quarry material: it is not clay but silt[6]. It has no plastic coherence and could never be formed by extrusion.

Link on the Scherrer's relation:
http://en.wikipedia.org/wiki/Scherrer_equation

The same Paul Scherrer who co-invented with Piet Debye an operating mode in radiocrystallography, known as the *Powder diffractions*, in 1915.

---

[6]**Silt**: diameters between 2 μm and 50 μm. No plastic cohesion.

8.6. The Goos-Hänchen effect in plane polarization and the Imbert-Fedorov effect in circular polarization.

Open-Eyes:
- The Goos-Hänchen effect in plane polarization and the Imbert-Fedorov effect in circular polarization are two new proofs of a notable width of each photon, and give floor values of it.
http://journals.aps.org/pr/abstract/10.1103/PhysRev.139.B1443 but only the abstract is free.
http://journals.aps.org/prl/abstract/10.1103/PhysRevLett.96.073903 Idem.
https://inspirehep.net/record/1203926?ln=fr provides an excellent global synthesis.
https://imphscience.wordpress.com/revisiting-reflection-ii/
provides a shorter sum up. The original from Fedorov in russian, then its english translation:
http://master.basnet.by/congress2011/symposium/spbi.pdf
We will only give an abstract, as the details are over the level of this initiation handbook. They are slight deviations to the law given by the geometric optics (for Euclid) to the total reflection from a material more refractive than the external milieu. Total reflections as for instance in Porro prism. In plane polarization, the emerging ray is slightly farther than predicted, as there was an evanescent wave outside. The discrepancy is about the wavelength, between ten times and a tenth. In circular polarization, the deviation is to the right or to the left from the incident plane, depending on the polarization. Here again, the discrepancy is about the wavelength, between ten times and a tenth. All the experimental results are given in terms of beams; a complete re-reading is necessary to draw from them the laws at the individual scale of the photon.

Curious:
- You speak, but you do not offer any figure. Who can follow you?

Open-Eyes:
- Please be patient: this piece of science is still in progress, and you have the figures in the quoted references. Two kinds of phenomena require each three-dimensional figures in almost all kinds of polarization (plane or circular or worse, mixed): the first one has to describe the fields in the beam or in the photon, and their evolution by the reflection. The other describes the evolution of the bounds and the axis of the beam. Now all the experiments were done at the scale of the entire beam, though we demand a law at the individual scale of the photon. This law describes the light in a transparent material, so with strong coupling with the electronic clouds of this transparent dielectric.

CHAPTER 9

The relativist frame, and its constraints on the microphysics

9.1. The relativity, is it over?

Curious:

- I just have read in Agoravox that t*he Relativity, it is over and obsolete,* and they will soon find damn bloody better. Great heavens! Did they pull out the teeth to the devil? On Usenet, you also have some weird invaders who swear that they have revolutionized all that, and that Albert Einstein was *nothing but a bum...* Some more details, please?

Open-Eyes:

- At the end of the 17th century, Isaac Newton had all the excuses to believe that an absolute time, this one of his god, could exist. He also imagined that an absolute space and an absolute frame existed, precisely those of his god *to him he had.* Three hundred and thirty years later, we have no more excuses. We have *forty-seventeen-plus-eleven* proofs that from one place to another, from one particle to another, the flows of time differ, and the possibilities to measure space differ also.

Curious:

- In these conditions, with proofs everywhere, the Relativity cannot be else than perfectly integrated into microphysics! Rather! In a hundred and thirteen years!

Open-Eyes:

- Precisely not! It is not yet integrated. A big theoretical impediment is the corpuscularist postulate they kept against all the experimental evidence. For a long time, I had underestimated the other big impediment: the anti-relativist postulate. Stubbornly, they still believe in the Newtonian macro-time, divine ubiquitous parameter. So all the heirs of the Göttingen-København sect, entangled in their anti-relativist postulate for ninety-one years, for 1927, persist to trip over their beards, and must invoke mysterious *"collapses"*, whose they never provide the physics. The last paper from Roland Omnès persists in invoking his mysterious *collapses* whose mysteries resist as much as the *holy trinity* remained inexplicable to Aurelius Augustinus, bishop of Hippone (354-430): **Scheme of a Derivation of Collapse from Quantum Dynamics**. In pdf:

1601.01214.pdf at the address http://arxiv.org/abs/1601.01214 Magnanimously, I do not recall the other impediments: tribal, territorial, rhetorical, even criminal, whose the harasser Marmot is a sample.

9.2. Let us begin by the Pound and Rebka experiment, in 1959 at Harvard

https://en.wikipedia.org/wiki/Pound%E2%80%93Rebka_experiment
Link to the original publication:
http://journals.aps.org/prl/pdf/10.1103/PhysRevLett.3.439
What was to prove was: $gh/c^2 = 2.5 \times 10^{-15}$.
More recent experiments by Pound and Snider:
http://journals.aps.org/prl/abstract/10.1103/PhysRevLett.13.539
You will find a complete course on the used General Relativity at
luth2.obspm.fr/IHP06/lectures/mester-vinet/IHP-2GravRedshift.pdf

Professor Castle-Holder:
- What made the experiment possible at all was the exceptional fineness of the gamma ray emitted or absorbed by the nucleus of the 57 iron, discovered by Rudolf Mössbauer (in the first time, he worked on iridium 191): relative half-width = 3.10^{-13}. If you have access to a university library, I recommend you the *Atomic Physics* by Eduard Shpolsky, Mir Publishers. In § 129 and 130, he gives an excellent exposé. First, the use of the H_β spectrum ray of the hydrogen to verify the second-order Doppler effect; next he presents the experimental constraints, notably to thermostat the emitter and the absorber **for the spectrum line at 14.40 keV of ^{57}Fe.**

Mir Publishers practically disappeared with the USSR. Most of their publications are now unavailable. Springer Verlag has purchased the whole of french and english translations and publishes some very few old library success (the Landau and Lifshitz, essentially), only paperback and at a Springer price, and dropped the remaining to the mice. What interested them was to eliminate a competitor. The remaining used books from Mir are sold at a collector price.

Open-Eyes:
- In that experiment, mounted in Harvard in 1959, they measured the effect of the difference of gravity potential on 21 m, on the flowing of time, therefore on the resonant frequency of the ^{57}Fe. What was to prove was $gh/c^2 = 2.5 \times 10^{-15}$. The difference of altitude, so small compared to the Earth radius, multiplied by the average gravity at that altitude gives a difference of gravity potential which we denote $\Delta\varphi$. Hence the new Mössbauer frequency:
$\nu' = \nu(1 + \Delta\varphi/c^2)$.
The experimenters mounted the source of γ at the center of the cone of a big boomer (like those that Altec Lansing made then) driven in low frequency. Then the matter was to measure at which phase the selective absorption by a

^{57}Fe filter occurred before the detector, to deduct the speed which by Doppler-Fizeau shift, exactly compensated the effect of the gravity potential. The experiment was performed in the two directions, up to down, and down to up, in the tower of the physics laboratory. Already convincing as a confirmation of the equations of the General Relativity, the experiment was done again, with a much-improved accuracy in 1964 by Pound and Snider. Many measures have also been done with atomic clocks installed in civil liner planes for long ways. They have fully confirmed the corrections predicted by the Special Relativity (speed) and the General Relativity (altitude). Sure, those are very small deviations, which demanded the best metrologic means, while a scenario involving an intersidereal rocket at relativistic speeds, as necessary for the *Langevin's paradox of the twins* demands an unattainable stock of propellant, impossible to gather on Earth. The building of the monstrous rocket too belongs to fiction. We have many confirms that the relativist laws are precisely verified, on accessible and real cases. The precision of the experiments by Pound and Rebka, next by Pound and Snider, were well improved since, with a hydrogen maser, with an accuracy of 10^{-4} of the relativist effect to prove.

Curious:
- I sum up what I understand: a difference of altitude of 21 m is enough, at the surface of the Earth, to measure the difference of flow of the time, of the macro-time precisely! But then, does it imply that among all the molecules of gas, which agitate in all directions at speeds about the sound speed, they all have a different and incompatible *"sense of timing"* (the micro-time)? So would be the inevitable consequence of the Doppler-Fizeau effect, though discovered in the 19th century? It is the first time somebody focuses my attention on that.

Professor Castle-Holder:
- That is why the characteristic spectral lines of a gas are enlarged by the thermal agitation, as well in emission and in absorption. We can use these widths of the spectral lines to evaluate the temperature of a gas, even at astronomical distances.

Open-Eyes:
- The application to the microphysics is not less surprising, in the eyes of most people. We have seen sooner the differences in energy of an electron (or of the whole atom, or molecule as well) according to the level it occupies, the ground level or the lowest it can occupy, versus an excited one. Hence results in a divergence in the micro-times attached to these electrons, predicted by the Relativity.

Curious:
- But if I see this through, you imply that for the most bound electron, its micro-time flows more slowly than for an electron less bound, or a free electron? Not speaking about an electron in a particle accelerator.

Open-Eyes:

- Perfect! You got the message. And the difference of Broglian frequencies between these states is exactly the frequency taken out by the emitted photon, or brought in by the absorbed photon. It was already described by Erwin Schrödinger[1], in his paper sent in September 1926 to the Physical Review; Alas, in his forgetful bad luck, Schrödinger had omitted to come back to the relativistic frame for this end of paper so his initial and final frequencies for the electron were very far from real Broglian frequencies, and deprived of any physical meaning. It was enough for his enemies to forget this part of his job, instead of (easily) correct it.

Professor Castle-Holder:
- These lacunae and hard lucks in the work of Schrödinger were corrected two years later, in 1928 by Paul Adrien Maurice Dirac (1902-1984). "*The Quantum Theory of the Electron*",
http://www.math.ucsd.edu/~nwallach/Dirac1928.pdf
or https://www.jstor.org/stable/94981
Dirac already knew a first relativistic solution, now known as Klein-Gordon's – already found by Schrödinger, but left on the roadside – but he saw it as inadequate for the electron; since, we know it fits for the spinless particles, such as the pion. Dirac decided to no more having a quadratic equation, but entirely of the first degree (such as in the Liapunov's[2] representation, for the non-harmonic oscillations).

$$\left(\beta mc^2 + c \left({\textstyle\sum_{n=1}^{3}} \alpha_n.p_n\right)\right)\psi(x,t) = i\hbar\frac{\partial\psi(x,t)}{\partial t}$$

However, the mathematical difficulty took a "*Great Leap Forward*": the coefficients α and β in Dirac are 4 x 4 matrices. Whichever the variant adopted to express them, we are usually in moral misery for interpreting them.
The p_1, p_2, and p_3 are the coordinates of the momentum as operators. m is the rest mass. Now, the Ψ function has four components and is named a

---

[1]Erwin Schrödinger. *An Undulatory Theory of the Mechanics of Atoms and Molecules*. Phys. Rev. 28, 1049 – Published 1 December 1926.
Abstract (http://journals.aps.org/pr/abstract/10.1103/PhysRev.28.1049)
The paper gives an account of the author's work on a new form of quantum theory.
§ 1. The Hamiltonian analogy between mechanics and optics.
§ 2. The analogy is to be extended to include real "physical" or "undulatory" mechanics instead of mere geometrical mechanics.
§ 3. The significance of wave-length; macro-mechanical and micro-mechanical problems.
§ 4. The wave-equation and its application to the hydrogen atom.
§ 5. The intrinsic reason for the appearance of discrete characteristic frequencies.
§ 6. Other problems; intensity of emitted light.
§ 7. The wave-equation derived from a Hamiltonian variation-principle; generalization to an arbitrary conservative system.
§ 8. The wave-function physically means and determines a continuous distribution of electricity in space, the fluctuations of which determine the radiation by the laws of ordinary electrodynamics.
§ 9. Non-conservative systems. Theory of dispersion and scattering and of the "transitions" between the "stationary states."
§ 10. The question of relativity and the action of a magnetic field. Incompleteness of that part of the theory.
[2]Alexandr Mikhaïlovitch Liapounov, russian mathematician, 1857-1918.

bispinor. The matrices $\boldsymbol{\alpha}$ and $\boldsymbol{\beta}$ are Hermitian, and their square is the unity matrix.

$\boldsymbol{\alpha^2} = \boldsymbol{\beta^2} = I_4$

Moreover, they anticommut (i and j distinct): $\boldsymbol{\alpha_i \, \alpha_j + \alpha_j \, \alpha_i} = 0$

$\boldsymbol{\alpha_i \, \beta + \beta \, \alpha_i} = 0$

They are a good sample of Clifford's algebra, created in 1878 by William Kingdon Clifford (1848-1879).

Open-Eyes:
- The first surprise coming from this mathematical invention by Dirac was that the spin of the electron came as a natural consequence. The second surprise came in 1930 and 1932 by the solution given by Erwin Schrödinger, who proved as a consequence, the **Zitterbewegung**, or the « *Schrödinger trembling motion* » : The microphysical velocity of an electron is always alternatively +**c** and -**c** in the direction of the macrophysical motion. The frequency of this alternation is the double of the Broglian frequency, so **2mc²/h**. But the distribution of the two signs in time depends on the group speed. The amplitude of the displacement during the alternation seems to be h/mc, which lefts us embarrassed in rightfully interpreting this.

Next, Schrödinger proved that the spatial equidistance Dirac-Schrödinger is the right one to explain the Compton scattering by the Bragg law of diffraction. I had to rediscover it in 2011, as the entire world ignored this discover from Schrödinger; only the Nobel lecture on Dirac in 1933 mentions it; fanatic radio silence anywhere else in the world.

http://www.deonto-ethics.org/quantic/index.php?
title=Calcul_diffusion_Compton_et_Zitterbewegung

Curious:
- Please give a numerical application, so that we can have an idea of the magnitudes.

Open-Eyes:
- I fear that you will regret your request, as the mathematical things will become harder. Many readers will skip the next sub-chapter, and directly jump to the next chapter.

9.3. Instrumentation and constraints

An experimental dissymmetry is fundamental: when you and your laboratory pretend to describe the speed, the time and the length of what you consider to be "*a mobile*", you need a speed measurement basis, with its instruments, which is connected to your laboratory. To measure the speed of the mobile, you need two clocks, at a fixed distance, which next will be able to communicate; these two distant clocks, synchronized in your frame constitute for you a length basis. If moreover, you will question your perspective seeing of the time of the mobile, you need the mobile has a clock, which indicates the proper time of the mobile. So what you see on the mobile and its internal clock is never a proper time, but only a perspective view. If moreover, you will gauge the fate of the proper length of the mobile, but viewed in a relativistic perspective, you need that the mobile also has a speed measurement basis. And if moreover, you wish to proceed with the inverse operation, where the mobile measures your laboratory, it needs to have **two** clocks, synchronized in its frame.

Here is a speed measurement basis, as described by the *Instructions Nautiques* (*Service hydrographique de la Marine*) :

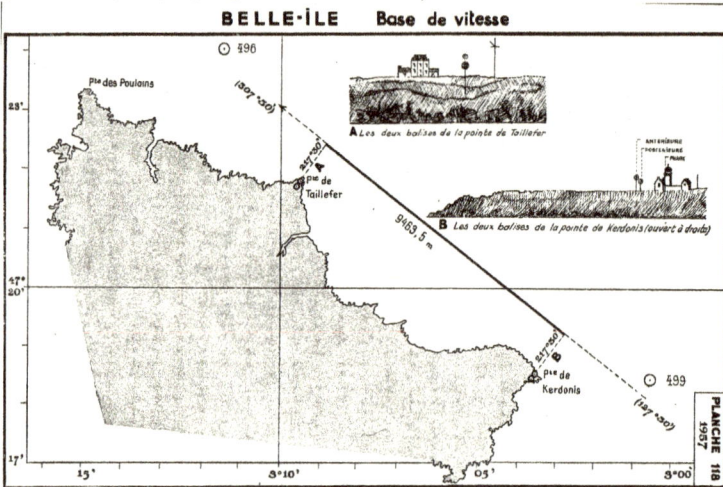

The main difference is that the chronometer is aboard your ship, to measure the interval of time between the two alignments, Kerdonis and Taillefer, though, for a laboratory which studies the trajectory of an elementary particle, the clocks are on the two end alignments. Application in microphysics: the clock proper to the mobile if it is a Fermion, is its intrinsic Dirac-Schrödinger $2mc^2/h$ in its average and smoothed frame, but only recently a team accessed to it, experimentally. Not any elementary particle carries a length basis.

Professor Castle-Holder:

- However, we have an astonishing and everyday verification of the relativistic perspective on the lengths: the magnetic effects of the electric currents, which are known since Œrsted in 1820, and which André-Marie Ampère was the first to give the laws.

The magnetic force is a relativistic correction in v^2/c^2 to the Coulomb force.

First the qualitative reasoning *with the hands*, which is enough when the two intensities are pointing in the same direction. Let us take two parallel wires A and B, where the same current i is flowing in the same direction.

We will draw them horizontal on the blackboard, with intensity flowing to the left. The lattice of copper ions in A sees the lattice of copper ions in B as motionless (in the frame of A). But it sees the conduction electrons of B with an average drift to the right at a speed of about few tens micrometers per second. So the relativistic correction applies to them, it "sees" them denser than the copper ions. So it is attracted by these charges "-" more than it is repulsed by the charges "+" of the copper lattice; and reciprocally, it attracts them (the "-"

charges).

And you repeat the reasoning on the "*seeing*" of the copper ions of B by the mean electrons of A.

The final result of this relativistic perspective: the wires A and B are attracted if the intensities point in the same direction. And what if the intensities are opposed? Repulsed? Now we need to carry the real calculation.

We take the metrological case, in principle:

Two conducting wires of infinite length, and from which we extract one meter, distant by one meter, and supporting one-ampere intensities.

i.dl = 1 A * 1 m = Q.v

For repartition of Q.v between Q and v we choose a reasonable electrotechnical value: $v = 10^{-4}$ m/s. Hence Q (by meter) = 10^4 C (ten thousand coulombs).

The contraction of the lengths, developing the radical only at the first order :

$1 - \frac{1}{2}\frac{v^2}{c^2}$.

Let **F** be the Coulomb force between all the copper ions of A and all the copper ions of B, repulsive. Between two punctual charges **Q** and **Q'** at the distance

R: $F = \frac{1}{4\pi\varepsilon_0}\frac{Q.Q'}{R^2}$

Between two wires, of negligible thickness, where **dl** is the element of length, of lineic charge λ, that is a charge d**Q**, at a distance R:

$dF = \frac{1}{2\pi\varepsilon_0}\frac{\lambda.dQ}{R}$

Summed on a meter of wire: $F = \frac{1}{2\pi\varepsilon_0}\frac{\lambda.Q}{R}$

$F = \frac{1}{2\pi\varepsilon_0}\frac{Q^2}{R.1m}$

And at a distance of one meter: $F = \frac{1}{2\pi\varepsilon_0}\frac{Q^2}{1m^2}$

Between the copper ions of A and the electrons of B: $-F.(1+\frac{1}{2}\frac{v^2}{c^2})$ (attractive).

Between the copper ions of B and the electrons of A: $-F.(1+\frac{1}{2}\frac{v^2}{c^2})$ (attractive).

Between the conduction electrons of A and those of B (speed 2v): $F.(1+\frac{4}{2}\frac{v^2}{c^2})$ (repulsive).

This term is new, in the case of the opposed intensities: $2\frac{v^2}{c^2}$

Total electromagnetic force, still at first-degree approximation: $F_e = F.\frac{v^2}{c^2}$.

While we had $-F.\frac{v^2}{c^2}$ with the intensities pointing in the same directions.

So we have the good signs. Have we the good dependence to intensities?

The force is proportional to the intensity in one conductor, and to the intensity in the other, so to i^2 if the two intensities are equal in absolute value.

Only remains to verify the correctness of the predicted value, with the good coefficient.

$|Fe| = F.\frac{v^2}{c^2}$ with $F = \frac{1}{2\pi\varepsilon_0}\frac{Q^2}{R.1m}$

$|Fe| = \frac{1}{2\pi\varepsilon_0.c^2}\frac{(v.Q)^2}{1m^2}$

Then $\varepsilon_0.c^2 = \mu_0$ and v.Q = i.L

$|Fe| = \frac{i^2}{2\pi.\mu_0}$

Or in the more general case of a length l of wires, at a distance **d**:

$|Fe| = \frac{i^2}{2\pi.\mu_0}\frac{l}{d}$

and by the legal definition of the ampere: $4\,\pi.\mu_0 = 10^{-7}$ H.m^{-1}. End of the demonstration.

Curious:
- I object to your demonstration of magic: you have used the speed of the average drift of the conduction electrons. While the Lorentz correction is in v^2/c^2, so nonlinear, and these electrons never stop to run at the Fermi speed. You have said it is about 1,570 km/s in the copper.

Professor Castle-Holder:
- This a very good question! I thank you for having posed it. Another way to pose your question is to ask whether the electric neutrality of a conductor is a myth, seen by a relativistic mirage. A third way would be to ask whether the Fermi speeds are not mythic. However, the macrophysical laws established in the 19th century are unassailable, very solid.

Curious:
- But you have not answered!

Open-Eyes:
- To simplify a problem in physics, the first rule is to look at the symmetries. The two conducting wires are the place of the Fermi electronic speeds, but they do not have a macroscopic effect. Switch off the current, in other words, zero the average drift of the electrons, and all the magnetic forces become null. Only remains the gravity, the elastic stresses, the thermal dilatations, and shrinkages. When a power transformer supplies a plant in full work, the noise it emits is intense: the coils vibrate, and the steel sheets of the magnetic circuit vibrate. When the work hours are over in the plant, no more much current at the secondary coil and the noise is now bearable: the primary coils feed mainly the losses in iron. Switch off the primary too, and the noise becomes null. We conclude that the Fermi speeds may interest the metallurgist, surely interest the physicist of the solid state, but do not concern the electrical engineer.
A personal point of view now: the relativistic explanation treated in the plane, the plane problems, without ever lumbering with a folkloric "*vector*" in an unuseful third dimension. The Relativity respects the physical symmetries; but it does not respect the traditions in "*cross product*" we owe to Oliver Heaviside (1888), a tradition which does not respect nor the physics, nor the mathematics.

9.4. The Lorentz transform

The Lorentz transform is nothing else than the Pythagoras relation, but in non-Euclidean space, with now a pseudo-Euclidean metrics; the conservation of the speed of light in all the frames is written by the conservation of the distance in the Minkowski metrics. Here we simplify the writing by taking x'Ox along the relative speed of the two frames.

$$ds^2 = dx^2 + d(ict)^2 = dx^2 - c^2\,dt^2 = -c^2\left(1 - \frac{v^2}{c^2}\right)dt^2$$

Hence the transformation of the time from one frame to the other:

$t_1 = \frac{t'_1 + \frac{v\Delta x'}{c^2}}{\sqrt{1 - \frac{v^2}{c^2}}}$ and $t_2 = \frac{t'_2 + \frac{v\Delta x'}{c^2}}{\sqrt{1 - \frac{v^2}{c^2}}}$ hence the difference:

$\Delta t = \frac{\Delta t'}{\sqrt{1 - \frac{v^2}{c^2}}}$

One may make the formulation more compact dy doing: $\beta = \frac{v}{c}$. it is a number, dimensionless.

Then the rapidity $\varphi = $ Artanh $\frac{v}{c} = $ Artanh β. This dimensionless quantity φ is often considered as a "hyperbolic angle".

$\gamma = \frac{1}{\sqrt{1 - \frac{v^2}{c^2}}} = \frac{1}{\sqrt{1 - \beta^2}} = \cosh(\varphi)$

In Relativity the speeds are not additive, but the rapidities φ remain additive. On may write $\Delta t = \gamma \, \Delta t'$

And the transformation of the lengths, in the direction of the relative speed:

$\Delta x' = \frac{\Delta x - v\Delta t}{\sqrt{1 - \frac{v^2}{c^2}}} = \frac{\Delta x}{\sqrt{1 - \frac{v^2}{c^2}}}$

The length shrinks, while the time dilates, seen from the observer who sees the mobile passing. The experimental dissymmetry remains fundamental: when you pretend to describe the speed, the time and the length of what is for you the mobile, you need a speed measurement basis bound to **your** laboratory.

Open-Eyes:
- There is a minus sign under the radical. So instead of the sines and cosines intervening in the isometries of our ordinary Euclidian space R^3 (that is the rotations, which are of $+1$ signature), in Relativity, we must use the hyperbolic trigonometry: hyperbolic sine, hyperbolic cosine, hyperbolic tangent. Next, we can use all the facilities of the linear algebra.

Professor Castle-Holder:
- Before the linear algebra in dimension 4, let us begin by an exercise: convert the 6 GeV of the synchrotron facility in Grenoble ESRF into wavelengths of the electrons in the beam.

9.5. Exercise of relativistic dynamics

Problem:
The electrons running in the main ring are accelerated and kept at the energy 6 GeV. Your mission is to give their wavelength in our fixed frame.

Formulas:

Energy $E = \frac{m.c^2}{\sqrt{1 - \frac{v^2}{c^2}}} = \gamma.m.c^2$

Momentum **p**: $m^2 c^4 = E^2 - p^2 c^2$,

hence $|\mathbf{p}| = E/c \sqrt{1 - (\frac{mc^2}{E})^2} = \gamma m v = \frac{v.E}{c^2}$

Therefore the momentum: $p^2 = [(6 \text{ GeV})^2 - (511 \text{ keV})^2] / c^2 = (E/c)^2$. $(1 - 7.25 \cdot 10^{-9})$.

Hence $\mathbf{p} = 6$ GeV $/ c = 961.3 \cdot 10^{-12}$ J $/ c = 3.206 \cdot 10^{-18}$ kg.m/s

Then we have already seen the de-Broglie law: $\lambda = \mathbf{h} / \mathbf{p} = \mathbf{206.6}$ **am** (attometers). It is very far from the human scale, and even of the atomic scale. Indeed, it is apt to yield nuclear reactions. Therefore, several weeks after stopping the ring, the metals of the apparatus remain radioactive, after side-effects of the 6 GeV electrons.

And what does become their intrinsic clock frequency, seen from our frame?
6 GeV / c = 3.206 . 10^{-18} kg.m/s
Divided by the rest mass of the electron: 03.52010. 10^{12} m/s = 11 741.8 **c**
Hence the rapidity φ (dimensionless number, in implicit unit **c**):
φ = Argsinh(11 741.8) = 10,06406
Hence the gamma factor: γ = cosh φ = **11 741.8** (at these high energies, the hyperbolic sine and cosine are almost equal).
Recall of the Broglian intrinsic frequency of the electron ($\frac{m \cdot c^2}{h}$ at rest): 1.23559 . 10^{20} Hz
Apparent slowing of the de-Broglie internal clock: $\frac{1}{\cosh(\varphi)} = \frac{1}{\gamma}$
Hence the apparent frequency, seen from the laboratory:
1.23559 . 10^{20} Hz / γ = **10.523001 . 10^{15} Hz**$_{laboratory}$
The other intrinsic frequency, the Dirac-Schrödinger one, which intervenes in the electromagnetic phenomena, such as the Compton scattering, is twice the Broglian one: 21.046003 . 10^{15} Hz$_{laboratory}$
The Broglian period in the frame of the laboratory, the inverse of the frequency: 95.02992 . 10^{-18} seconds per cycle.
And no, you cannot multiply this apparent period by the apparent speed to obtain the wavelength. It would be false!

Curious:
- So, according to you, it would be enough for me to review the lesson to soon become a distinguished relativist?

Professor Castle-Holder:
- Anyway, now you have a compendium of relativistic formulas, and two samples of the use. Further, you will find another example, the experimental measurement of the Dirac-Schrödinger frequency in the frame of the Linear Accelerator of Saclay (ALS).

9.6. The Lorentz transform, diagonalized.

Open-Eyes:
- I will worsen my case, by showing that if we respect the discipline of the algebraist and search the proper directions according to which the matrix for the Lorentz transform is diagonal, the result is heavily humiliating for our anthropocentrism: the proper directions are all on the light cone. Now, we will never be on the light cone, we... Forever, we will remain improper observers: we have a mass, and all our instruments have, too.

9.6.1. All the proper directions are on the light cone. Let give the index **0** to the time coordinate, so it leaves unchanged all we have already done

with the spatial coordinates 1, 2, and 3 in dimension 3. The habit of writing **c** = 1 is dangerous, let us do so nevertheless, for the brevity in notations.

The special feature of the Minkowski space, is that for all events related to some electromagnetic wave, or for any massless wave, the competent metrics seen from macroscopic frames with mass, is pseudo-Euclidian:
$ds^2 = dt^2 - dx^2 - dy^2 - dz^2 = (dt; dx; dy; dz) .g . {}^t(dt; dx; dy; dz)$
where the physical unit is omitted, m^2 or s^2. And **g** is the metric tensor.
We cannot deduct anything about the metrics proper to the photon:
how it could "*see*" its spreading in a few periods (femtosecond laser) to many milliards periods.
Coordinates of the metric tensor in one of our human frames:

$$\mathbf{g} = \begin{pmatrix} 1 & 0 & 0 & 0 \\ 0 & -1 & 0 & 0 \\ 0 & 0 & -1 & 0 \\ 0 & 0 & 0 & -1 \end{pmatrix}$$

Any propagation of the photon kind, with a null mass, has the feature that its proper time is null.

In addition to the known isometries in the spatial sub-space, we have three new elementary Minkowskian isometries (denoted M-isometries). Let write the matrix describing a uniform translation along the x-axis with the speed **c.tanh(φ)**:

$$\mathbf{R_{01}}(\varphi) = \begin{pmatrix} \cosh\varphi & \sinh\varphi & 0 & 0 \\ \sinh\varphi & \cosh\varphi & 0 & 0 \\ 0 & 0 & 1 & 0 \\ 0 & 0 & 0 & 1 \end{pmatrix}$$ which may be diagonalized into:

$$\begin{pmatrix} e^\varphi & 0 & 0 & 0 \\ 0 & e^\varphi & 0 & 0 \\ 0 & 0 & 1 & 0 \\ 0 & 0 & 0 & 1 \end{pmatrix}$$ on the basis appropriate to this direction of propagation.

The matrix of this appropriate basis: $\frac{1}{\sqrt{2}}\begin{pmatrix} 1 & 1 & 0 & 0 \\ 1 & -1 & 0 & 0 \\ 0 & 0 & \sqrt{2} & 0 \\ 0 & 0 & 0 & \sqrt{2} \end{pmatrix}$

which we can rewrite so: $\begin{pmatrix} \frac{1}{\sqrt{2}} & \frac{1}{\sqrt{2}} & 0 & 0 \\ \frac{1}{\sqrt{2}} & -\frac{1}{\sqrt{2}} & 0 & 0 \\ 0 & 0 & 1 & 0 \\ 0 & 0 & 0 & 1 \end{pmatrix}$, independent of the swiftness φ.

Beware! Do not multiply those proper values e^φ and $e^{-\varphi}$ by **c**, hoping to obtain celerities: they are only proportions of each snapshot propagation; the majority in the direction of the macroscopic drift or propagation of the particle, the minority in the backward direction, each at speed **c**. Let us take the example of the electrons in a cathodic tube of a color TV set, accelerated under 24 kV. The ratio (total energy / rest energy) is $ch(\varphi) = 1.0469668$. The argument φ

equals Argch(1.0469668) = 0.305230. The speed of the electrons seen in the frame of the television set is 88,784,827 m/s, as macroscopic average. It is composed of a direct luminic move of duration proportional to e^φ= 1.35703, and a retro-luminic move of duration proportional to $e^{-\varphi}$ = 0.73690.

This is in contrast with the phase velocity of the Broglian wave on the propagation axis, obtained by an inversion whose radius is the light velocity. For these electrons, the phase velocity of their wave is largely supraluminic: $\mathbf{c^2/v}$ = $\mathbf{c.coth(\varphi)}$ = 3.37662 .c

While the Broglian wavelength of the electron is then 8.193 pm, or 8193 fm.

It is noticeable that so, the basis of diagonalization is entirely real. But its two basis vectors are M-isotropic, as both on the light-cone:

$$\frac{1}{\sqrt{2}} \cdot \begin{pmatrix} 1 \\ 1 \\ 0 \\ 0 \end{pmatrix} \text{ and } \frac{1}{\sqrt{2}} \cdot \begin{pmatrix} 1 \\ -1 \\ 0 \\ 0 \end{pmatrix} \quad \text{Their M-norms are null.}$$

In the same spatial direction, one is orthochronous; the other is antichronous. An alternative way of seeing is to consider that both are orthochronous, or antichronous as well, but with opposite propagations. Up to now, only one orientation was privileged, without experimental argumentation. These two vectors of a proper base are intrinsic to this x'Ox spatial propagation and are independent of the speed of propagation. Both on the light cone, these intrinsic axes are the asymptotes of two branches of equilateral hyperboles containing all the events at usual (human) finite distance.

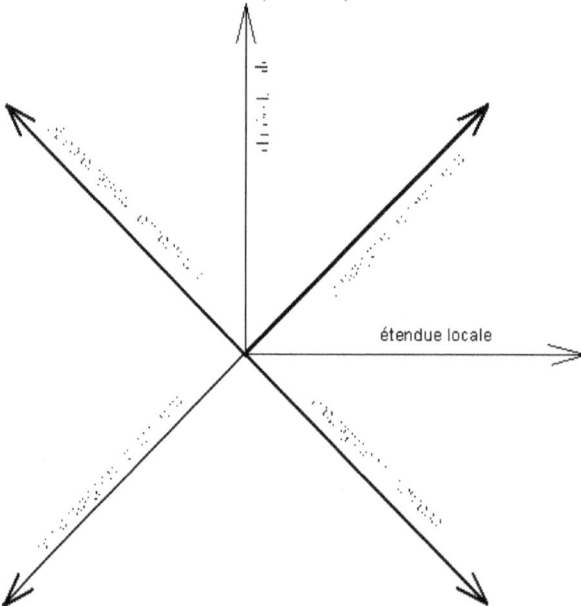

étendue locale

Figure 9.2. This figure is provisionally incomplete, without the branches of hyperboles.

By a Taylor expansion, we obtain the differential operator, so the generator:

$$\mathbf{dR_{01}} = d\varphi. \begin{pmatrix} 0 & 1 & 0 & 0 \\ 1 & 0 & 0 & 0 \\ 0 & 0 & 0 & 0 \\ 0 & 0 & 0 & 0 \end{pmatrix}, \text{ which is diagonalizable on the same basis into}$$

$$\begin{pmatrix} 1 & 0 & 0 & 0 \\ 0 & -1 & 0 & 0 \\ 0 & 0 & 0 & 0 \\ 0 & 0 & 0 & 0 \end{pmatrix} \text{ multiplied by } \mathbf{d\varphi} \text{ } (\mathbf{d\varphi}, \varphi \text{ real, for the proper isometries)}.$$

These three pairs of proper vectoroids (only one has been developed) belong to the light-cone or M-isotropic cone, where every vector in it is of null M-length in the meaning of the Minkowski pseudo-Euclidian norm.

We conclude that the three other prolongations of the gyrors (in $\mathbf{R^3}$ space) are completed by three other base operators which are symmetric, forming a base for the directions of propagation of an electromagnetic wave:

$$\mathbf{J_{01}} = \begin{pmatrix} 0 & 1 & 0 & 0 \\ 1 & 0 & 0 & 0 \\ 0 & 0 & 0 & 0 \\ 0 & 0 & 0 & 0 \end{pmatrix}, \mathbf{J_{02}} = \begin{pmatrix} 0 & 0 & 1 & 0 \\ 0 & 0 & 0 & 0 \\ 1 & 0 & 0 & 0 \\ 0 & 0 & 0 & 0 \end{pmatrix}, \mathbf{J_{03}} = \begin{pmatrix} 0 & 0 & 0 & 1 \\ 0 & 0 & 0 & 0 \\ 0 & 0 & 0 & 0 \\ 1 & 0 & 0 & 0 \end{pmatrix}.$$

As with the rotations in the sub-space $\mathbf{R^3}$, we continue to state the exponentiation relation: $\mathbf{R_{01}(\varphi)} = \exp(\mathbf{J_{01}}.\varphi)$.

Professor Castle-Holder:
- Haha? As with the rotations in the sub-space $\mathbf{R^3}$? Please develop! It would be nice if you could prove the necessity of the spin.
I see you were going to forget to say to the readers that the exponentiation of a matrix is developed in a Taylor series:
$$\varphi J \mapsto e^{\varphi J} = \sum \frac{\varphi^k}{k!} J^k$$

Open-Eyes:

9.6.2. The rotations in a Euclidean space of dimension 2.

For any rotation matrix, of angle ϑ as: $R = \begin{pmatrix} \cos{(\theta)} & -\sin{(\theta)} \\ \sin{(\theta)} & \cos{(\theta)} \end{pmatrix}$, it may be written as the matricial exponential $\exp(\vartheta J)$ of the gyror J $d\vartheta$. An infinitesimal rotation is written as $1 + J$ $d\vartheta$. The gyror J, as genérator of the rotations is the coefficient of the infinitesimal angle $d\vartheta$, in an infinitesimal rotation, $1 + J$ $d\vartheta$.

$$J = \left(\frac{\partial R(\theta)}{\partial\theta}\right) = \begin{pmatrix} 0 & -1 \\ 1 & 0 \end{pmatrix}.$$

The characteristic polynomial of R is: $(\lambda^2 + 1 - 2\lambda \cos{\vartheta})$, whose roots λ = $\cos{\vartheta} \pm i.\sin{\vartheta}$, are the eigenvalues of $\mathbf{R}$. But the particular case of null angle ϑ (mod π), where where $\mathbf{R}$ is reduced to $\pm$ identity in all bases, R may only be diagonalized on a complex basis, whose eigenvectors are on **isotropic**

directions. So, on the eigenbasis $\frac{1}{\sqrt{2}}\begin{pmatrix} 1 \\ i \end{pmatrix}$, $\frac{1}{\sqrt{2}}\begin{pmatrix} i \\ 1 \end{pmatrix}$, J takes a diagonal form:
$\begin{pmatrix} -i & 0 \\ 0 & i \end{pmatrix}$, and R takes the form: $\begin{pmatrix} e^{i\vartheta} & 0 \\ 0 & e^{-i\vartheta} \end{pmatrix}$.

Its first-order invariant, or trace : $R^i_i = 2\cos\vartheta$ (Einstein notation).
Its second-order invariant is the determinant: $R^1_1 .R^2_2 - R^2_1 .R^1_2 = 1$.
In dimension two, the rotation of right angle, and of positive sign, is not discernible of its generator J.
These isotropic eigendirections are directly related to the factorization of the Euclidean metrics on the corps of the complex numbers : $ds^2 = dx^2 + dy^2 = (dx + idy)(dx-idy)$. Here it is expressed on a canonical and orthonormal basis, so the metric tensor is unitary. Omitted physical unit: m^2.
We will not expose the decomposing of the rotations into a couple of reflections.

9.6.3. The rotations in a Euclidean space of dimension 3.

Any rotation may be expressed on a proper basis, whose a direction is aligned with its invariant straight line, and its matrix takes a simple form by blocks. Here are three forms :

$$R_{xy}(\vartheta) = \begin{pmatrix} \cos(\theta) & -\sin(\theta) & 0 \\ \sin(\theta) & \cos(\theta) & 0 \\ 0 & 0 & 1 \end{pmatrix}. \quad R_{zx}(\vartheta) = \begin{pmatrix} \cos(\theta) & 0 & \sin(\theta) \\ 0 & 1 & 0 \\ -\sin(\theta) & 0 & \cos(\theta) \end{pmatrix}. \quad R_{yz}(\vartheta)$$

$$= \begin{pmatrix} 1 & 0 & 0 \\ 0 & \cos(\theta) & -\sin(\theta) \\ 0 & \sin(\theta) & \cos(\theta) \end{pmatrix}.$$

The exponentiation relation is similar: $J_{xy} d\vartheta$ is the generator of the rotation $R_{xy}(\vartheta)$.

Defining the gyror J_{xy} by: $J_{xy} = \left(\frac{\partial R(\theta)}{\partial \theta}\right) = \begin{pmatrix} 0 & -1 & 0 \\ 1 & 0 & 0 \\ 0 & 0 & 0 \end{pmatrix}$, then: $R_{xy}(\vartheta) = $
$\exp(\vartheta J_{xy})$.

The gyror is not more identical to a right-angle rotation, but is the composed of a right angle rotation, by the orthogonal projector on the stable equiplane of R. This projector has for kernel (pre-image of zero) the invariant equiline of the rotation **R**. Any rotation commutes with the projector associated to its stable sub-space.
The diagonalization is in prolongation of the one already seen in dimension 2: on the complex basis whose matrix is

$\frac{1}{\sqrt{2}}\begin{pmatrix} 1 & i & 0 \\ i & 1 & 0 \\ 0 & 0 & \sqrt{2} \end{pmatrix}$, J_{xy} takes the form $\begin{pmatrix} i & 0 & 0 \\ 0 & -i & 0 \\ 0 & 0 & 0 \end{pmatrix}$, and $R_{xy}(\vartheta)$ takes the

diagonal form $\begin{pmatrix} e^{i\theta} & 0 & 0 \\ 0 & e^{-i\theta} & 0 \\ 0 & 0 & 1 \end{pmatrix}$.

This basis is used as a "*standard basis*" in quantum mechanics.
When the dimension is over 2, the rotations are no more commutative. However, two matrices of infinitesimal rotations, by infinitesimal angle $d\vartheta$, their products,

on the left or on the right only differ by a second-order term in $d^2\vartheta$. That is why the gyrors are useful everywhere in mechanics and in electromagnetism: angular speed, angular moment, torque, magnetic field, magnetic moment, etc. We will not treat the factorizations of the metric form (by quaternions or Pauli spinorials).

9.6.4. Rewriting the metrics on the appropriate basis.
We wish to re-express ds^2 on the basis proper to the propagation, here noted x, and still with the writing convention $c = 1$.

"ds" is the infinitesimal distance, and "ds^2" is it square.

On the basis proper to the direction dx/dt, the metric tensor g takes this form:

$$\begin{pmatrix} 0 & 1 & 0 & 0 \\ 1 & 0 & 0 & 0 \\ 0 & 0 & -1 & 0 \\ 0 & 0 & 0 & -1 \end{pmatrix}$$

Let e and f designate the new orthogonal basis vectors (conserving the physical unit), respectively e the orthochronous (with "e" like Einstein, who believed only in strictly orthochronous causality), and f the antichronous (with "f" like Feynman, who dared to draw an antichronous causality in his diagrams). Let o and a designate the coordinates on these new basis vectors, respectively "o" for orthochronous and "a" for antichronous. Their handling is now symmetric. Let us write the M-distance: $ds^2 = 2do.da - dy^2 - dz^2$. On the direction of propagation where dy and dz are null, only remains the factorization: $ds^2 = 2do.da$ (the physical unit is still omitted). The continuous transform between this proper form of the metric tensor and the known form in a massic basis (such as the basis of the laboratory) is a complex Lorentzian, of argument $i\frac{\pi}{4}$ (pure imaginary *swiftness*). While with real coefficients Lorentz transform (for a real *swiftness*) is only competent to pass from a massic basis to another massic basis. It leaves invariant the metric tensor and remains unable to reach a proper and biluminic basis. With abusive anthropocentrism, they precipitated to forget the proper basis and the proper metrics, under the accusation of being "*unphysical*", to translate into "**not anthropocentric**".

Curious:
- Is this proper basis, that you say "*biluminic*", accessible to human experimentation?

Open-Eyes:
- Inaccessible to human experimentation, but inescapable for the theoretician. Unless he (she) prefers to do monkey business. You will have to raise the same objection in § 10.8, when we will re-examine the Compton scattering of an X photon by a loosely-bound electron, in the frame of the center of inertia, which is unpredictable by the experimenter.

Curious:
- You are rough, you! You pretend for yourself to be adamant on the experimental discipline, but claim the privilege to use concepts that are inaccessible

to experimentation!

Open-Eyes:
- We should even resurrect Antonio Gramsci, to have him explain the historical dialectics of this, in his inimitable way. Please acknowledge me that I am playing cards on the table: I announce what is by nature inaccessible to experimentation and what is only relevant to the tribal rumors in some tribal tribe.

Professor Castle-Holder:
- But now, Mr. Open-Eyes, what is, according to you, the right microphysics? Up to now, I had let you surreptitiously introduce your modifications. Now, go for it!

Open-Eyes:
- Just before, have a look at the colorful grievances of a colleague at Jussieu, against the popularization of science, and against the popularizing authors.

9.7. Is popularization the enemy of science?

9.7.1. Why these grumbles? Popularizing is a difficult task, whose results are rarely convincing. We give the speech to a researcher who is disgusted with any popularization. In my sense, he goes too far, anyway he has the speech now. I only translate him.

Precedent posting on Usenet, forum fr.sci.physique, the 4 December 2004:
> *What I know is that the question of scientific popularization is not an idle question. Unless I am mistaken, it is even a dimension which is included in the missions of the researchers of the CNRS, for instance.*

Another **Professor Castle-Holder**:
- Yes, and it is a blatant scandal. Some researchers in the CNRS forge career only in popularization, are professional of popularization, and find ways to sing their praises, and be promoted with that; or with the administration or any twaddle so that the only one things that do not count are the scientific results. The popularization has always been a perfectly unuseful activity, so I will not blame the Bogdanoff to have written silliness in their book, as all the popularization books contain silliness from the beginning to the end. It is already difficult enough not to write blunders in a serious book or paper, but coming to say something wise and understandable by the mythical "man of culture", is radically impossible.

Norbert R. (popularizer in astronomy):
- Is that humor, or do you really think so?

Another **Professor Castle-Holder**:
- I am so silly, I really think so. The one who wishes to learn Quantum Mechanics just takes a good book, like the Dirac's or the Feynman's and will sooner learn something than by listening to the haywiring of a popularizer. Each time I was naive enough to open La Recherche, I did not understand anything when it was a domain I did not know and found mistakes everywhere when it was in a domain I already knew. ... By the way, the research takes on water from everywhere.

[pseudonym]
- Do you also class the missions of teaching by the teaching researchers as twaddle too?

Another **Professor Castle-Holder**:
- We were talking about the CNRS. The teachers of the University have an essential mission, which is teaching. I feel reticent to the multiple missions, which often lead to that nothing is correctly done.

End of citations.

Open-Eyes:
- I shorten here. The whole discussion is at
https://groups.google.com/forum/?hl=fr#!topic/fr.sci.physique/
7EuUwzJxbbY[101-125]
On other threads of discussion, he is still more bitter in his indictment against
any popularization, accusing it of favoring the laziness and the presumptuous-
ness of the public. Alas, many are the examples which justify his pessimism.
Now, we are of an opposite opinion so that I will go on.

A problem: this acrimony against the popularization is specific to the Quantic
Mechanics, it is itself a symptom which tickles the ears of the clinician psychol-
ogist. Nothing such exists in any specialties of the biology, nor in geosciences,
nor in chemistry, nor in astronomy, nor even in mechanics; however, the rationale mechanics is also a very mathematized discipline... On the contrary,
there are similar reticences and tricks by the psychoanalysts, the Freudians who
also practice voluntary obscurities, a deceptive parlance; a skittish sect...

I have heard an associate professor vehemently demanding, in a workshop in
Lyon 1, that the courses of physics should be forbidden *"to the weird spirits"*.
Oh, his *black sheep*, it was myself: once, only once, I had asked him an unexpected question which disoriented him. Distraught, he replied that I *"should
read books"*. I had them, the books, I had read them, and I was dissatisfied
with their faults in methods, their contradictory and weird assertions. These
faults of method which remain unnoticed by most of the young students, *jump
to the eyes* of an experienced research engineer. As for the young students
who perceive the faults of method and the contradictions specific to the sect,
disgusted, they change of discipline, and turn themselves into more sane works.
Year after year, as I amassed documentation on the history of this discipline,
I had more and more evidence that this acrimony against any popularization
comes from the fact that this tribal tribe (this sect if you prefer) has many
shameful secrets to hide, including some skeletons in the closets. In clinical
psychology, we know the pathologies driven by one or more family secrets,
whose toxicity persists down the generations. Here too, it is a collective psychi-
atric case. On the professional side, they are prisoners of some artifacts which
hide them the reality: the fairy tales instead of semantics and physical axioms,
that their *Great Ancestors* bequeathed them, following their winning coup in
1927.

CHAPTER 10

But what is the correct microphysics?

Open-Eyes:
- We expose the renovated quantum physics, with now the head upright and the feet down, that is with **transactions** between emitters and absorbers. We will present original developments, whose we argue the implications are largely worth the toil. In a few words, the believing in a whimsical corpuscle in place of the photon, the elementary unit of light, is ended: it is a successful electromagnetic transaction between an emitter, an absorber, **and the space or optical devices in between**. Even if the photon was emitted fourteen milliards (human) years ago, even if it will meet its absorber in **sixty-five milliards (human) years. Sure it is humiliating for our egocentrism; does it matter? The photon remains the unit of light** in the sense it has one emitter and one absorber, **but never ceases to be an electromagnetic wave, ruled by the physical optics of Fresnel** (1819) and the Maxwell equations (1873), modified for bosons since. Not any kind of "*corpuscles*", nor "*corpuscular aspects*", nor "*wave-corpuscle duality*" have no more any validity in transactional physics; only remains the undulatory nature, often quantified (but not always). Above, we had stated that we put an end to the confusion between **individual** wave and collective of waves. The macroscopical physics only had to know the collectives.

At the sources of the transactional physics are two very despised and censured discoveries: these two intrinsic frequencies discovered in 1923 by Louis de Broglie, and in 1930 by Erwin Schrödinger, respectively $123.56 \cdot 10^{18}$ Hz and $247.12 \cdot 10^{18}$ Hz for the electron (two hundred and forty-seven milliards of milliards cycles per second). When you take them into account, you can spare monstrous volumes of calculations, which were considered as the *nec plus ultra* of the physics for about sixty years. Indeed, when you take into account the intrinsic Broglian frequency, you observe that the transfer of a photon or any other "*particle*" occupy only a thin and stiff beam, without spreading everywhere, as Feynman and his so many followers have repeated. Intrinsic frequencies for any particle with mass. The width of this Fermat spindle and its angle of the tangent cone (at both ends) have many applications in astronomy and instrumentation. Lastly, we predict that the yield of the scientific teaching will be greatly improved when the contradictions of the hegemonic Copenhaguist semantics will be abandoned. It was frozen in 1927, and we observe it is a millstone. We do not more put the randomness at the same place as the Copenhaguists: from the de-Broglie-Dirac ground noise emerge some successful transactions, but its influence on the transfer (of a photon, an electron, a neutrino, etc.) during the transfer is minimal and negligible. On the contrary, most of the Copenhaguists, even Nobel laureates, pretend without proofs zigzagging and squirmy trajectories, exorbitant from any physical laws.

The fact that we take into account the Broglian ground noise, the compulsory frame of handshaking the transactions, is a major innovation, a breaking one.

10.1. *"Perhaps to Jupiter and back"*? How far will dare Stephen Hawking?

Here we will let the anti-transactionists develop freely their errings and contradictions, where they hold all the powers. The discussion was in english, and the curious man was an english-speaking man. You will see that the violence is the ultimate refuge of the incompetence.

PhysicsForum-Curious:
- I am just inhaling *"The Grand Design"* and am stuck in the chapter on the *"buckyballs"* double slit experiment. The authors say that in case of the experiment, a particle may take any possible way (*"perhaps to Jupiter and back"*), which then Feynman depicts as adding vectors to a result vector (as I understand). However, I wonder how this can be real, as the buckyball (or photon) has a definite speed s (or c) on the result vector path. But in case the particle takes the path to *"Jupiter and back"* the length of the path it has taken cannot fit the speed of the particle on the result vector, resulting in the (presumably false) supposition, that it had a speed greater than **s** (or **c**).

Open-Eyes (aside):
- Here the *curious* denotes by the letter "**v**" the mean speed of the center of inertia of the particle with mass, such as an electron, as taught and verified in macroscopic physics. The transactional physicist recalls that Jupiter is at 43 light-minutes from here, between 35 and 51 light-minutes depending on the year and the season. Hence it is obvious that when Hawking popularizes, he infringes the Relativity.

American Professor Marmot
(A manager of PhysicsForum, anti-transactionnist):
- *I believe the book I read about QM stated that almost all of these different paths cancel each other out in probabilities so that something like that doesn't happen*

Open-Eyes (aside):
- This beginning of the conversation is at
https://www.physicsforums.com/threads/
feynman-paths-and-double-slit-experiment.513139/
No one of the anti-transactionists will notice the contradictory wording: *"the result vector path"*.
For the curious reader, we will insert here three figures from the *Feynman's Lectures on Physics*, lectured at Caltech, which we have rejected his anti-<atomic limit> postulate:

Recall: Confusionist postulate.
To deny the atomic limit in undulatory, prescribe to confuse all kind of "waves", each individual wave (quantic wave) with any collective of waves, and these collectives with gravity waves or elastic waves in a collective of matter, then

mathematically unify all these kinds: the individual waves (quantic ones), the collective, and the waves in a material collectivity. The Born-Heisenberg copenhaguism is founded on this trick, and it is so for ninety years. A hegemonic swindle.

PhysicsForum-Curious:
- So, the particles that travel to Jupiter interfere altogether with themselves, but they don't travel at a speed greater than **s** (or **c**). Right? [Edit] After reflecting that I assume that any path which would result in a speed greater than **s** (or **c**) might be eliminated, which then leads to the conclusion that only the direct paths remain [?] But in this case, there wouldn't be an interference, would it?

American Professor Marmot:
- *No, I mean that the possible paths interfere with each other similar to the interference pattern on the double slit. Where they destructively interfere the particle(s) have a much less chance to take that path, if they have one at all.*

Open-Eyes (aside):
- I draw your attention to the facts that the anti-transactionist uses the reasoning from the physical optics (Fresnel, 1819), to conclude on the statistics of the passing of small balls. So is the contradiction in power for 1927. I let them go on.

Figure 10.1.

Fig. 1-1. Expérience d'interférence avec
des balles de fusil.

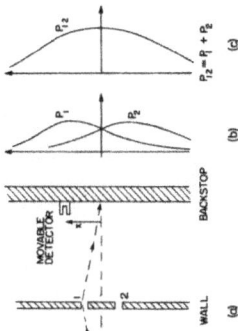

Wave source = Source d'ondes
Wall = Paroi
Detector = Détecteur
Absorber = Absorbeur

Fig. 1-2. Expérience d'interférence avec
des ondes produites dans l'eau.

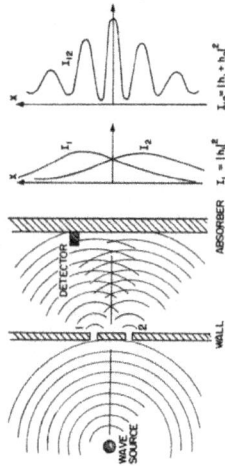

Électron gun = Canon à électrons
Wall = Paroi
Detector = Détecteur
Backstop = Plaque d'arrêt

Fig. 1-3. Expérience d'interférence avec
des électrons.

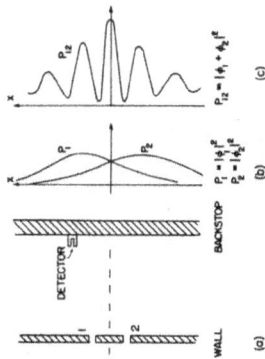

PhysicsForum-Curious:

- I am not sure if I fully understood your answer. Does that mean, that there may be some particles traveling to "*Jupiter and back*" according to the interference pattern?

PhysicsForum-Open-Eyes:
- You are just dreaming with dreams, not more. The theorist may compute such an aberrant path, but the conclusion will not yield another result than null. So why to exhibit magic sentences when the real result remain null? When a photon is emitted by an emitter, its history begins, as seen from our laboratory. When a photon is absorbed by an absorber, its history ends, as seen from our laboratory. Meanwhile, it is tight by the laws of physical optics (the Maxwell equations), as long as lasts the synchronous transfer from emitter to absorber. This leaves not so much room for theoretical and magic fantasies. More strange is the noise before any transaction succeeds, but alas beyond the reach of most experiments. However, the hope is not null: someones claim that some radioactive decays depend on external conditions. Stay tuned for eventual confirm.

PhysicsForum-Curious:
- ... in the current chapter basically the different histories of a particle are taken as a basic argument in order to explain that the universe has infinite histories (and perhaps infinite futures). And the authors say that understanding that is very important for understanding the chapters afterwards. Otherwise, the book (from Hawking and Mlodinow) would be science fiction?

American Professor Marmot:
- *From my quick reading of this thread, this has more to do with trying to understand Feynman's path integral. Unless someone has some extraordinary capability with written communication, trying to illustrate this principle is almost impossible on a public forum such as this. So maybe a readable source that introduces what a Feynman path integral is might be useful. Try this one: http://scitation.aip.org/getpdf/servlet/GetPDFServlet? filetype=pdf&id=CPHYE2000012000002000190000001&idtype=cvips [Broken] Zz.*

PhysicsForum-Open-Eyes:
- This famous work of Feynman was reinventing the wheel, but less practical, with heaps of unuseful mathematical fatigue. Why? Just because of american arrogance: what is not published in english is thought not being, for american physicists. So Feynman simply ignored the periodic character of any quanton that has a mass, its two intrinsic frequencies: Broglie frequency for all of them: $m.c^2/h$, proved in 1924 (published in french). Dirac-Schrödinger electromagnetic frequency for fermions, such as the electron: $2.m.c^2/h$, proved in 1930 (published in german). This fact (the Broglie intrinsic frequency) drastically reduces the alternative paths to mathematically explore, as they very very quickly become unphysical: the interferences become destructive.
EDIT. Err, there is a frequency in the text cited above, but is not the good

one, it is much lesser, not relativistic, not intrinsic:

This fundamental and underived postulate tells us that the frequency f with which the electron stopwatch rotates as it explores each path is given by the expression: f=(KE−PE)/h.

So with this inappropriate tool, Feynman explores much much broader paths than necessary, much much broader than the real physical paths.

American Professor Marmot:
- Sounds like you just have a problem with Feynman and his work, not that it is incorrect or not. ...

Objections by **Open-Eyes**:
It is to notice that no one of the anti-transactionists has ever noticed these key-words, quite new for them: *"synchronous transfer from emitter to absorber, ground noise, transaction"*. As of now, we practice a new physics, very different from their, and they still have not perceived: too arrogant in their posture of *" We the-initiated-who-are-knowing"*, too imbued of their intrinsic superiority by the pack upon the remaining of the world, despising *"the plebeians-who-do-not-know"*. The anti-transactionists prefer to take refuge in attacking against the person, mercilessly, and in repression, mercilessly too.

The reader may verify himself in the original discussion the rise of the violence and dishonesty against the science-venturer, soon banned:
https://www.physicsforums.com/threads/
feynman-paths-and-double-slit-experiment.513139

Indeed, violence is the ultimate refuge of the incompetence.

What is grave in this quarrel, is that the experimental situation is well known, we practice it in all the cathodic oscilloscopes, all the radar or TV cathodic tubes, all the electron microscopes, in the "magic eye" monitoring the tuning of our radio sets when I was a child, in all the engraving machines for the integrated circuits, in the Castaing electron microprobes, the SEM, etc. Not any of these apparatus could work with the fairy tale of Hawking and Mlodinow. Here is a device for the demonstration in the classroom of the principle of the cathodic oscilloscope, used here to measure the ratio q/m of the electron, its deviation by an electrostatic field.

Figure 10.2.

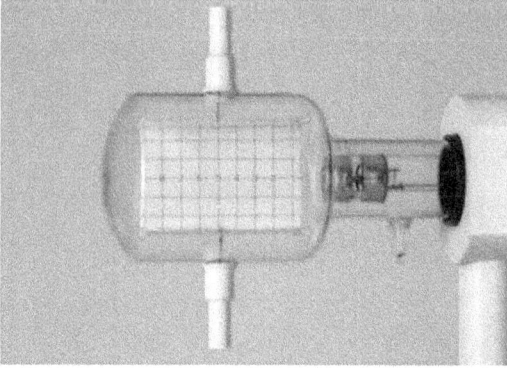

In all the handbooks of the final class of Lycée, you have a similar photo of this device, but switched on. Regarding the deviation of the cathodic rays by a (constant) magnetic field, here it is:

Figure 10.3.

The electron beam is deflected in the shape of a circular arc, following the same sense of rotation than the electric current in the Helmholtz coils. Here more details on the electron gun in the bulb filled with rarefied dihydrogen.

fig 10.4

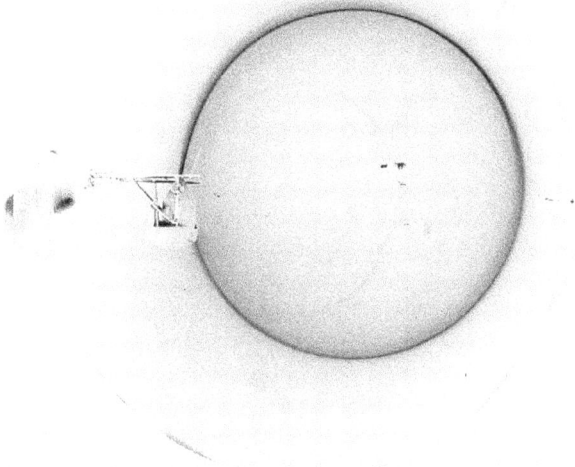

In electron microscopes, the lenses are magnetic, as the builder can obtain less optical distortions than with electrostatic lenses. Their fields are oblique regarding the axis of propagation, while in the photo above, the initial speed is in the equiplane of the magnetic field (the plane of the page), to obtain a circular trajectory and not a helix. In an electron microscope, where the fields are heterogeneous along the axis, the magnetic lenses produce nearly helicoidal trajectories, but on convergent generatrix. Therefore, when you change the magnification or the focusing, the image rotates on the screen. It does not disturb much the biologists (maybe but the embryologists ? They need chemical gradients), who are the biggest market for the powerful electron microscopy, but it is more annoying for the metallurgists, who often have to study rolling or forging textures, not to have a goniometer in the microscope.

Further, in the very inflamed discussion (the violence is the ultimate refuge of the incompetence), one of the caciques started yelling against the proofs of electron diffraction on a crystal or a crystalline powder (methods either from von Laue, either from Debye[1] and Scherrer).

---

[1]**Peter Joseph Wilhelm Debye** (born **Petrus Josephus Wilhelmus Debije** 24 March 1884 in Maastricht - 2 November 1966 in Ithaca, New York, U.S.A.) was a dutch physicist and chemist. Nobel laureate in chemistry in 1936.

Here is an equipment for Debye-Scherrer diffractograms with electrons, which is sold by Leybold Didactic for our classrooms:

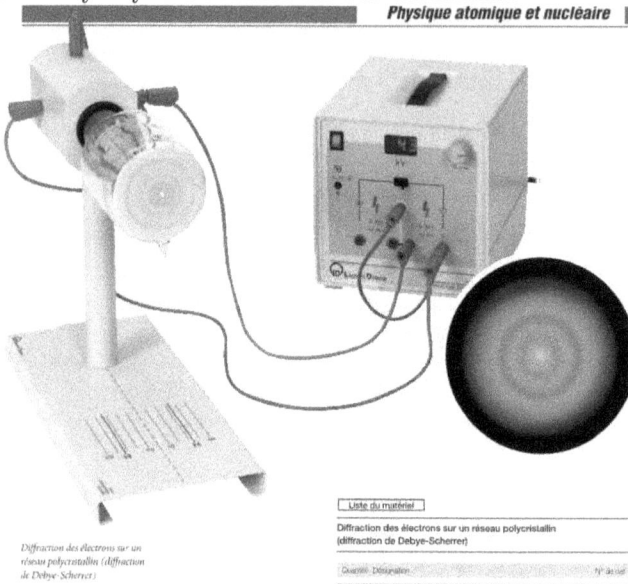

Figure 10.5.
The target is crossed through, and is polycrystalline, presumably with all the orientations. The electron beam is too wide for fine crystallographic measures, as for classroom use, they needed to illuminate an appreciable area of the thin plane target. What is minimized here are the cost, the danger, the bulkiness, compared to a professional X-ray radiocrystallographic equipement.

You see that they use higher voltages to obtain more Bragg reflections on a small screen. The angular resolution is poor. We obtain far better results with K_α of a metallic anticathode. Now the question arises: what causes such a poor angular resolution? The crystallites in the metal or graphite sheet may be too small. Each incident electron may be too small. However, it must be at least three to four interatomic distances long, and five to six interatomic distances wide, otherwise not any diffractogram could be seen at all. Too much speed dispersion in the electrons beam. Too much angular dispersion in the electrons beam. To know more about the length of coherence of an electron in a beam, one must make the electron interfere with itself. So in Aharonov-Bohm type of experiments, the speed and the accelerating voltage are much lower. As far as I know, the so proved length of coherence is in the magnitude of ten wavelengths, at least.

Going up with the quotations.
American Professor Marmot:
- *I'm not sure what you wrote in that post has anything to do with the topic. Still, since you're talking about the "real world", how about my checking if you*

also are aware of such a thing. What you described above are not clear examples where QM of any kind (be it path integral, straight forward QM, etc..) is applicable. When dealing with free electrons, and when electron-electron correlations are negligible, many of these phenomena resort back to the CLASSICAL description! Don't believe me? Look in beam dynamics of particle accelerators. Particle tracking codes used such as PARMELA considers each electrons as classical particles! And CRT? Yes, even that! Look at the codes used to model photomultipliers, especially the electron trajectory through the MCPs or the dynodes! I'm not sure why these examples are being used here, and how exactly they are applicable. All I can see is a bunch of things taken out of context that were never based on QM's description in the first place! And for your information, I routinely use SEMs, work at a particle accelerator, and am involved in a photodetector project to design and build new detectors. Would those qualify as "real world" experience? Zz.

Reference
https://www.physicsforums.com/threads/
feynman-paths-and-double-slit-experiment.513139/page-2

Open-Eyes:
- Yeah... Here are electron diffractograms done in an electron microscope by transmission. They are Laue diffractograms (Max von Laue, 1879-1960), obtained by bringing the crystallite on the axis of the beam, diaphragming on it, then by changing the focusing on infinite and wobulating the accelerating potential for modulating the wavelength of the electrons. The Laue method requires a continuous spectrum of incoming wavelengths.

Figure 10.6.

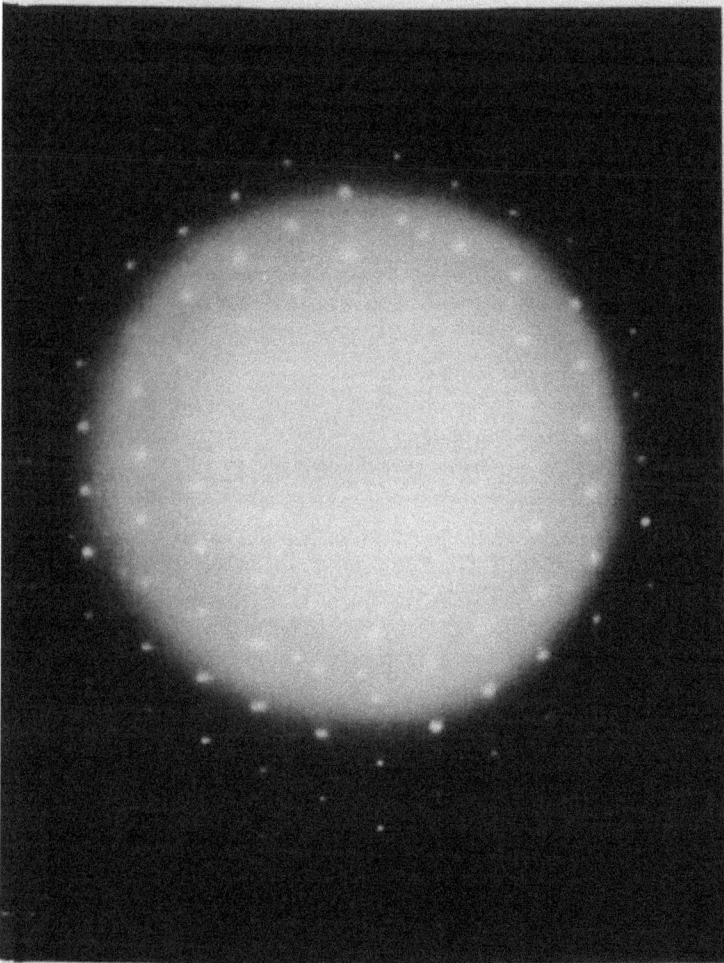

Fig. 12. — Cliché de diffraction électronique du produit synthétique

Figure 10.7.

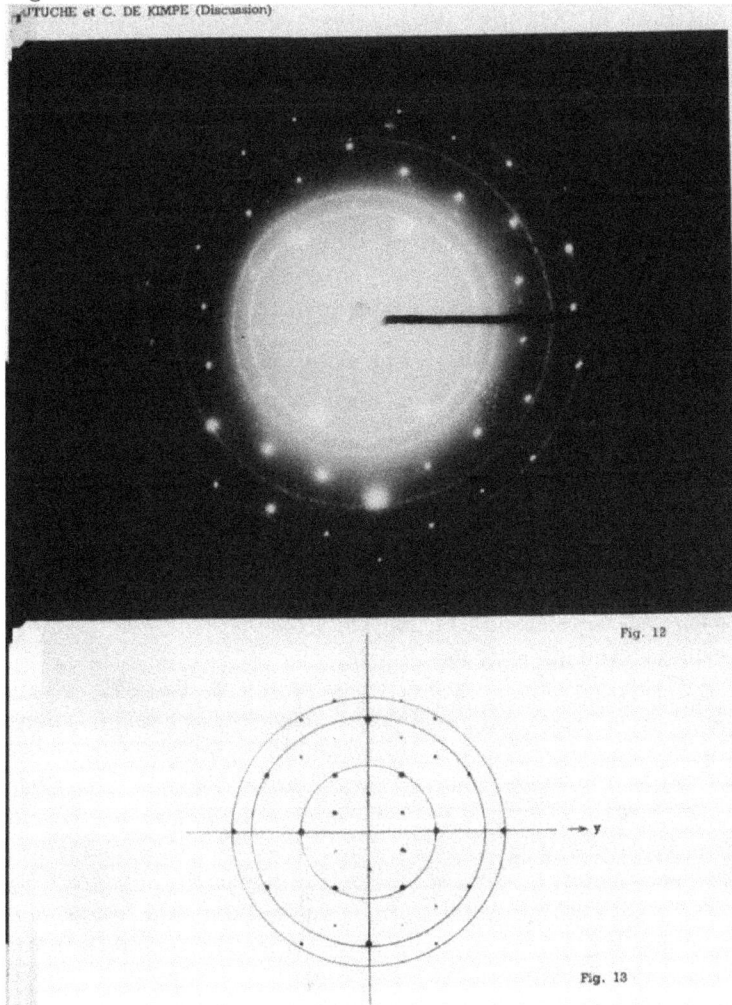

You can read that these diffractograms were done by Gastuche and De Kimpe, here on clays, of nearly hexagonal symmetry.

At the INSTN we had obtained similar diagrams on carbide inclusions (cubic symmetry) in a thin plate of steel.

Those anti-transactionists could never prove that these diffractions grounded on the undulatory features of the electron could ever be "classical". Therefore those tribal chiefs of the pack soon banned the man who did not believe in their catechism. Things are more simple when you prohibit any word from who does not share the beliefs of the chiefs of the pack.

The war is a confrontation of wills. Just ask our soldiers who had to fight in the Adrar of Ifoghas! In principle, science is just the contrary: cooperation of intelligences, across the geographical distances, across the cultural distances,

across the beliefs and affectivities, across the generations. In principle, in principle... But over there in that tribe (locally physicsforums.com) and lots of others, the confrontation of wills and territorial instincts, the war persist to replace the scientific debate. Now it lasts for December 1926, the confrontation of the most combative against the less combative, Niels Bohr against Erwin Schrödinger. Read the narrative by Werner Heisenberg:
http://citoyens.deontolog.org/index.php/topic,1141.0.html

The telling is by Werner Heisenberg himself, though he had much to earn in defeating Erwin Schrödinger, either by a loyal or by a disloyal fight.
Source: Franco Selleri. *Le grand débat de la physique quantique*. Champs Flammarion, Paris 1986. Page 96.
This telling is confirmed from second hand by Emilio Segrè, in *Les physiciens modernes et leurs découvertes*. Fayard, Paris 1984 for the french translation.
Selleri quotes the original source with the letter from Heisenberg: S. Rozenthal, ed. Niels Bohr. North-Holland, Amsterdam, 1968.

Quotation: Schrödinger had to face a difficult fight in Copenhagen. Bohr invited him to give a lecture at the end of 1926 "*and asked him to, not only give a lecture on his undulatory mechanics, but also to stay in Copenhagen enough time for discussing of interpretation of Quantum Mechanics*".
Heisenberg describes so the intensity of the fight:
"*Though Bohr was really an obliging and attentive person, he was able in such discussions about the epistemological questions he considered as absolutely vital, to insist in a fanatic manner, and with a terrifying inflexibility on the complete clarity of all the arguments. After hours of wrestling, he still could not resign himself in front of Schrödinger, to admit that his interpretation was insufficient, and even unable to explain the law of Planck. Each attempt form Schrödinger to evoke this embarrassing point was refuted, slowly, point by point, in painstaking and endless discussions. Probably because of this overworking, after some days Schrödinger fell ill and had to stay in bed in Bohr house. But even then it was difficult to keep Bohr far from the bed of Schrödinger...*"
And Heisenberg concludes:
"*Finally, Schrödinger went out of Copenhagen rather demoralized, when at the Bohr institute we felt we got rid of the interpretation given by Schrödinger to the quantic theory, which hastily used classical undulatory theories as models.*"
So were treated the fundamental questions, and so a small-group became hegemonic: by sheer violence.

10.2. When two luminaries mess around

When two luminaries forget to clutch the mathematical formalism, and reveal the ground of their ideas, err, well, the show is worth the voyage!

It is the book "*Soyez savants, devenez prophètes*", from Georges Charpak and Roland Omnès, in a repetitive series by the publisher Odile Jacob. Never we will enough outline the responsibility of the publisher in the matter of scientific popularization. Too often, the publisher behaves like a rogue, disrespectful towards the public, and does not more than serving again what has already well sold.

In his (her) defense, however, popularizing well is a difficult craft. From just the background of the journalist or the publisher, not to be bluffed by the authoritative argument is difficult. We have already seen some colleagues who storm at any popularization, as it exonerates the public from the necessary toil and work. They go too far, but the problem of the right place for the popularization remains.

For the sake of the rhythm, we will postpone to Appendix J the case of Bernard d'Espagnat and his xx versions of the same *veiled real*, a file yet burdened by Jean Staune and the John Templeton Foundation. The microphysical reality does not care about our moods, about our poignant feeling of *cruel uncertainty*... By an authoritative argument which nobody had the guts to roast, Niels Bohr and Eugen Wigner have played a dirty trick to the posterity when they put the human observer and his moods in the middle of the microphysical picture. And is yet taught! There is no physics in it, just transfero-transferential auto-theory, a flight behind empty words.

Charpak and Omnès had good scores of selling, so Odile Jacob serves them again, and it is appalling of dishonesty and incompetence, said as an understatement, this "*Soyez savants, devenez prophètes*", from Georges Charpak and Roland Omnès.

Mostly, they are out of their field of competence. Sure they have the right to take this risk. We all take risks, we people of hard sciences, when we treat of the history of sciences and their insertion in the affairs of the realms: we are not historians, nor sociologists, we do not have enough time to search all the original documents and to submit them to comparative critics. Are we wrong to take such big risks? No, because the professional historians do not have our competences to all understand in the history of sciences. Therefore the cooperation and the interprofessional dialogue are indispensable.
And there, did these luminaries submit their work to the supervision of a historian who could interrupt them, and ask them to review their text? No. They enjoyed themselves at two, to compose their fairy tales, and pretended to themselves that they would play a saving social role. The duty of the publisher was to shout Breakneck, but she did not.

Professor Marmot:
- Stop! It is so rare to see luminaries condescending to popularize and so rare to see publishers taking the risk of publishing a popularization! I forbid you to criticize that!

Open-Eyes:

- And at least in their specialty, the quantic physics? It is appalling too. Here is forensic evidence, their figure at their page 87:
Figure 10.8.

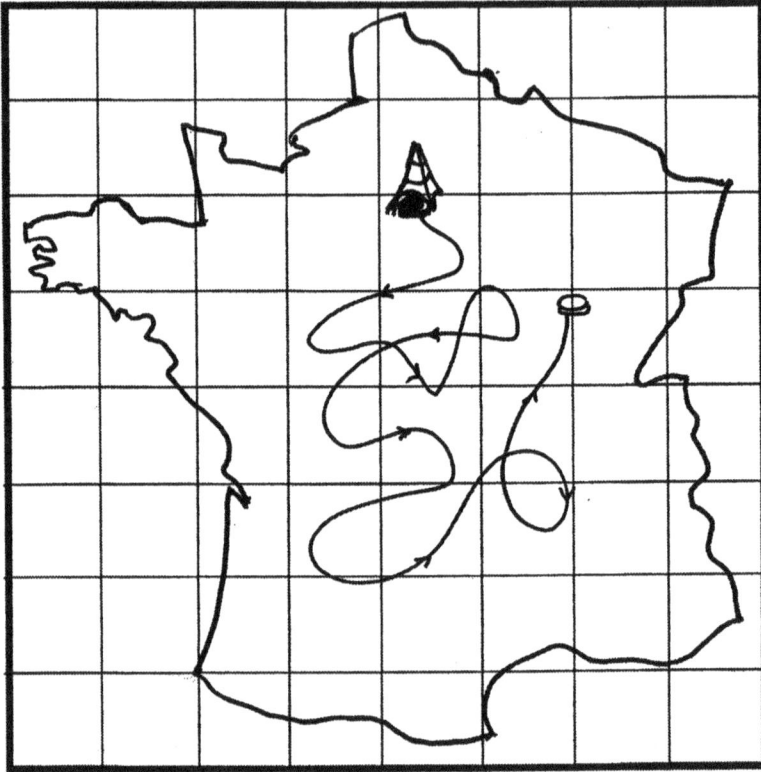

Le mouvement erratique d'un clone de particule.

And the rest of the chapter runs along the same lines. Sure, one could argue that they were betrayed by the illustrator, just as Olaf Magnus was betrayed by his illustrator who never saw in Italy the skis of the Sami and the Swedish. Next comes the following:
Figure 10.9.

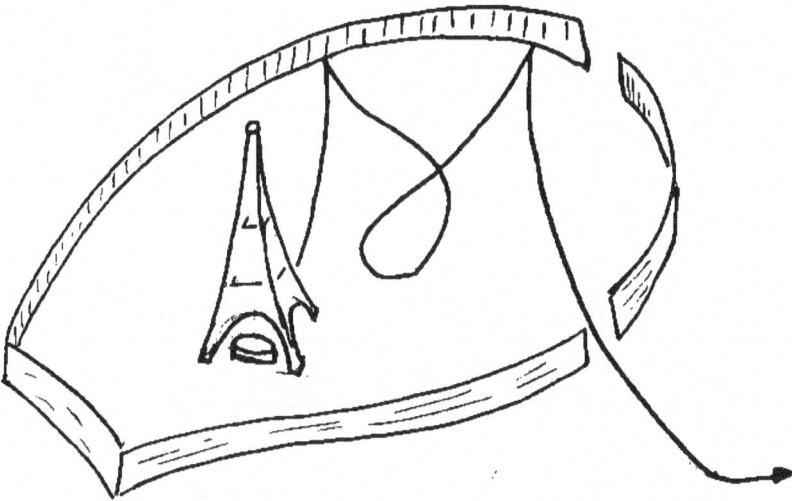

Le mur murant l'image de Paris ne laisse que deux ouvertures pour le passage d'une particule (on voit ici le mouvement d'un clone).

And their text, worth its weight of guano:

Quotation:
The particle is let go, this time with a speed, and the clones spread again, bump on the wall, rebound several times until they go out by one of the doors, and scatter in zigzag in the room.
Now you had at home in your living room, the counter-experience: the electron-gun of your cathodic televisor. If the physics of the electrons was as wriggling as explained by these luminaries, no one televisor could work, no one cathodic oscilloscope could work, no one electron microscope could work, no one engraving machine which engraves the microprocessors and all the integrated circuits of the electronic industry could work, no one accelerator of particles, nor the ESRF synchrotron could work, no one radar cathodic display could work, etc. Maybe one could save the triodes, tetrodes, and pentodes, maybe, maybe one could save the X-ray generating tubes, to which we owe so much in medicine, and all the radiocrystallography, but only with much luck, and by altering the geometry of the anticathodes, and all the design of the collimation of X-ray beam would be different, etc.

Professor Marmot:
- A scandal! That small villager from nothing who dares to say that luminaries write bullshits!

Open-Eyes:
- But why these luminaries told you such bullshits? Because they feel safe that you are unable to guffaw of their trickery. They think they will never be caught

with their hands in the jam jar. Their scientific virtue is as frolicsome as the virtue of Dorabella and Fiordiligi: it depends on the others regard and on the **"what people might say?"**. *Cosi fan tutti!*
"*Yes, may one object, but out of the box, their electrons fly in a straight line in conformity to well-know optics for Newton. Only in the mystic box, they have a mystic and goblin behavior!*". So, like before Galileo and Kepler, we are again burdened with two incompatible physics: a terrestrial physics, which is knowable by experiments, and a celestial metaphysics, accessible only to the theologists. Please admire the progress!

Please admire the other victory of the theology: Out of the box, the electrons are *corpuscles* but obey the laws of optics, including interferences... Yes but it is mystic, so again it proves the superiority of the theologian on the layman.

When they proceed with calculations in the frame of their craft, these two luminaries use the standard formalism, which - puff! - remains undulatory and deterministic. But when the matter is to impress and deceive the public, the students, or the laymen, the fairy tales come back immediately: the "particle" reverts to a corpuscle, with defined trajectories, but just to make hazardous, this trajectory wriggles like a worm for being the longest possible.
Charpak and Omnès explain that it was such way they understood Feynman and the principle of least action. Now in 1924, a certain Louis de Broglie united the least-action principle (Hamilton, about 1834), in mechanics, with the Fermat's principle in optics: if any "*particle*" is undulatory, then the path of least action is also the isophase with all its close first-order neighbors, which **all arrive in phase at destination**, at least at first order. Exception to this simplified statement: when two or a few more paths can give an offset in phase of an integer number of periods then these (few) paths can contribute **together** to the transfer of the photon or fermionic particle (or helium atom, etc.), each one with a finite and fuzzy width of path. Since Young and Fresnel, this is named **interferences**.
Obviously, Charpak and Omnès had forgotten the contribution by de Broglie, eighty years before, maybe too new for them. Oh yeah, for the Solvay meeting in 1927, in theoretical physics only remained the winners and the defeated, and as de Broglie and Schrödinger were defeated, their results were thrown into the *Memory Hole* by the winning pack. The Schrödinger equation is carefully un-Schrödinger-ized: the periodic term in the solution is erased at the beginning of the handbooks, just after appeared once on just one line. In stride, Charpak and Omnès forgot all the contributions to the physical optics done in the 19th century: Thomas Young and Augustin Fresnel, hop! Into the *Memory Hole*!

It is not all their fault: Feynman also had all forgotten. When I was a young student, I was fascinated in 1964-1965 by these all-new Feynman volumes. Like all the others, I was fascinated by the special lecture on the minimum of action. I am no more a youngster, and the fault *jumped to my eyes*: this principle of minimum of action, when so presented, remains a mathematical miracle.

But it is obvious when translated into coherence of phase, via the optics of the Broglian waves. This optical evidence flows from Christiaan Huyghens and Pierre de Fermat in the 17th century.

I insist for the beginners: "*quantic*" means periodic, undulatory and transactional, but in hiding it at most. It is coded so to avoid you could understand too easily something so simple. Why that secret coding? To maintain watertight the frontier between "*We the initiated who know*", and "*You the laymen who do not know*". The narcissism of the pack has its reasons the reason does not share.

The researcher at Jussieu who earlier stormed at the popularization, likes to object: "*Oh! But I know a high-level physicist who does not practice the confusions you denounce. So nobody teaches this confusion!*". But now we have the printed proof that even high-level physicists, one of which was Nobel laureate, practice and teach the confusions I denounce for many years already. So low-level teachers, ahem! I don't tell you...

Curious:
- Are you sure it was not an isolated aberration, by two oldest old men?

Open-Eyes:
- Well, we caught the first guilty. Alas, it was Feynman himself. The original paper of 1948, "*Space-Time Approach to Non-Relativistic Quantum Mechanics*" is at pages 321 to 341 of the gathering by Julian Schwinger "*Selected Papers on Quantum Electrodynamics*", Dover Ed. Bad news: the magnifying glasses are indispensable to read. It is reprinted really small. The hypotheses Feynman used are not explicit, and are buried deep under the Lagrangian formalism. Indeed the merit of Taylor, Vokos and O'Meara is precisely to have them put in evidence; only then their unrealism is obvious.

Quotation:
This fundamental and underived postulate tells us that the frequency f with which the electron stopwatch rotates as it explores each path is given by the expression: f =$\frac{kE-pE}{h}$.
Now this frequency, implicit by Feynman, explicit by these authors – thanks to them – is totally fictitious, immensely variable, and milliards of times slower than the real, intrinsic frequency. And Feynman, as an internee in the group-think coming from the copenhaguists pack, strongly believed that the electron wave was only fictitious, just a magic trick for calculation; corpuscularists, they believed in corpuscles, just endowed with magic powers. A fictitious and un-realistic frequency for a supposed-fictitious wave. The result is that all the paths that Feynman and his readers could imagine were far too slack and un-stringent, are exorbitant from any physical law, and their calculation had to embrace gigantic spaces for a null result. Not surprising they had to struggle with heaps of diverging integrals, though condemned to give zero.

Harasser Marmot:
- All that because you are jealous of the success of Feynman on the wives.

Open-Eyes:
- *In fine*, the wrigglings of Charpak and Omnès are only the symptoms, exaggerated beyond the absurd, of a collective mental illness. A real electron, a real neutron have much more stringent properties than those postulated by this chapel of authors. Here are two illustrations extracted from the Greiner:

Figure 10.10.

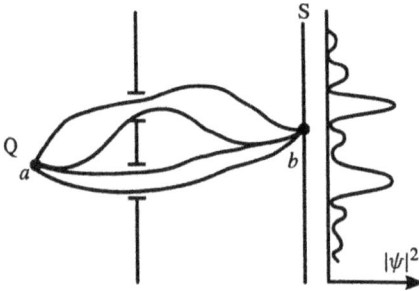

Walter Greiner. Quantum Mechanics, special chapters. Springer Verlag 1989. Chapitre 13.1 Action Functional in Classical Mechanics and Schrödinger's Wave Mechanics. Greiner too does not explicit the fictitious frequency used by Feynman. Here are drawn wrigglings instead of trajectories.
Figure 10.11.

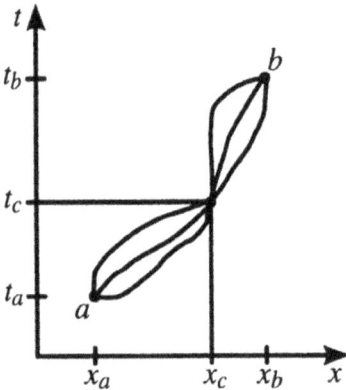

Fig. 13.4. Two successive events (x_c, t_c) and (x_b, t_b) and the corresponding paths

This total unrealism flows from the initial choice by Feynman of a "fictitious wave" with fictitious frequency, where he confused phase velocity and group speed. But well, he was raised in a corpuscularist tribe... To Joseph Louis Lagrange, who worked in the 18th century, we will forgive to have elaborated a non-relativistic formalism. Richard Feynman is less forgivable to have returned to the Lagrangian formalism, therefore corpuscularist and non-relativist, which gave him a frequency and consequently Huyghens and Fermat constraints fully unrealistic, so much in opposition to experience. A consequence to have been raised by the corpuscularists.

Harasser Marmot:
- All that because you are just jealous of the sexual success of Feynman at Cornell! How do you dare to contest a Nobel Laureate?

Professor Castle-Holder:
- No clear ideas without an exercise, scrupulously quantitative. We will take an alpha radiation, emitted by a decaying heavy isotope. Which one do you prefer? A long-period isotope of uranium, thorium, or radium?

Curious:
- Let us take the thorium.

Professor Castle-Holder:
- The 232 thorium is the longest-half-life isotope: $1.41 \cdot 10^{10}$ years. At the end of fourteen milliards years, still the half remains. Now the implosion of a supernova which synthesized the nuclei of heavy atoms found in the Solar System occurred about five milliards years ago. Therefore, the big majority of the thorium synthesized and ejected then is still there. It emits an α, or nucleus of helium 4, whose energy is 3.994 MeV in 77% of the occurrences, of slightly less energy in 23%. We write the nuclear reaction: $^{232}_{90}Th \rightarrow ^{228}_{88}Ra + ^{4}_{2}\alpha +$ 4.08 MeV
The α does not take out all the liberated energy: first, the recoil of the atom may be not negligible, next the new nucleus of 228 radium may be excited and will de-energize by emitting a gamma ray. Moreover, this new radium atom is now ionized and must lose two electrons and reorganize its electron cortege. First, we will calculate the flight speed of the α, with supposing it is not relativistic.
1 AMU (Atomic Mass Unit) = 931.48125 MeV = $1.6605656 \cdot 10^{-27}$ kg.
Helium mass: 4.00260 AMU = $6.64658 \cdot 10^{-27}$ kg.
α mass: the mass of the neutral helium minus two electrons = $6.64658 \cdot 10^{-27}$ kg $- 2 \times 9.1093897 \cdot 10^{-31}$ kg $= 6.64476 \cdot 10^{-27}$ kg. We will not quibble on the bonding energy of the helium atom. The first exploration is with considering the flight as not relativist: we suppose that its new energy is kinetic and computable by the Newtonian mechanics. 3.994 MeV = $3.994 \cdot 10^{6} \times 1.6020 \cdot 10^{-19}$ J $= 6.398 \cdot 10^{-13}$ J. Hence the flight speed of the alpha nucleus: 13,877,060 m/s, that is 4 % of the light velocity. We keep the non-relativistic approximation.

Open-Eyes:
- Now, what is the Broglian frequency of this α? $\mathbf{mc^2/h}$, here $9.012886 \cdot 10^{23}$ Hz.
Remind the intermediate universal result: $\mathbf{c^2/h}$ is $1.35639 \cdot 10^{50}$ kg^{-1}.s^{-1}.
We calculate the initial momentum: $6.64476 \cdot 10^{-27}$ kg $\cdot$ 13,877,060 m/s $= 9.2210 \cdot 10^{-20}$ kg.m/s.
Hence the Broglian wavelength $\mathbf{h/p}$ of this α:

6.6260755 . 10^{-34} joule.seconde/cycle / 9.2210 . 10^{-20} kg.m/s = 7.1821 . 10^{-15} m/cycle = **7.1821 fm/cycle**.

As the momentum will decrease in a braking medium, the wavelength increases as the inverse. That is very little by crossing a bubble or fog chamber, much more in an anti-radiation armor until the complete stop. The wavelength of the emerging α is similar to the diameter of the strong interaction of the emitting nucleus: a rather coherent physics. All the energies of emerging α are about the same.

Professor Castle-Holder:

- And with good reason! It is mostly the electrostatic repulsion between protons: the two protons departing with the alpha nucleus, and the protons in the remaining nucleus. The nature of the emitting nucleus and its excess of internal energy intervene mostly on the statistical probability of individuation of an alpha sub-nucleus will individualize the time of crossing through the barrier of strong interaction. The effect on the energy of the emitted alpha is a minority.

Open-Eyes:

- Thanks! And now, reader my friend, try to imagine how this alpha, oscillating at 9.012886 . 10^{23} Hz and with a propagation always perpendicular to the wavefronts spaced by 7,1821 fm could sort something to have a non-stiff, flabby and wriggling trajectory as in the drawings of Walter Greiner, Roland Omnès, and Georges Charpak, as seen above. The laws of propagation rule it as very stiff. What the above calculation did not provide are the length and the width of this alpha wave. We need more investigations, preferably experimental. Here is an image of trajectories and two reactions, obtained in a bubble chamber: Figure 10.12.

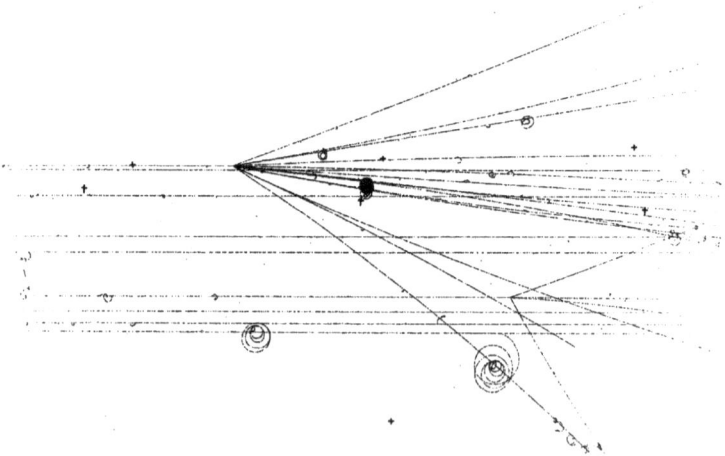

Let us see the phase celerity of this alpha: V = c^2/v = $(299{,}792{,}580 \text{ m/s})^2$ / 13,877,060 m/s = 9,476,558,510 m/s, that is 603 times more fast then the group speed. I had already observed that when wake waves of a ship come to the shore, the phase velocity differs from the group velocity: the crests arise at the back of the group and vanish at the front. For an α ray, this discrepancy is much bigger.

Curious:
- But you do not know the real length of a wave train of an α?

Open-Eyes:
- Yes, we almost do, as we know the experimental indefinition on the kinetic energy 1/4,000. Now it is quadratic with respect to the speed, hence the indefinition on the speed and the momentum: 1/8,000. Thereon, we have to take into account our ignorance of the experimental protocol, whether the conclusion was statistic or individual, and presume the published value could be the average of more dispersed, even poorly defined ones. With prudence, we divide by four to obtain the individual indefinition; hence we deduct that the magnitude of the length of the wave train is about two thousand wavelengths, that is at least 14 pm = 0.14 Å. Not so far the radius of a hydrogen atom.

Professor Castle-Holder:
- And about the duration of an α wave train according to these hypotheses we divide the length 14 pm by the group speed, 13,877,060 m/s, it yields **1 as** (one attosecond: 10^{-18} s). It is still an approximation.

Professor Marmot:
- Your calculations are completely implausible! When an alpha nucleus comes at the boundary of the nuclear potential well and tilts outside, it is very small, and can never become as long as you pretend.

Open-Eyes:
- Sure, in 1928 the historic reasoning from George Gamow relied on a corpuscularist presupposition, and since, you take this presupposition for granted as an absolute. Now in September 1926, Erwin Schrödinger gave the right model, that the community of the physicists did not notice: a photon is emitted by the beat of the Broglian frequencies of the final and the initial states, in the emitting atom or molecule. And the same beat at the resonating spectral absorption. For the heavy atom emitting an alpha, there is also a beat between the final and the initial state, both very stable, or with very long lasting metastable states, so this beat has a long duration, rated at the nuclear scale; we have given its magnitude: the attosecond. The final and the initial state are of long stability, so well defined in energy; therefore their difference, though relatively less well defined, is still however well defined. The experimental paradox to endure is that the known precision is better on the part of the emitted alpha than on the total energy <alpha + radium 228>. A vexation to take with calm.

Professor Castle-Holder:
- I fear that you assume that the Broglian waves are enough to rule all the physics of the nucleus.

Open-Eyes:
- No! Certainly not enough! I only pretend it is weird to discard their reality, only because Schrödinger and de Broglie were defeated in Brussels in 1927, by the Göttingen-København pack. A hegemonic tribe nowadays. This victory acquired by the violence of a pack is not a receivable scientific argument.

Professor Castle-Holder:
- Do not forget to bring experimental proofs, at least some indices.

Open-Eyes:
- To obtain that an α ionizes molecules along its path, it is preferable it has dimensions, mainly in long, comparable to the diameter of the liquid dihydrogen (eventually butane) molecule which filled the bubble chamber. How many molecules can an α ionize, before its kinetic energy is exhausted? Let us take the energy of ionization of a hydrogen atom: $E = 13.53$ eV. To squander 3.994 MeV, you need about 295,196 ionizations. It corresponds roughly to the fineness of the photographs taken in these bubble chambers.

Curious:
- And may we imagine that these about three hundred thousand shocks deviate the trajectory appreciably, as drawn by Greiner, Charpak, and Omnès?

Open-Eyes:
- Not at all like they drew, with continuous and artistic curvings. It is not *bocce*, the provençal boules, with their "*carreau*" shooting, and the laws of the elastic shock are not applicable: all ionizations are inelastic. The inelastic molecular shocks towards right or left, high or low are equiprobable, are unable to deviate the alpha on coordinate and artistic curves. Moreover, there is no equipartition of the momentum. The initial capital is $9.2210 \cdot 10^{-20}$ kg.m/s, a vectorial quantity, is not easy to squander in all directions, as the ionized molecules have only their ionization energy as a handle, plus a tiny kinetic energy given to the plucked electron.

Professor Marmot:
- You're done for with your contradictions! Before, you said that a particle has only one emitter and only one absorber, **and now you have 300,000 absorbers-re-emitters**!

Open-Eyes:
- Well, you have made your point: the destiny of an alpha, and so on with an e⁻ or an e⁺ in the matter, are very different from a photon in the vacuum. What is exactly the eventual transitivity of each reaction at each ionization apex? Yes, there is a quasi-conservation of the frequency, and locally of the phase at each of these apexes. While globally, on the whole run, it is indeed a decoherence. No one of the ionization of a molecule by the fast and ionizing particle may be considered as reversible in time: To hope to accelerate a particle by deionizing

a molecule, it is unfeasible in a repetitive way, and already quasi-impossible even once.

Curious:
- Mr. Open-Eyes, you got distracted by your subject, so that I will sum up myself. You do not believe anymore in the corpuscles; you pretend they never exist (but maybe the atomic nuclei, we will see later), but that **only exist waves, progressive or stationary**. Is this right?
You do not more believe in the standard arrow of time, of the macro-time, at least at the quantic scale. At the quantic scale, you only see equations, symmetric regarding the micro-time.
Therefore, simultaneously you are a relativist, and you are not, as you admit upstream actions which are not admitted in Relativity. You give to the absorbers an importance that your competitors ignore or deny. In your vision, **the absorber is exactly as causal as the emitter is, for any photon or any transferred particle**.

Open-Eyes:
- Your summing up is excellent. Quite a pleasure to have a so curious Mr. Curious, and whose patience is high enough to be at the necessary level to all understand! From 1905 to 1923, the Relativity is a strictly macrophysical theory. We had to wait for de Broglie in 1923-1924, and above all Dirac in 1928 to have its integration into quantic microphysics. Thanks to Dirac, became evident the retrochronous components, with negative frequencies and negative energies, considered with so much reluctance and defiance by the remaining of the tribal community.
Another difference is fundamental: we do not place the random in the same place. The Copenhaguists put it during the trajectory. In their dreams, the quantic particle escapes any physical law as long as you do not *observe* it, but by miracle, on the great numbers, its statistics returns under some laws. It is magical!
In my sense, we must add that the macro-time is only a statistical emergence, it flows in the same direction as the entropy does; now the entropy is a statistical emergence too. Along the 19th and 20th centuries, they took the habit to consider the macro-time as a fundamental parameter, while this silent assumption is not yet grounded. The macro-time has no causal power in microphysics.

Curious:
- However, you have not yet succeeded in explaining the "*spin*", which remains a big mystery to me. You promised to explain further the ferromagnetic and ferrimagnetic materials. And you must explain more clearly what are the atomic and molecular organizations: you have shot down in flames the planetary model, but without explaining what instead.

Open-Eyes:

- Not so simple to give you right today the good formulas, and the good diagrams, as all our external sources are affected by the same sin: they coded with a soldering iron the theory of Max Born in all their calculations; they disguised the solution given by Schrödinger under its Hermitian square (defined positive), to only give *"the probability of presence of the corpuscle"*. For 1927, the ideation from Max Born and Werner Heisenberg became the mandatory Newspeak of the theoretical physics. We must redo everything to draw things as they should be drawn, with evidence of changes of sign in the stationary waves, whose the copenhaguist deny the interest, a major interest anyway. Alas, all redoing, more quickly said than done!

However, an experiment was done, with the molecule of dinitrogen N_2: when it is excited, the changes of sign of the phase are in evidence.

10.3. The experience has sorted: the domains of phases of the orbitals are observed

The end of the recreation: the experiment has sorted, *so the experiment must be wrong!*

References: Itatani and al. Nature 432,867, 2004.S. Haessler and al. Nature, Quoted by Les Dossiers de la Recherche, n° 38, February 2010, pages 63 and 64: "*Des flashes toujours plus courts*". Page 63, the figure n° 2 shows a dinitrogen excited molecule, with domains of positive and negative phase for the outermost electrons of the molecule.

Figure 10.13.

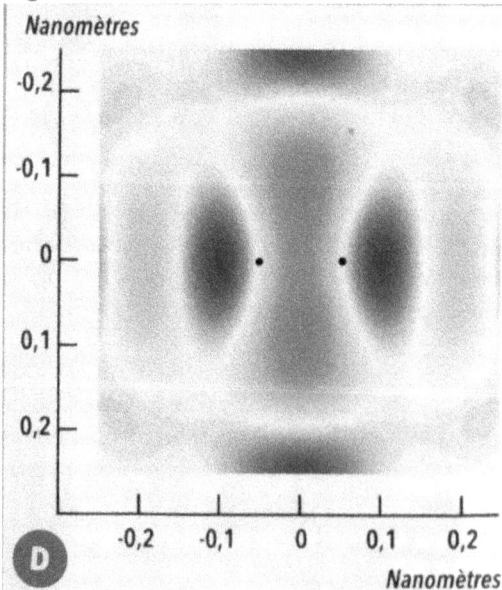

Figure 10.14.

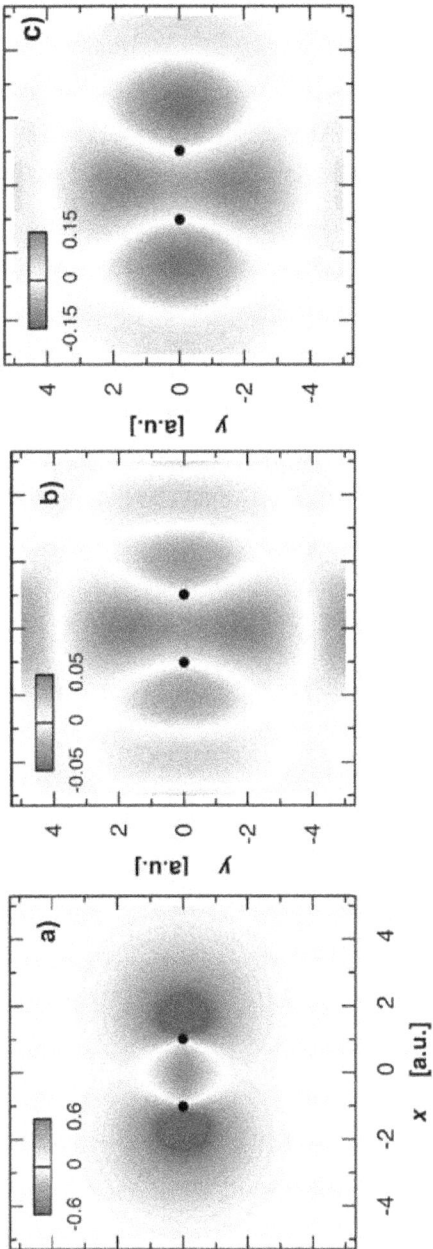

References:
http://tel.archives-ouvertes.fr/docs/00/44/01/90/PDF/
thesis_DrStefanHaessler.pdf
http://iramis.cea.fr/spam/MEC/ast_visu.php?
num=101&keyw=Atto%20Physique&lang=fr ,
http://iramis.cea.fr/spam/Phocea/Vie_des_labos/Ast/ast.php?

t=fait_marquant&id_ast=1550 ,
http://iramis.cea.fr/Phocea/file.php?file=Ast/1550/
CP-Photographie_electron__3_-vCNRS.pdf ,
Orbital HOMO of the dinitrogen molecule N_2,
a) exact calculation (Hartree-Fock), fundamental state.
b) experimental reconstructed by the tomographic technique (re-collision of the electron wave packet in laser field),
c) theoretically reconstructed in the conditions of b). The reconstructed orbital b) presents the characteristic structure, in amplitude and sign, of the exact HOMO.

http://iramis.cea.fr/Images/astImg/1550_4.jpg
Credit: Nature Physics.
Authors: S. Haessler, J. Caillat, W. Boutu, C. Giovanetti-Teixeira, T. Ruchon, T. Auguste, Z. Diveki, P. Breger, A. Maquet, B. Carré, R. Taïeb & P. Salières,
To complete the legend, we precise that in the figure a) N_2 is in the fundamental state, while b) and c) N_2 is in an excited state.
Of course, the message is overwhelmed with the usual claptrap in *"probability of presence"* under the threat of being excluded from the club.
But it is too late! A researcher (Michel Talon) had objected that as soon as we cannot more integrate by hand the solutions of the equation of Schrödinger (modernized by Pauli, next Dirac), the frontiers between signs of phases ceased to exist, and now, his hypothesis is refuted and ruined. These frontiers are there for topological reasons, and nothing can abolish them. The experiment has confirmed.

Conclusion: The one who was right against his winning enemies of 1926 and later years, remains Erwin Schrödinger. His equation (modernized by Pauli, next Dirac) describes the electronic density of the electronic wave, here a stationary wave, and not the messy complications into *"probabilities of the apparition of the goblin corpuscle"*.

Researcher Castle-Holder:
- Ah no! Impossible! Nobody can obtain the phase and the sign of the phase experimentally! How did they do?

Open-Eyes:
- The chemists do not spit on the phase, on what you spit. This, plus the spin give them so much, to evaluate and calculate the chemical bonds, the bonding and anti-bonding orbitals, in the molecules, as well as crystals and glasses.

Curious:
- I feel that you are hesitating to add something, after this very technical message.

Open-Eyes:

- Indeed, I have to add. The copenhaguists as this researcher at Jussieu are so suspicious on the operating protocol which shows the phases he denies, asked how did they do. He pretends that no such things as a phase and its sign can exist, pretending the phase is not an *"observable"*, and cannot have any experimental importance.

The real discussion is at

http://deontologic.org/quantic/index.php?title=

Les_surfaces_infranchissables_au_%22corpuscule%22_pr%C3%A9tendu

As for them the solutions of Schrödinger is only good to be squared to give the *"probability of presence of the corpuscle"*, and as the geometry of orbital turn of the planetary model from the years 1911-1913 is invalidated, they come to pretend that electron-corpuscle zigzags everywhere in the assigned domain, and so statistically performs exactly the assigned density. In the fundamental states, it is not so bad, as the domain is connected. But in excited states with n > 1 the domains of phase are separated in zones of opposed phase, not connected, so no electron amplitude on these frontiers. But how the goblin electron-corpuscle manages to cross these frontiers of no *"presence"* at all? Can you spend the half of your time in the bed of your concubine Zeinab and the other half in the bed of your concubine Zobeid without spending a fraction of time in the lobby between their two rooms? Well, yes, you are a prophet gifted with supernatural powers...

Curious:
- I understand that you must have many enemies when you object so!

Open-Eyes:
- More than you think. One of these harassers, a dead loss in physics, much conceited and despotic, accuses to be *"Jealous of the sexual performance of the prophet"* (or of the fifth evangelist? We are getting lost...). he is himself a *faggot*, scatological obsessed.

This is not all. I pretend that their ideation of a crazy electron-corpuscle, crazying around the atom or the molecule is incompatible with the fineness and stability of the spectrum lines the spectroscopists study from the 19th century. The copenhaguist theoreticians have never obtained nor analyzed a spectrum. It is a heavy analysis, that is practiced only by the analytical chemists, but the copenhaguists never practiced analytical chemistry. Six thousand spectral lines of iron are listed, and on each film we impress, we reserve one half for the spectrum of iron for calibration of the film, on the side of the half for the unknown material. A goblin and crazy electron-corpuscle would forbid this liability, fineness, and accuracy of the spectral lines: it would never stop to muddle the spectrum.

10.4. *"Nobody understands the Quantum Mechanics"*. Here is the process he used to

Open-Eyes:

- *"Nobody understands the Quantum Mechanics"*, it is so fashionable to repeat this avowal from Richard Feynman often, to accuse of a *sin of pride* those who are not satisfied by this obscurity.

Here is the process used to understand nothing, and you have seen the figures at the beginning of this chapter.

Reference: **"The Feynman Lectures on Physics"**, T.3 **"Quantum Mechanics"**, chapter 1.

Feynman pretends to explain the behavior of the electrons by the bullets from a gun. Let's see.

At the Lubianka, the usual threat during the questionings was the *"nine grams of lead"*. It will be enough with a bullet of five grams.

That is five moles of nucleons, or three millions of milliards of milliards of nucleons. Five milliards and five hundreds of thousands of millions of milliards of milliards of times heavier than an electron. Are you sure it is representative? When the bullet crashes on the armor or the target, there are milliards of milliards of milliards of milliards of milliards of quantic reactions. Representative? Are you sure?

An electron has only one quantic reaction at its emission, only one quantic reaction at its absorption.

After the macroscopic bullet, Feynman takes either a macroscopic beam of light, either the gravity waves in a waves tank. Can the water and its waves be representative of an electron?

If so, what could be the intrinsic frequency of the water, where we produce waves in the tank? Please explain what could be the absorption reaction which puts the propagation to an end! And moreover, please explain how the emitter and **the absorbers of the wave** will be submitted to the constraints of stationarity of the electron wave, therefore quantified by discrete states like an atom or a molecule are.

Throughout the lecture by this historic pedagogue, not the slightest idea, nor quantitative, nor qualitative, of what distinguishes our macroscopical world from the quantic world. Proceeding so, nothing surprising in the result *"Nobody understands the Quantum Mechanics"*: they took the right means to obtain such a result. It is much deplorable, and it exacts a shameful and exorbitant cost, in loss of educational yield of science teaching.

Exact sciences... You said *"exact"*? Bizarre! Bizarre!

Curious:
- Say! Mr. Open-Eyes! You seem to have accumulated strong grievances during the last twenty years, as you have not yet developed what you do well, only what the *"Copenhaguists"* do wrong!

Open-Eyes:
- You are right, and it is high time I explain my discovery, the Fermat spindles, though their exact geometry remains unachieved. At least an approximation

by above exists and has well withstood the trial of weathering. The Fermat spindles do all that was to do.

10.5. In propagation, the Fermat spindles

Open-Eyes:
- I recall the Fermat's principle (1657), its explanation by Augustin Fresnel, next by William R. Hamilton, and the exploitation of the results of Hamilton by Louis de Broglie.
Fermat had sum up the laws of the reflection and refraction (Snellius, Descartes): any light path is of stationary optical length, most often minimal, compared to its immediate neighbors. Elaborating the physical optics (1819), with wavelength and transversal polarization, Fresnel clarified: *"Evidently! The light propagates at right angles to its wavefronts - or iso-phase surfaces"*. A real path respects the wavefronts.

About 1834, Hamilton elaborated a formalism of the mechanics, which makes it strikingly looking like the optics; the only difference is that the wavefronts are replaced by the iso-action surfaces, where the considered "action" is the circulation of the momentum. In 1924, Louis de Broglie resolved the mystery of this resemblance, by discovering that these iso-action surfaces in the Hamiltonian Mechanics are precisely wavefronts, where the wave is his Broglian one, of intrinsic frequency $\mathbf{mc^2/h}$ for any particle of mass **m**.

Curious:
- But? But waves of what?

Open-Eyes:
- We had already tens of hours of sterile disputes, for instance with Florent M. who adamantly wanted to impose a hyper-restrictive definition of "wave", imposing that no matter would be carried, and moreover, that one could monitor in macrophysics a macrophysical oscillating quantity, and as he pretended all would be macrophysical, we should measure the oscillating thing... You must accept that **an electron wave is the electron itself**, it is material, it carries a mass, an electric charge, a spin, a magnetic moment; similarly accept that a neutron wave, so appreciated in the fundamental physics of the crystals, is the neutron, and nothing else, it carries a spin, a magnetic moment, a mass and some hadronic features. You must accept that an electron never has another existence that would be non-undulatory; even when stationary around an atom or a molecule, the electron remains undulatory. You must accept that never any instrumentation will give you some information of the kind $\mathbf{a\ sin(\omega t)}$ on the path of the electron: the electron is indeed elementary, while neither we, neither our instruments are. When we pose equations and their solutions, we are steps farther beyond the fineness our instruments access. It is a forever curse, and we have to do with: we are not at the scale of the individual electron, so we must innovate in our theoretician bag of tools.

Curious:
- But now? Your process?

Open-Eyes:
- I put, facing and parallel, two dipolar electric antennas, such as the carbon monoxide molecule described earlier, or such as an oscillating atom between a spherical orbital and an oblong orbital. One antenna is said emitting, the other absorbing. The whole energy emitted by one is absorbed by the other, and their distance is much larger than the wavelength. Therefore, when no other interferential devices meddle in between, the region of space where the energy is transferred by is limited by the maximum dephasing at the end: less than a quarter of a period. Hence, a maximum width all along the transmission spindle. The next task is to calculate this maximal width in a perfect vacuum, and its evolution along the abscissa. Outside this spindle, the transmitted power is null, or nearly-null. When interferential devices are there, we have to consider several branches of paths, and phases differing by an integer number of periods; it means several Fermat spindles, as many as integer phases in the interference.

Provisory, in the absence of the exact calculation, it will be enough with the approximation by excess, found in May 1998. One guesses that in the vacuum or a perfectly homogenous medium, the concurrent paths are arcs of circle, in other words, are of constant curvature. It underestimates the diameter near the ends but overestimates it at mid-journey. Next, we settle for the second order in the Taylor expansion, which is to approximate the arc of circle by a parabola, and so to worsen the overestimating of the maximum diameter in the middle of the spindle. Seventeen years during, I kept a bad opinion of this approximation by arcs of circle for the Fermat spindles. For astronomical purposes, maybe I was too pessimistic. Let us say that beyond 20 diameters of the source or the absorber, we are in a far field, and the approximation seems to become liable, at least in astronomy.

Figure 10.15.

Schematized for placing the dimensions:
Figure 10.16.

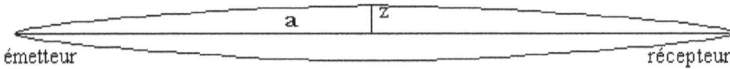

The condition of the Fermat spindles is: $2\,\alpha\,R - 2\,R\,\sin(\alpha) < \frac{\lambda}{4}$.

Limited to the first order: $\alpha^3 < \frac{\lambda}{4R}$.

Now $\mathbf{a} = \mathbf{R.sin(\alpha)}$, reducible at the first order to: $\mathbf{a} = \mathbf{R.\alpha}$.

So we can eliminate the radius R, and keeping the first not null term of the Taylor expansion, it remains: $\mathbf{z^2 = 3/16\ a.\lambda}$ where λ is the wavelength.

We take the square root: $z = \sqrt{3a\lambda}/\,4$.

Expressed in function of the wavelength: $\frac{z}{\lambda} = \sqrt{\frac{3a}{16\lambda}}$.

We also need the half-angle at the end of the tangent cone (near the emission and absorption reactions, but farther than twenty diameters), or angle of Fermat: $\alpha = \sqrt{\frac{3\lambda}{4a}}$.

Numerical examples:
A. We take the infrared photon resonating with the longitudinal vibration of the carbon monoxide molecule, of wavelength 4.6 μm.
Take a total path of 20 cm between the lamp and the molecule, so a = 0.10 m.
a . λ = 4.6 . 10^{-7} m².
$\sqrt{3a\lambda}$ = 1.17 mm.
Dividing by 2 to obtain the maximal diameter of the Fermat spindle: 0.59 mm.
Not negligible; we obtain all the previously known results since Young and Fresnel, on the diffraction on an edge, a hole, or an obstacle.

B. An optical disposition is designed to simulate a stationary wave in front of a mirror. Let us take λ = 500 nm, **a** = 1 m.
What is the width of the stationary wave in front of the mirror, at the individual scale of one photon?
We must add two hypotheses:
Diameters of both the reaction of emission and absorption = 1 μm.
Angle of (small) tilt of the emitter and the absorber, from the normal = arcsin(0.01) = 34'.
$1 - \cos(\arcsin(0.01)) = 5 . 10^{-5}$.
$\frac{z}{\lambda} = 2\sqrt{\frac{3a}{16\lambda}} = 1{,}224$. It is the number of wavelength in the width of the spindle, in addition of the sum of the emitter and absorber widths.
This diameter in metric units: $(1{,}224 + 2)\ \lambda$ = 0.62 mm.
So with this angle of tilt, the footprints of the wavefronts creep laterally, in the centrifuge direction.
How many wavelengths in the diameter of this track? $1{,}226 * 5 . 10^{-5}$ = 0.06.

C. We will calculate the mirror, pretended to tremble, necessary to experiment the assertions from Elitzur and Vaidman.
http://www.agoravox.fr/culture-loisirs/culture/article/
contrafactualite-penrose-elitzur-155565
First step: enlargement of the Fermat spindle on that distance:
radius z = $\sqrt{3a\lambda}/\,4$
where we will count 50 cm from emitter to absorber, included the re-focusing optics. That is **a = 0.25 m**.
Let us take the ray: λ = 0.5 μm.
3 . a . λ = 375 nm²
$\sqrt{3a\lambda}$= 612 μm*radius **z** = 153 μm = 0.153 mm. ==> Diameter: 0.306 mm.
On this, we must add the mean of the physical radii of the emitter and the absorber, plus the uni-axis oblonging of the diameter of the spindle by the half-reflecting mirror. Let us set the influence of the half-reflecting to 4λ = 2 μm, negligible compared to the belly of the Fermat spindle.

So the Fermat spindle is wide about 0.4 to 0.6 mm near the mirror.

The total reflection by a mirror without diffraction demands a mirror definitely wider than the photon, to obtain that the moving of the electrons under electromagnetic influence remains far from the edges. Hence a mirror of 1 mm cross 1.4 mm, minimum. The affair should be better (or not as poor) in the near UV: if we divide the wavelength by two, the momentum should be multiplied by four, and the belly of the Fermat spindle divide by 2, so a smaller mirror, provided that the size of the source and the sensible zone of the absorber could be reduced, too. But even so, the contrafactual hoax remains impossible to test by an experiment: the mirror remains far too heavy by eight or nine orders of magnitude, to be "*shakered*" by one photon.

Conclusion: we do not practice anymore the same physics as the copenhaguists sorcerers do. We, we can dimension this experimental apparatus, as they cannot, and do not see their error.

10.6. The geometry of the Fermat spindles, applied in astronomy

The astronomy provides some powerful tests for the validity of the geometry of the Fermat spindles. So it is a priority to take risks in this domain.

10.6.1. long bases interferential astronomy and Hanbury Brown and Twiss effect: bosonic features of the photons.

When you calculate the diameters of the Fermat spindles on astronomical distances, you find diameters that become nearly astronomical. So it is obvious why the Very Long Baseline Interferometry (VLBI) and Very Large Array Interferometry work: though the absorbers are distant from several tens of kilometers, these photons coming into them had all the time to synchronize during their long shared journeys, during which they largely shared their width of propagation; width that could exceed the diameter of a star, even an astronomical unit.

Let us take the nearest star, which one can inflict you sunburns. The melanin in your skin is a copolymer, whose molecular weight is poorly known, but high. No principle mistake but a plausible assumption if we assign to it a diameter of 300 nm, and studying the propagation of a 300 nm wavelength photon from the Sun.

The Sun is on average at 149.6 millions of km from here. We neglect the radius of the Earth, but subtract $\frac{3}{4}$ of the Sun radius, and it remains $1.492 \cdot 10^{11}$m. At 6 µm of the melanin molecule, we are in a good approximation of a far field, and the angle of the tangent cone is about 0.8 µrad. Under the assumption that the emitter is of small size, the maximum width of the belly of the Fermat spindle is about 192 m. The bosonic character of the solar photons is perfectly justified.

Now let us take the light coming from Andromeda Great Galaxy (M31 in Messier catalog), which is at 2.2 millions of light-year. A short digression about the age of this incoming light: 2.2 M years ago, it was here the upper Pliocene, beginning of the Donau glaciation.

2.2 M lyr $= 20 \cdot 10^{21}$ m.

Radiation at 0.55 µm $= 5.5 \cdot 10^{-7}$ m

$3 \cdot a \cdot \lambda = 33 \cdot 10^{16}$ m²

$\sqrt{3a\lambda} = 4.54 \cdot 10^{7}$ m. Hence a diameter of the spindle of 22,700 km, about a twelfth of light-second, or a small third of the altitude of a geostationary satellite. It remains very small compared to the diameter of a star: less than half of the diameter of the Sun.

Let us see now much farther objects. We know at least one quasar, whose optical image is split by a galaxy acting as a gravitational lens. OK, a gravitational lens for the ensemble of the incoming photons. But are there individual photons which are split by this galaxy? So we must calculate the normal width of these individual Fermat spindles from this quasar 0957+561. Image at
https://upload.wikimedia.org/wikipedia/commons/
9/9d/QSO_B0957%2B0561.jpg

The distances are estimated by the redshift, according to the hypothesis from Hubble. So the quasar would be at 8.7 milliards of light-years (twice elder than our Solar System) and the aligned galaxy acting as a gravitational lens at 3.7 Gly. More details at
https://en.wikipedia.org/wiki/Twin_Quasar. For this example, our question is idle, as has been proved during the last thirty years that the time lag between the two optical paths and images is about 417 days. To have a chance to observe some interference, this difference of path should not exceed about ten thousand wavelengths or periods, and moreover, nothing exists for the photons that could split them individually as a negative thread does for the electrons. However, we persist in studying the other conditions. So we assume no interposed optics, nor gravitational, nor any else. Choosing the wavelength: 500 nm.
8.7 Gly = 82 . 10^{24} m, hence:
3 . a . λ = 62 . 10^{18} m². And $\sqrt{3a\lambda}$ = 7.9 Gm, and a diameter of 3.9 Gm.
It is 2.5 % of an astronomical unit; at the galactical scale, it is microscopic.
However, for the efficiency of the Very Long Baseline Interferometry, and for the synchronization in phase and frequency of the photons from the same origin, it is valuable.
For the correlation in intensity on a basis of 6 meters at the reception, see Hanbury Brown and Twiss:
https://en.wikipedia.org/wiki/Hanbury_Brown_and_Twiss_effect. It has allowed them to measure the apparent diameter of Sirius. Encyclopedia at
https://en.wikipedia.org/wiki/Astronomical_interferometer.

These facts prove a retrochronous causality from the crossed medium to the emitter. It never looks like the sequential causality as inherited from the artillery: The explosive charge explodes, the gun shoots the shell, **then** the air and the wind intervene in the course, **then** the shell explodes and disperses its shrapnels. In microphysics, the conditions of propagation and interaction between photons in space partly rule the emissions.
So we retrieve this retroaction from the laser medium to the individual atomic emitters, predicted by the Einstein coefficients in 1916. No one of the potentially emitting atoms ever individually *"decides"* when, nor at which precise frequency, nor on which duration, nor with which phase it emits a photon: it is a collective affair, with resonance with the cavity between the two end mirrors. On the distance after getting out of the cavity, the collateral bosonic effects gets looser, and the final destinies of the photons are largely independent on far targets. For instance, on the Moon, not all the photons are reflected by the mirrors, a majority is absorbed by the lunar regolith.

Professor Castle-Holder:
- Pity! Your reasoning is wrong on a point: though it is coherent with the width of the photons according to the width of your Fermat spindles, however, nothing can oblige these photons to become long during the journey. And in the visible domain, their length of coherence has a ceiling about one meter; maybe the total length of a photon in the visible could be about two meters.

To let the photons to find a travel companion with which to *boson* together, it lefts a really short length of interaction, and a very short time lag between them. This objection is not critical, but we have to refine the model from the experimental characteristics of the bunching of the photons. I presume, but it should be confirmed, that the bright stars give a good bunching, while the rarefied interstellar gas does not, because of a rarefied emission.

Open-Eyes:
- Yes, it is still a work to do. However, our simplified reasoning imagined photons with only strictly parallel paths. This implicit hypothesis is not realistic on the star scale. If the initial trajectories are slightly crossing, each photon will have chances to cross an eventual twin with a very small angle. And their bosonic interaction will cluster their momentum vectors and frequencies. In other words, they are building a common destiny and a quasi-common destination. This is how, in a substantial part, they arrive in packs on our instruments.

Open-Eyes:
- After reflection, I think that your objection is quite judicious: it implies that the bosonic attraction intervenes on the emitters and the absorbers, to incite them to transact the emissions the most synchronized, the most in bundles. To dare to consider this perspective is a revolutionary act.

10.6.2. Geometry of the Fermat spindles, and scintillation of the stars.
10.6.2.1. *Scintillation and turbulence.*
Curious:
- Please begin by defining what this **scintillation** of the stars.

Professor Castle-Holder:
- It is one of the effects of the atmospheric turbulence, the only one that you can detect with the naked eye, the night. With a clear sky, mostly at the beginning of the night, when the ground is still hot, the light from the stars seem to blink, irregularly: their brightness varies quickly. In the extreme cases of turbulence, some bright star near the horizon such as Capella, oscillate in colors too; even Jupiter may blink to our eyes when it is low on the horizon, and the turbulence is strong.

Curious:
- Low on the horizon? Why?

Professor Castle-Holder:
- The most the light path is near the horizon, the longest it crosses a dense atmosphere. This turbulence may have two causes: thermal columns from the hottest spots on the ground compensated elsewhere by down-flowings. And wind turbulences caused by obstacles and relief. And the all can combine. A similar aberration by convection occurs when the Sun heats an asphalt road:

the heated air rises in irregular patterns, and with low-angle light, seems a liquid and agitated road. According to its temperature and density, the optical index of the air changes, and that in moving convection cells, so the light deviation fluctuates.

Open-Eyes:
- The other big effects of the turbulence are obvious through a spotting scope or a telescope. In my previous flat, I learned to adjust my scope by aiming a neon sign on top of a building, 3 km further. Depending on the wind, the hot exhaust of a plant in between could blur the target. At a lower distance, about a kilometer, when I tried to decipher a shop panel, aside from a door, it could be impossible at the hot and sunny hours, when the convection was active, then the air in between calmed only at sunset, when the sunlight came from a low angle. In the night, the wind turbulence could shred the boundaries of the moon into ten to fifteen fugitive crenels. An idea of the rhythm? About 2 to 3 Hz, but without any regularity. I tried to spot the North Star, with the camera at the focus of the scope: it wriggles, this star! Or at least its image on the zooming screen, with a magnification about 480. One may infer that if instead of taking the whole image but diaphragming on few pixels on the edge of the moon, or the edge of the agitated North Star, these pixels would record a scintillation.
We still have to understand why our naked-eye sight of a star looks like the pixels on the edge of an agitated image.

Professor Castle-Holder:
- Not so fuss! We know that in a telescope, the scintillation disappears. It is enough to enlarge the entry pupil to weaken or suppress the scintillation. But it does not suppress the fast and erratic movements of the fragments of the image. So in astronomy, the adaptative optics brought relief and far better accuracy, very appreciated. Now it is necessary to query photographic documents of this scintillation. Images were once at
http://www.je-comprends-enfin.fr/index.php?/Notions-sur-la-lumiere/
pourquoi-les-etoiles-scintillent-elles-et-pas-les-planetes/id-menu-73.html
Problem: Since, this domain name has been purchased by a quite different activity, and the real authors of these remarkable images are unknown.
Their explanations in words of « *molecules* » are false, but now nobody can read them.

Figure 10.17.

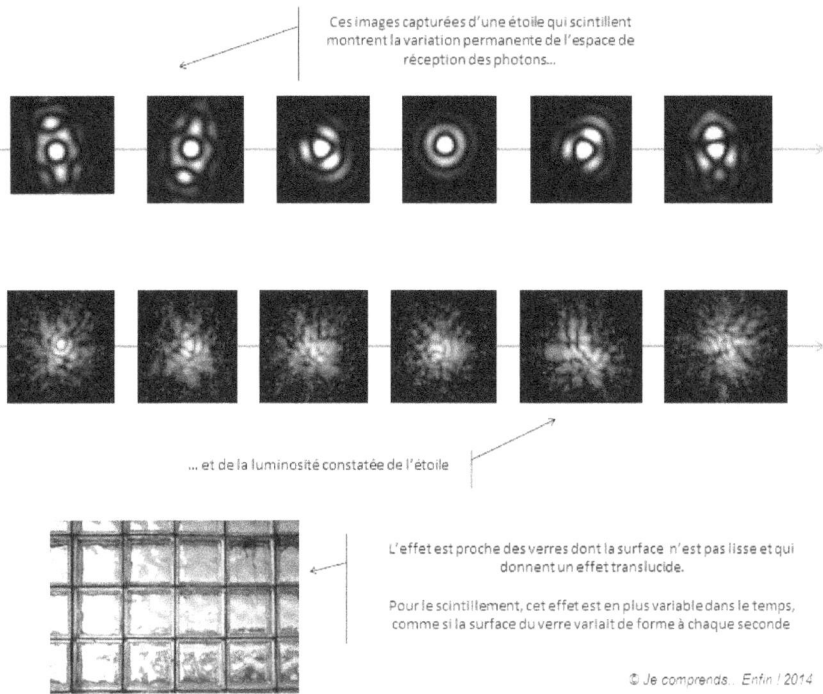

Ces images capturées d'une étoile qui scintillent montrent la variation permanente de l'espace de réception des photons...

... et de la luminosité constatée de l'étoile

L'effet est proche des verres dont la surface n'est pas lisse et qui donnent un effet translucide.

Pour le scintillement, cet effet est en plus variable dans le temps, comme si la surface du verre variait de forme à chaque seconde

© Je comprends... Enfin ! 2014

The proof by the adaptative optics:
Figure 10.18.

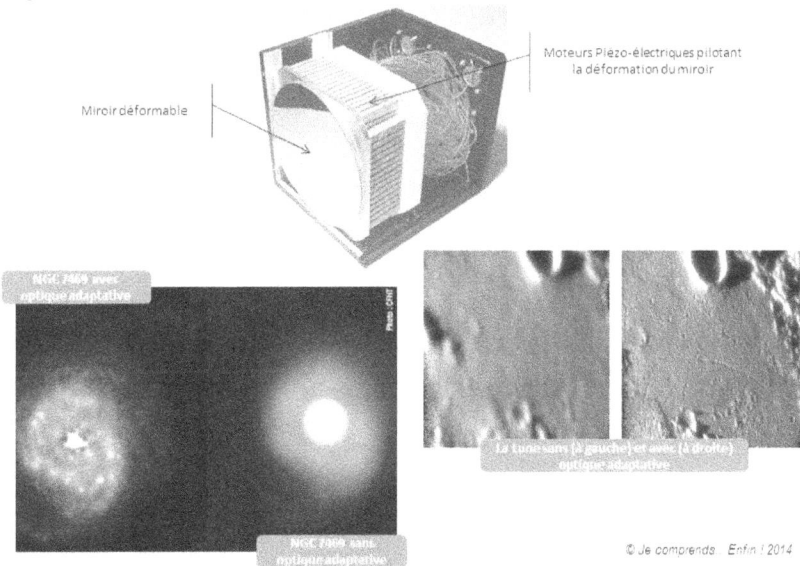

Moteurs Piézo-électriques pilotant la déformation du miroir

Miroir déformable

NGC 7469 avec optique adaptative

NGC 7469 sans optique adaptative

La lune sans (à gauche) et avec (à droite) optique adaptative

© Je comprends... Enfin ! 2014

http://www.je-comprends-enfin.fr/images/stories/restreinte/big/
2014-NDC-RELATIVITE-RESTREINTE-0009.jpg
Nor the optics nor the sensor are known. And an interesting video at
https://intra-science.anaisequey.com/physique/categories-phys/34-
astronomie/316-etoiles-scintillation#r%C3%A9ponse-avanc%C3%A9e

Again, the pupil, the optics, and the sensor are not said, alas.

Open-Eyes:
- But not so simple with the human vision! We have two kinds of pupils to consider. In the literature, the modeling we find is at the scale of a crowd of photons and a crowd of absorbers. So they construct a long geometrical cone holding on the edges of the surface of the star and on the pupil of the eye. Example:
http://elib.dlr.de/7341/1/LASE2004-5338-29_Perlot_ApAv_measur_
HANDOUT.pdf
Now an opsin and its retinal work at the scale of the individual photon. What is new in TQP (Transactional Quantum Physics) is that we work at the scale of the individual wave, photon, electron, so on. So we are concerned not only by the entry pupil in front of the lens but also by the network of photosensitive molecules in the retina. Our sclera of primates is opaque and absorbing: a descent of diurnal mammals for about 58 My – say since the common ancestor to us and the tarsier (Tarsius tarsier).

But the sclera of the *Laurasiatherians* who are familiar to us proves an excellent reflection, so hardly 10 % absorption at first crossing through the retina, and 10 % at crossing back (plus about 5 % absorption at the reflection on the tapetum lucidum). *Laurasiatherians*: the cohort of placenta mammals whose ancestors 95 million years ago lived on the North continent or Laurasia, comprising present North America, Greenland, and Eurasia. In opposition to the Gondwana on the south, from where parted South America, Africa, Antarctic, India, Madagascar, and Australia. Our ancestors of the primates, the Dermoptera, the tupaias (treeshrews), the lagomorphs and the rodents, in one word the Euarchontoglires, were on the Gondwana.
Figure 10.19.

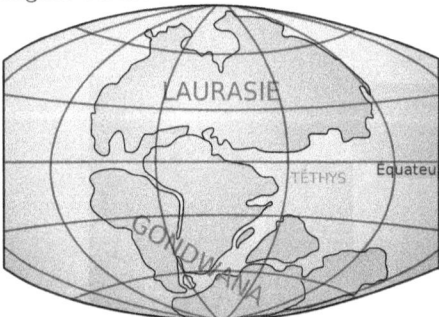

TRIAS
Il y a 200 millions d'années

Professor Castle-Holder:
- Objection! The Laurasiatherians do not have all the same structure of *tapetum lucidum*, and moreover, also the majority of the strepsirrhines (the prosimians but the tarsiers) have a *tapetum lucidum* behind the retina. With the *Carnivora*, the rodents and the cetaceans, the tapetum lucidum is made of cells containing patterns of very refringent crystals, acting as reflex reflectors. With

the bovids, the equidae, and the sheep, the *tapetum lucidum* is made of extra-cellular fibers. They generally have in common to be iridescent.

Open-Eyes:
- Our knowledge is too short about the transient or durable character of this transparency of the retinal pigments. The familiar experience of the flashlight in the eyes of our domestic animals may give a misleading idea on the capture efficiency of the pigment when it was in the dark, especially at its resonance frequency. Until its first captured photon, the opsin was really dark or "*black*" (pink in reality, hence its name rhodopsin, "*black*" at its resonance frequency, 604 THz), but afterward, from the picoseconds after the capture until the photosensitive cell depolarizes and re-synthetizes the association opsin-retinal in all the opsins dissociated by the light, it remains transparent, and the animal is blinded. The familiar experience shows that the blind and transparent state may last several tens of minutes.

Curious:
- But do you speak about cones or sticks? Sticks I presume?

Open-Eyes:
- However, when the amateur astronomer observed color scintillation, he brought a certitude and a question: the certitude that the illumination from Capella is enough to excite cones. Next comes the question: is the retinal image small and mobile enough to shoot sometimes a cone specialized in one color and sometimes another for another color? Or was it already a chromatic separation before arrival at the cornea? My preference is for the first hypothesis: the main image of the star oscillates on several different cones.

Professor Castle-Holder:
- If we open an encyclopedia, they build a cone holding at the arrival on our retina, and at the starting on the edges of the heavenly body, planet or star. Do you agree?

Open-Eyes:
- Not so much, for two reasons: They think crowds, crowds of emitters, crowds of photons, crowds of absorbers in the retina. It is not false but it has to be confronted to the individual geometry of the photon which comes to a photosensitive cell (and more restrictive: molecule) and is absorbed by it. We had a similar question with the big mirrors such as at the Mount Palomar: does the 5 m diameter photon exist? Does the photon having the diameter of the pupil at the pupil, 3 to 7 mm, exist? And what about the diameter of the daylight photon on the pupilar slit of a feline? While these optical systems with a wide opening have some defects, some astigmatism, geometrical aberrations, a focal that depends on the distance of the entering ray from the axis of the system... We must take account of the Abbe conditions, even imprecise: will converge on an opsin whose depth is small, only photons which have crossed the cornea and

the lens under a small aperture. Alas, we can only conjecture the efficient photonic aperture. Provisionally, let us take its radius as 0.5 mm, that is a disk of
1 mm diameter as the efficient-pupil-for-the-receiving-opsin. Versus 3 to 7 mm
of anatomic pupil for crowds of photons for a crowd of cones and sticks. Later
we will criticize this provisional value, and when we will study the rainbow, we
will have to reduce it.

Figure 10.20.

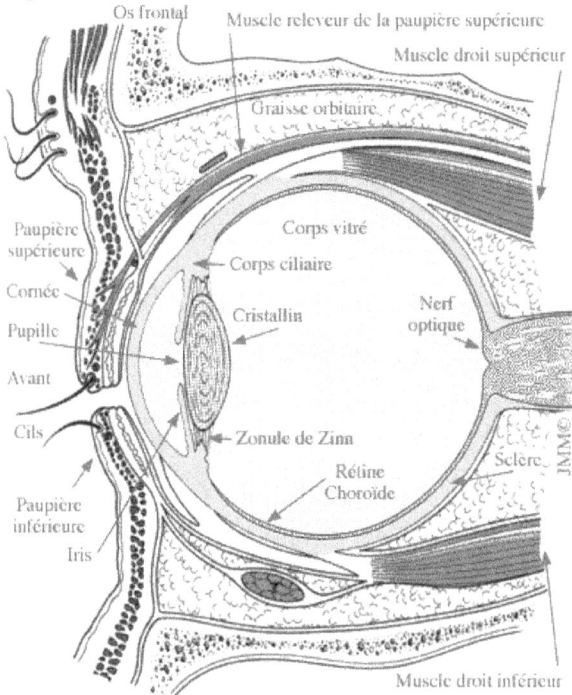

Curious:
- And do you have proofs of sub-pupil, limited by the quality the ocular or
astronomical optics? And my next question: what is the importance of the
depth of field in the image space? If we look at the shapes of the cones and
sticks, they seem adapted to spread in depth convergences. Attention: the
photosensitive disks are the nearest to the opaque membrane, the sclera; the
light they receive has already crossed the layers of neurons. The eyes of the
cephalopods have the inverse disposition: photosensitive cells in front, neuronal
interconnection behind.
Figure 10.21.

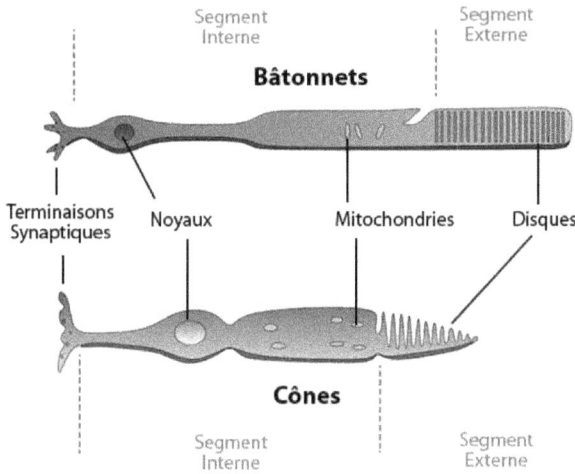

Figure 10.22.

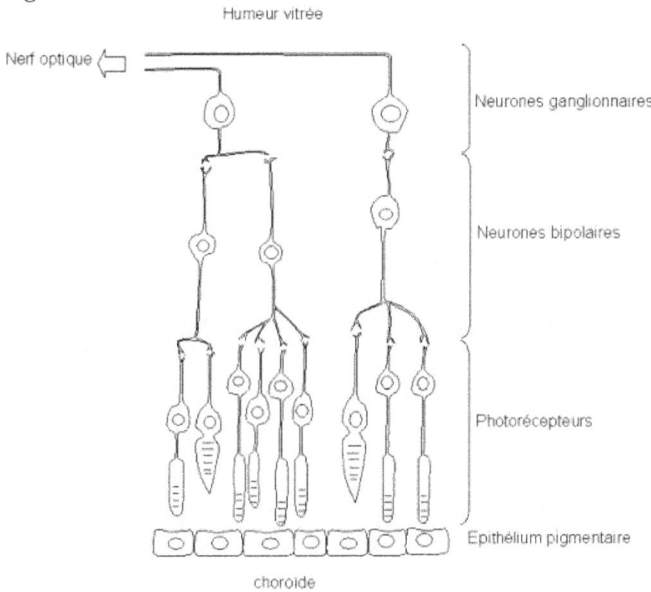

Open-Eyes:
- Sub-pupil, proof? This is the task to perform now.
What is the most significant of the angular characteristics of a Fermat spindle, viewed from the absorber?
We may think to $2z/a$, the maximal diameter of the spindle, drawn from the absorber, but it is not an observable. From the theorem of the inscribed angles, we deduce it is also the Fermat angle of the complete arc, and the angle from the axis to the tangent cone. In the approximation of Fermat by an arc of circle of small angle, we do not need more for the use in astronomy.
But what becomes this theoretical Fermat local cone when the light from a star passes through the atmosphere? We aim to explain the scintillation of the

stars when their light passes through our pupils, and why nor Saturn nor the Andromeda galaxy scintillate, while no star scintillates in a modest telescope, say a 130 mm one.

We will calculate for one type of receptor: the 11-cis-retinal in our rhodopsins in the sticks in our retinas.

The 11-cis-retinal of our rhodopsins in the sticks has a long axis about 18 Å and about 5 to 10 Å of short axes. So at 36 nm from the retina, in the vitreous humor, and still far from the pupil, it is a far field.

For each photon coming from a star, one can calculate the angle α of the tangent cone, to the axis of the cone. It is precisely the angle at the center of the half-arc, from an apex to its belly. The primitive idea for this calculation was to elaborate a quantitative theory for the scintillation.

Professor Marmot:
- And you failed?

Open-Eyes:
- « *Point n'est besoin d'espérer pour entreprendre, ni de réussir pour persévérer* »
One need not hope in order to undertake, nor succeed in order to persevere.

William the Silent.

$$\alpha = \sqrt{\frac{3\lambda}{4a}}.$$

We calculate this angle of the Fermat cone for Sirius at 8.6 lyr and main wavelength 480 nm: 8.6 lyr = 81.4 . 10^{15} m.

$\alpha^2 < 8.85 . 10^{-24} \Rightarrow \alpha < 2.97$ prad. Less than three picoradians.

Under this angle, 0.5 mm of radius of pupil should converge at 68,100,000 m in the vacuum, about 0.561 light-second, about 43 % of the distance to the Moon.

Professor Castle-Holder:
- No! Breakneck! The retina is deep in the eye, through an optical system like a strong magnifier, whose vergence is about 58.6 to 70.6 diopters for a normal eye. Your cone angle is correct only outside the eye, but we have to calculate what it becomes in the vitreous humor to know the width of the absorbing molecule, on the image side, seen near the cornea.
Figure 10.23.

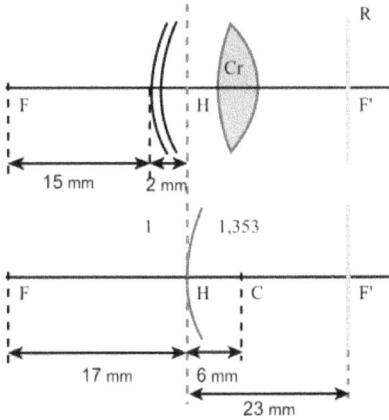

Open-Eyes:

- 55.6 diopters correspond to 17 mm focal in the air, but 23 mm in the eye, through vitreous humor of index 1.336. I would deduce the pupilar image of the stellar object by the eye system. Alas, the result is meaningless.

Curious:

- You are chasing your tails now, you physicists. You lack too many lessons in the physiology of seeing. I have inquired, and I can teach you that: The opsin is a transmembrane protein, it closes an ionic channel, but indirectly, via two successive transmitters. There are about a hundred thousand to a hundred and fifty thousand opsins per disk of stick or cone.

Figure 10.24.

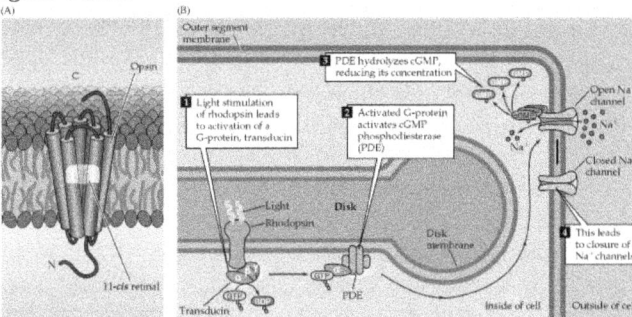

The sticks are sensitive to light by a pigment, the rhodopsin (rhodo: pink, opsis: seeing) whose feature is to absorb photons, with a resonance in the blue-green, about 498 nm (602 THz). There are about 10^8 (a hundred million) molecules of rhodopsin in the external segment of a stick, about 10^5 (a hundred thousand) per disk. One may find higher figures, but I could not find their proofs.

Figure 10.25.

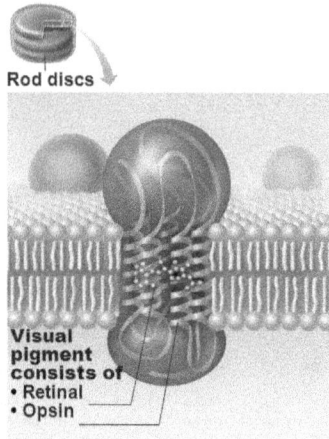

Rod discs

**Visual
pigment
consists of**
• Retinal
• Opsin

**(b) Rhodopsin, the visual pigment in rods, is embedded in
the membrane that forms discs in the outer segment.**

Figure 15.15b

Each molecule of rhodopsin, as the other photo-pigments, is composed of two
elements: a glycoprotein (the opsin) and small lipid derived from the A vitamin, the cis-retinal. Only the retinal directly interacts with light. The retinal
can exist in two shapes, and each intervenes for a phase of the visual cycle. In
the dark, the retinal is in a bent shape, called cis-retinal. When it captures a
photon, it takes the unbent shape and splits form the opsin; this new form is
called trans-retinal. To be reused, the trans-retinal must be re-isomerized into
cis-form. This is done out of the stick in the scleral epithelium. This photo-
isomerization initiates the cascade of biochemical events which produces the
electric signal, the receptor potential through the membrane of the photoreceptor. After a lengthy exposition to intense light, the most of the rhodopsin has
been dissociated, and the trans-retinal molecules are in transit to the epithelium
to be restored into cis-retinal; in the meantime, the sticks are less sensitive to
light. After exposition to a very intense light, when the rhodopsin is dissociated,
it takes almost one hour to re-synthesize the bulk of the rhodopsin
(http://slideplayer.fr/slide/501777/#).
There are about a thousand active disks in a stick, about 25 nm wide, 10 nm
thick. The disks are formed by membrane folding near the internal body, at a
rhythm of one to four per hour (varying along the day). At the end of life, they
break off and are phagocyted by the epithelial pigmented cells.

Open-Eyes:
- I suspect that you fell into a polysemic trap: "disk" takes two meanings in
the texts you used. Sometimes it designates an internal disk, which holds the
transmembrane opsin molecules (but I do not discern what its metabolism is),
and sometimes it designates the part of the cell, containing the internal disk,

in contact with the extracellular liquid (the vitreous humor), and which hyper-polarizes negatively by the closure of the sodium channels.

10.6.2.2. *Box: Colors.* -
https://www.bioinformatics.org/oeil-couleur/dossier/photoreception.html#
COURANT-OBSCURITE
Each kind of photoreceptor is more sensible in a part of the spectrum. The dissociation and the regeneration of the pigments of the cones seem similar to those of the sticks but in a far faster way.
The current of dark First, look at what happens in a cell in the complete dark. The photoreceptors work on the contrary to the normal neurons: the cones are depolarized when in the dark. It implies that a permanent current crosses the photoreceptive cells, and the receptor potentiall is $-$ 40 mV from the external fluid. This potential is created by inequality of charges between the extracellular and the intracellular media. But what is the origin of this current? Indeed, in the cones as well in the sticks, the current is created by the move of cations, predominantly Na+ for the cones, but also a current of Ca^{2+} and Mg^{2+}. These ions come from the vitreous humor. To let them through the plasmic membrane, necessarily the cationic pores must remain open, letting the cations come from the external medium. These pores are held open by a cyclic nucleotide, the cyclic guanosine monophosphate (cGMP), acting on the internal face of the plasmic membrane. Its concentration must be large enough to maintain open the pores unless the current weakens. The cations entered in the cell are next evacuated by the internal segment by sodium pump for Na^+ cations, by exchange for the Ca^{2+} and K^+. So, in the dark, thanks to the cGMP, the photoreceptive cells carry a flow of cations.
Figure 10.26. cGMP

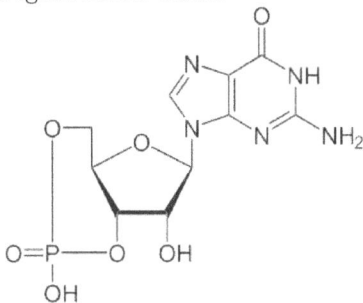

The hyperpolarization of the cone
. Figure 10.27

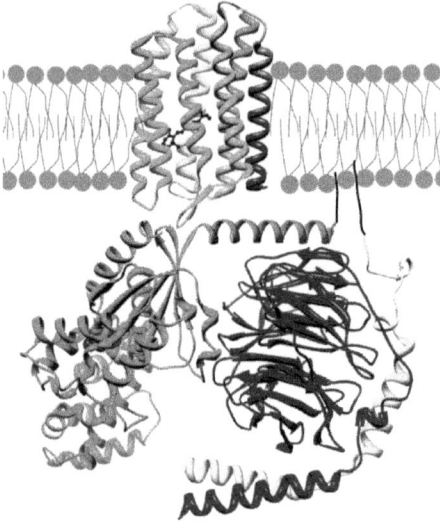

Iodopsin and transducin
. As seen above, the incoming photon changes the conformation of the iodopsin, so changes its function. Now begins the transduction. Now activated, the iodopsin undergoes a series of intermediate dissociations. The metaiodopsin activates the transducin, protein of the G group[2], which mediates the activation. It activates the phosphodiesterase, which hydrolyzes the cGMP. So the concentration of cGMP in the cell falls quickly, which carries the closing of the cations channels. The closing of the cations channels brings a rise in resistance of the cellular membrane. So the current crossing the photoreceptor drops, even ceases: it is a hyperpolarization. From a rest value of -40 mV, the receptor potential drops to – 80 mV, with a higher concentration of cations outside.
Now, we have a nervous signal: the receptor potential propagates along the plasmic membrane to the synaptic end, where the secretion of the glutamate transmitter is then inhibited. The captured photon has triggered potential energy stored in the cell, that transforms a luminous signal into a nervous signal. Of course, no kind of *"transformation of luminous energy into nervous energy"*. The amplification in energy is essential in this transduction. The energy of a photon is several magnitudes smaller than the one of a nervous signal. The amplification is in two steps: first, an activated opsin molecule can activate hundreds of transducin molecules. Next, when hydrolysis of cGMP occurs, one hydrolyzed molecule of cGMP drives the closing of 10^6 sodium channels (a figure to be verified). Now the receptors do not necessarily receive several photons together; giving a transmissible message by a synapse depends on the amplification in hyperpolarization. The cones are far less sensitive than the sticks and can not give a hyperpolarization signal by only one photon: they need at least a dozen in a limited time.

---

[2]Group of proteins implicated in the reception of external signals (light, olfaction, etc.)

Indeed, accumulation of signal occurs in the cones in the form of accumulation of partial hyperpolarizations, following the decompositions of the iodopsins by captures of photons.
End of the box.

Professor Castle-Holder:
- We retain that in a living eye, the metabolic activities are intense, so terribly fragile to perturbations and alimentary and vitaminic lacks. Not mentioning the parasites, like the *Onchocerca volvulus* transmitted by the Simulia (black fly).

Open-Eyes:
- You stimulate a quasi-personal memory: when the war in Algeria intensified, the doctor Mohammed Benabid, physician in Bordj Bou Arreridj, sent his wife and their three boys in refuge in Grenoble, care of friends there. Next, the FLN guerrilla captured him, as they were in great need of a medical doctor in the maquis. Captured next by the French Army, Benabid faked to read in his cell, to hide from the army that the malnutrition in the maquis made him blind (for a time).

Open-Eyes:
- But how do you know that?

Open-Eyes:
- The colonel Georges Buis intervened for the liberation of Benabid, and he was house arrested in Grenoble. Soon my father became a friend of the Benabid family, and he clandestinely drove Benabid in Geneva, to meet Ferhat Abbas. Their second son Jean-Claude was a fellow in my class.

Professor Castle-Holder:
- Time to add that a stick is never alone in detection: first, they are coupled in parallel by the neuronal wiring, from several tens around the optical axis, out of the fovea, to several hundred in the far peripheral. And next, the photosensitive molecules seem far to line at 100 % the retinal surface, leaving insensitive space between them. It is why most of the nocturnal species, especially all the descent of the Laurasiatherians – for instance, the sheep and the wolves – wear a mirror behind the retina, to double the photonic yield, and double their chance to detect in the night: the *tapetum lucidum*. I suspect that the diameter of the retinal molecule may not be a significant size. We must pay attention to the chance to hit a cis-retinal molecule or not, as an absorber in the retina.

Open-Eyes:
- And to the chance to hit or not our pupil, through the atmospheric turbulence.

10.6.2.3. *Scintillation, conclusion.* Considering the collective character of the reception on a cone, via some tens of hundreds of millions of opsins, or on a collective of several tens or hundreds sticks to trigger a neuronal signal towards the lateral geniculate nuclei, it is absurd for us to focus on the scale of the individual photon for treating the scintillation of the stars, though I got accustomed to doing so. For the experimentation, the individual-photon scale is out of reach. Towards which cis-retinal converges which photon remains a futile question: considering only one stick, replenished in cis-retinal by enough time in the dark and a healthy metabolism, at least a milliard cis-retinal molecules are in concurrence to get the photon.

However, the **astigmatism** of the cornea poses quite a bigger challenge – indeed any other vergence defect of the eye poses the same challenge. In a Newtonian time and a Newtonian causality, astigmatism just forbids the convergence of a photon on only one opsin, at some precise depth in the retina. So the logic demands a retro-causality acting from the absorbing opsin to the geometry of the Fermat spindle, via a function of *"attractivity of the absorber"* that I did not suspect or imagined when I began working the geometry of the Fermat spindle, at spring 1998.

In macroscopic optics with crowds of photons on crowds of absorbers, no fuss: on an axis of astigmatism, the convergence is less far than on the perpendicular axis, and between these two partial convergences, the image is fuzzy.
Well! But how to ensure the convergence on one retinal in one opsin in one disk for one photon, individual wave? Under the Newtonian laws which compose the background of our culture, it is merely impossible. But it remains impossible to perform the quantic reaction of a cis-retinal molecule, without its absorbing of the whole photon.
As soon as we accept the existence of the absorbers in microphysics, we are compelled to consider how the astigmatism of the eye brings the modification of the external geometry of the Fermat spindle for each incoming photon, makes it adjust incident angles into the eye according to an elliptic profile. In other words, you have to start from the absorber and calculate the geometry of the anti-photon emitted, with negative energy, propagating backward in macro-time, then calculate which profile of tangent cone is implied incoming on the astigmatic cornea to give in the eye the perfect convergence on a cis-retinal.
And this, statistically on each of the photons which will give a precise quantic reaction, either on a retinal photopigment, or on any other absorber.

Will it be experimentally proved? I have no idea. By principle, no direct verification is possible. And up to now, no indirect verification is known.

Let us come back to the **geometrical conditions of the scintillation**. We have to compare the geometry of a cone from the celestial body to our pupil, to the one of an individual photon.
Sirius is at 8.6 lyr = 81.36 Pm; a representative wavelength is 480 nm.
Radius: 1.711 solar radius = 1.711 x 696,000 km = 1,190,856 km approx.
⇒ angular diameter: 29.27 nanoradians.

From another source: angular diameter: 5.936 milliseconds of arc = 28.8 nrad.
Keeping 29 nrad is quite enough. 14.5 nrad for the half angle of the cone.

Compare with the Fermat tangent cone, according to $\alpha = \sqrt{\frac{3\lambda}{4a}}$.

We calculate this angle of the Fermat cone for Sirius at $\mathbf{2a} = 81.36$ Pm, and main wavelength 480 nm:

$\alpha^2 < 8.85 \cdot 10^{-24} \Rightarrow \alpha < 2.97$ prad. Less than three picoradians.

We compare 2.97 prad (Fermat cone) to 14.5 nrad (geometrical cone from the star), which is 4,920 times bigger.

So the transactional quantic theory adds nothing to the theory of the scintillation of the stars under the atmospheric turbulence: the classic calculations already done by the astronomers are quite enough, at least in the visible range.

Aldebaran is at 68 lyr, a representative wavelength is 750 nm (in the red).

When it culminates at midnight, Saturn is at 1,300 Gm. It looks pale, and we choose 550 nm as a representative wavelength.

So: $\alpha^2 < 3.17 \cdot 10^{-16} \Rightarrow \alpha < 17.8$ nrad. Over-width of the cone of the photon at 3000 m: 1 mm (+ sub-pupilar diameter still unknown).

Polar radius: 55,225 km

Equatorial radius: 60,268 km

We choose the biggest. So the angle of vision of the equatorial radius: 46.4 μrad.

It is 2,607 times bigger than the angle of the Fermat cone. And more than three thousand bigger than the apparent radius of Sirius.

At 3000 m of atmosphere, the cone is 2 x 139 mm wide. At such an atmospheric distance, the real diameter of the human pupil does not play any significant role.

So, as long as the turbulence does not move more than about one decimeter the real cone from the planet to our eye, its image may shake, but the lost photons have substitutes. So the naked eye receives a constant illumination from Saturn. No flicker (unless the turbulence is extreme).

In conclusion, our conception of the geometry of the Fermat spindles for the photons has not brought anything new to the theorization of the scintillation of the stars. But it has overcome this obstacle without damage: it is compatible with the experimental facts.

Curious:
- Can you, broadly, evaluate the over-estimation of the diameter of the Fermat spindle, brought by you provisory approximation by constant curvature?

Open-Eyes:
- I was expecting an over-estimation factor between 1.2 and 2. The estimation improves at far distances $\mathbf{a}/\boldsymbol{\lambda}$. Now we are more familiar with the astronomical applications, I trust more its quality in the far field.

In the next edition, we will study a possible ceiling of these diameters, coming from the properties of the rainbows. Many factual data are still to be collected.

Professor Marmot:
- Nineteen pages for this subchapter, and not any result! You are a mere crank!

Curious:
- You forget that in the journey, Mr. Open-Eyes has opened a new question: how a photon can **converge** onto one cis-retinal molecule, despite the defects of convergence or stigmatism of the eye. So he made his day in favor of his retro-causality from the absorber to the geometry of the individual beam. Do you object?

Open-Eyes:
- Astigmatism of 1 diopter is not rare among humans. Focusing on infinite, it yields a difference of 400 µm in the image planes, well above the thickness of the sensitive layer in the retina, which is about 7 to 28 µm. Now, in these photosensitive disks, one cis-retinal molecule captures one photon. For a stick in darkness, just one photon is enough to toggle the stick into hyperpolarization. More than ten photons in limited time are necessary for toggling a cone. Anyway, each photon is absorbed by only one absorbing molecule, 1.8 nm long. It implies that the absorber has governed the fine geometry of the Fermat spindle already in the air, before the cornea, to obtain this perfect convergence. This is in total divergence from the Newtonian time and Newtonian causality, which pretends the sequence: The emitter emits light, THEN light travels, THEN it meets the cornea, THEN it is refracted by the lens, THEN it meets the retina, etc. This Newtonian scheme is incompatible with the properties of the human eye.

In microphysics, the Newtonian macro-time has no validity. The individual waves dwell in their micro-times.

10.7. Radiation and polarization of the individual waves

10.7.1. Sizes of the crystallites and Scherrer Law. .

Open-Eyes:
- Radiocrystallography works with waves, and the optics of Fresnel, 1819. Not with corpuscles.

Curious:
- Have you more proofs against the corpuscles?

Open-Eyes:
- Many indeed. All the radiocrystallography was done again with electrons or neutrons instead of X-rays. Below is the geometry of the Bragg law:

Figure 10.28.

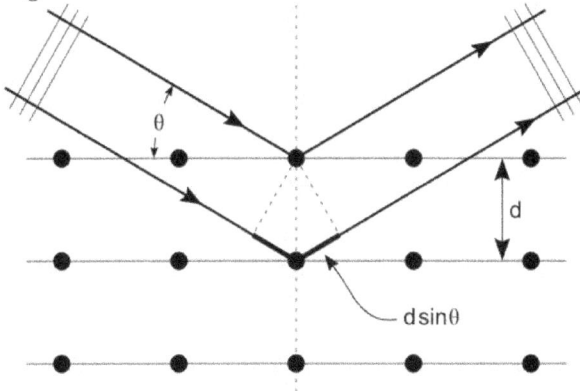

The ray reflected by the second atomic plane is late by a wavelength from the one reflected by the first atomic plane; two wavelengths late for the reflection on the third atomic plane, etc. Hence the Bragg law: $n\lambda = \mathbf{2\ d\ sin(\vartheta)}$.

For the simplicity of the drawing, here we draw a simple cubic lattice, a lattice that does not exist in nature, as not enough compact. However, in a crystal of halite NaCl, the anions Cl^- and the cations Na^+ alternately occupy the nodes of a simple cubic lattice, it is the halite lattice.

And now the application of the Bragg law in electron radiocrystallography, for showing a classroom, on a thin polycrystalline material, interposed on the run of cathodic rays, approximately monochromatic (the speed of the electrons is governed by the acceleration potential, but submitted to the initial thermal noise of the emitting cathode).

Figure 10.5 already given.

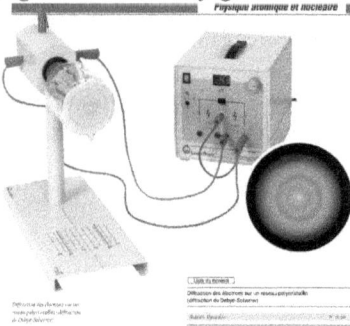

Compared to an X-ray apparatus we use in a laboratory of metallurgy (with an anticathode of molybdenum) or mineralogy (with an anticathode of copper or cobalt), here the precision is pitiful, but the cost, the danger, and the volume are not at the same scale, too. We also know how to obtain a Laue electron diffractogram with the big electron microscope (the one which occupies three floors of the building), by diaphragming on the inclusion that intrigues us in a thin plate of metal, wobulating the accelerating tension, and changing the focus to infinity. A historic paper from Albert Einstein in 1916 proved the perfect directivity of each emission of a photon by an atom in gas at thermal equilibrium. Though it is incompatible with the extremely low directivity that the emitting atom could yield without the absorbing atom. Quantentheorie der Strahlung (On the Quantum Theory of Radiation) Mitteilungen der Physikalischen Gesellschaft, Zürich, 16, 47–62.

But the copenhaguist theorists had no experience in radiocrystallography, which developed later; only the metallurgists, the mineralogists, and the solid-state physicists have concern and use of the radiocrystallography. And the radar and the microwave links, are not in the concerns of the copenhaguist theorists either.

Figure 10.29.

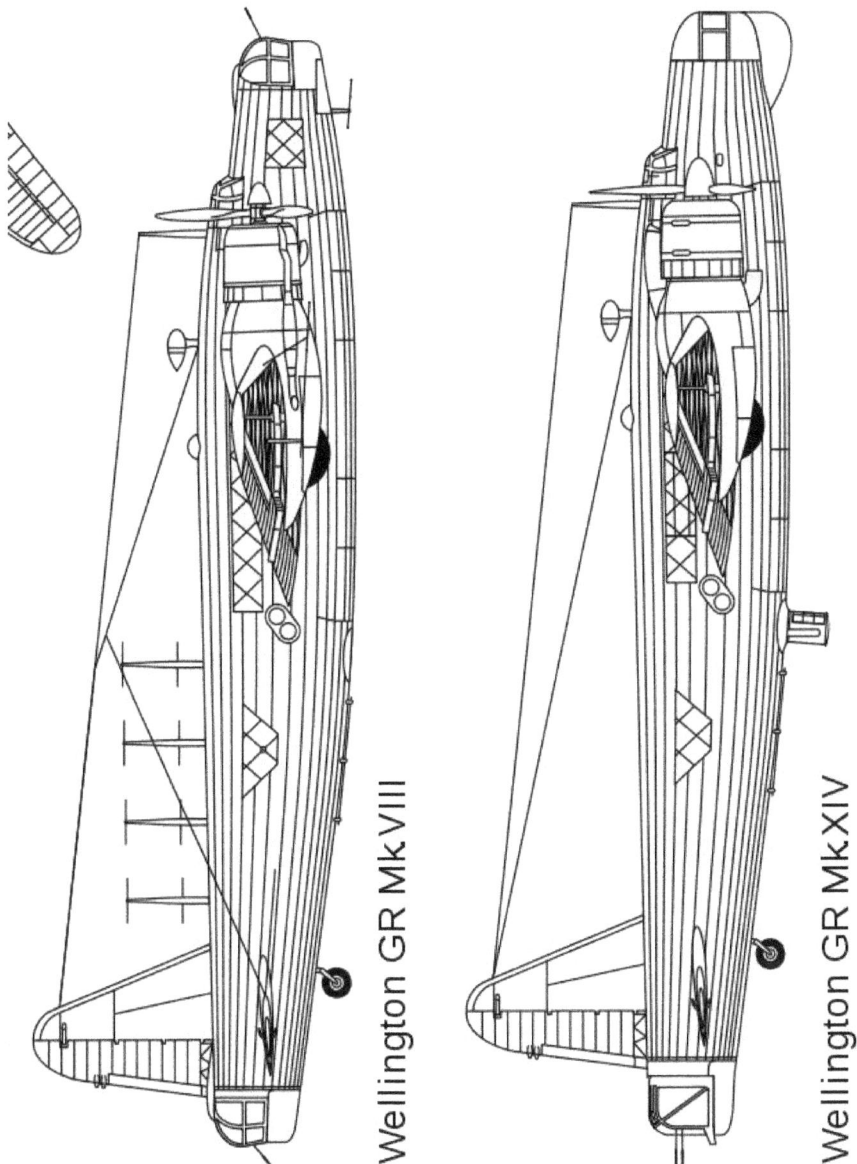

Wellington GR MkVIII

Wellington GR MkXIV

Just compare the difficulty obtaining a useful directivity with a radar, when you have only access to a 1.7 m wavelength, and what the same plane (for detecting and hunting the submarines) became when the allied got radars with 9.1 cm wavelength, where the directivity was driven by parabolic antennas, with diameters more than fifteen times the wavelength. The directivity of a photon emitted by an atom in gas is incompatible with the axiom *"the emitter, only the emitter, nothing else than the emitter"*: a small atom is very far from

232 OF CORRECT MICROPHYSICS

the antenna dimensions necessary to direct a photon. Only the association emitter-absorber can obtain the directivity proved by Einstein in 1916, in a synchronous transfer. This is the ground of the transactional reformulation of quantum physics.

Professor Marmot:
- You *drown the fish*, now! Crystallography and metallurgy are only kitchen plays for low peasants! They are not hermetic mathematics; on the contrary, everybody can understand, so they are not true sciences.

Open-Eyes:
- I am still far from having exploited all that the radiocrystallography brings as proofs against corpuscles. In sedimentometry, the fineness or width of the rays of each mineral gives a measure of the fineness of the crystallites: the broader the line, the smaller the crystallite. It is the Scherrer relation, published in 1918: P. Scherrer, "*Bestimmung der Grösse und der inneren Struktur von Kolloidteilchen mittels Röntgenstrahlen*", Nachr. Ges. Wiss. Göttingen 26 (1918) pp 98–100. The broadening begins to be visible for crystallites under 1 µm. This law is incompatible with any corpuscular representation of the X-rays, and generally of the electromagnetic radiations.

Application: compare the widths of the rays from a clay from Saint-Jacut du Menez (lake kaolinite and glauconite, with excellent plasticity), with the fineness of the minerals in the alluvial-loessal silt, taken in the alluvial plain, 40 km south of Qazvin (Iran), probably at 35°50'11" N, 50°07'34" E. Altitude 1,200 m, a silt that never had the properties of a clay.

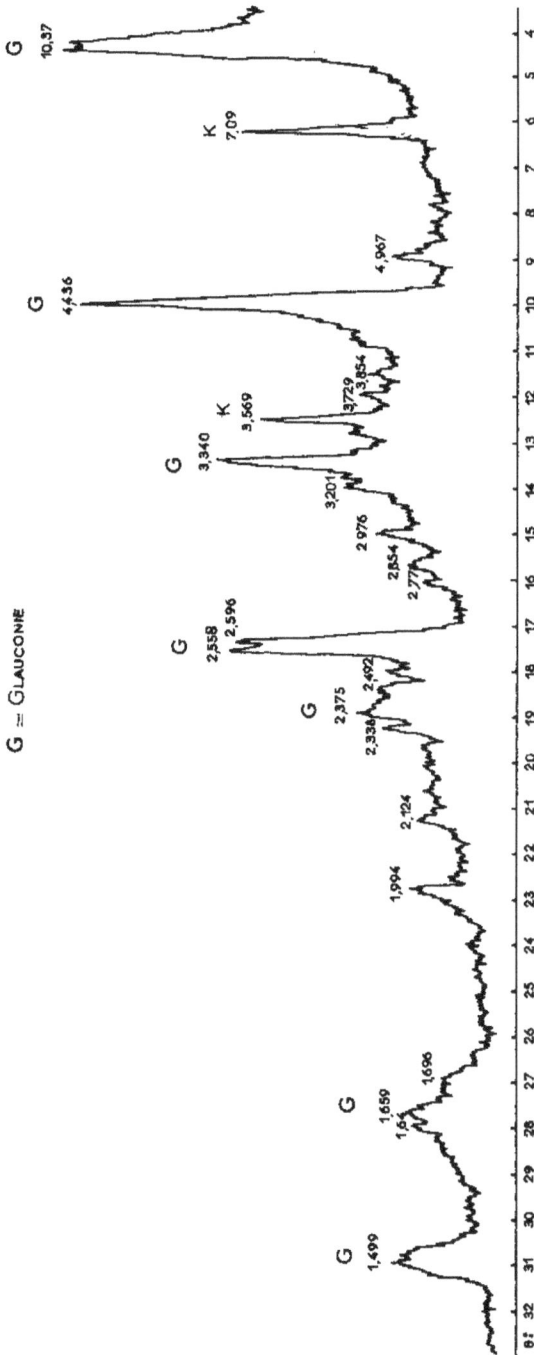

GLAUCONIE' DE S! JACUT

K = Kaolinite
G = GLAUCONIE

Figure 10.30.

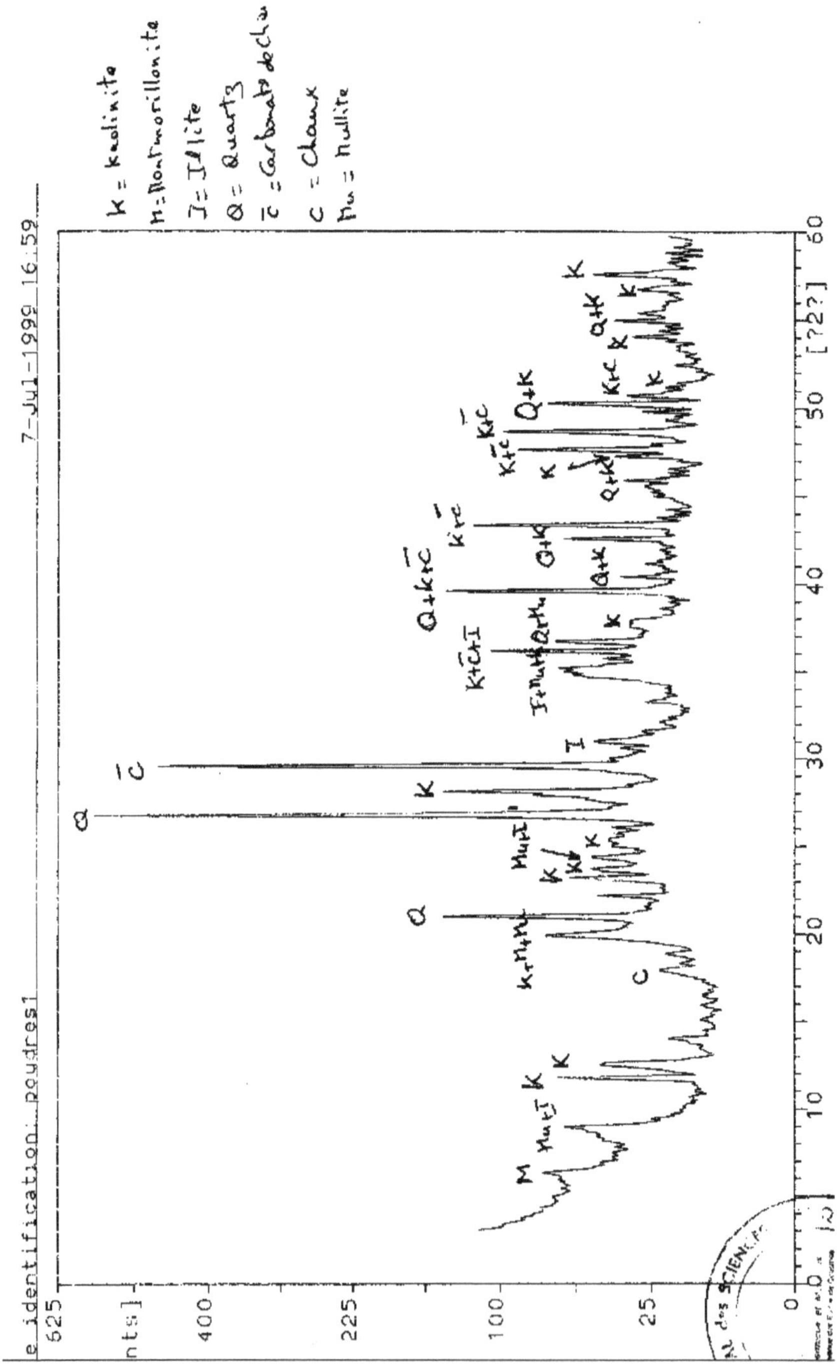

Figure 10.31.
Don't be astonished that the material of this quarry that Mr. Michel Laquerbe

made buy by his Iranian ex-colleague and swindled, never was clay, never had the required plasticity, never was extrudable; the plant could never produce. Here are the photos taken by Mr. Jean-Louis Gleizes, the engineer in charge of starting the facility.

Figure 10.32 : The silt quarry.

Next Mr. Gleizes made dig three meters deeper. Then they found a more plastic material, but you will see on the next images that it was still far from the goal. Sure, by soil leaching, the fine elements flow deeper in the soil profile, but in that altitude plain, it is a climate with moderate rainfall with low leaching. At the piedmont of the mountain at Takestan, the economic resource is fruits, especially the table grape.

Figure 10.33. : The extruded product has no cohesion, tears immediately.

Not extrudable! No plastic cohesion.

Figure 10.34.

Curious:
- I have not all understood. What is the link?

Open-Eyes:
- If the fineness of the reflections on the atomic planes depends so much of the fineness of the crystallites, it is because the width of each photon where it

meets the crystal is of the same magnitude than the width of the crystallites, and the reflection under the Bragg law depends on the photon meets much good crystal and few edges. A micrometer is about a normal diameter for very well crystallized kaolinite, but it is tens to hundreds of times bigger than a plastic clay micelle. While it is small compared to the grains of silt, here the silt from the plain of Qazvin, near Bo'hin. Another discriminating measure, which in a laboratory of the Ponts et Chaussées they achieve in twenty minutes, with just a pipette, a beaker, and a magnetic stirrer, is the measure of anionic surface per mass, by adsorption of methylene blue. But Laquerbe also ignored that. He ignored that the plasticity of the clays comes from the huge area of the fine micelles of clay (macro-anions) surrounded by water and hydrated alkaline cations: between 20 m^2/g and 650 m^2/g. The ceramists know that; the scientists of the soils, geologic risks, foundations, and civil engineering know that, but not Laquerbe, who taught in construction and civil engineering, but only the concrete, without knowing anything in foundations nor hydromechanics of the soils. Though he only had to step one floor above, in the same building on the campus of Cesson-Sévigné.

10.7.2. Parenthese in granulometry. .

Professor Castle-Holder:
- Objection! Our pupil only faked to understand when you spoke granulometry, opposing clay to silt. Only the ceramists, or the laboratory technician of the civil engineering, or sedimentologists or pedologists are in knowledge of the granulometric realities. A celebrity as Laurent Nottale built his career on a delirium, his *"scale invariance"*; it gives a measure of the ignorance of his public, though believed as scholars. You must give the reader a viaticum in granulometry.

Open-Eyes:
- Another industry must have precise understanding in granulometry: those who make fillers for the polymers, rubbers, and paints. Your tires would be dangerous without these men. Foundry sands too, have to meet precise granulometric criteria to have the required cohesion. Here is a table extracted from André Vatan, **Manuel de Sédimentologie**, Éditions Technip:

L. Cayeux	C.K. Wentworth	J. Bourcart
Blocs : 20 cm	Boulder: 256 mm	Cailloux ou ballast :
Galets : 5 cm	Cobble: 256 to 64 mm	1 m à 2 mm
Graviers : 5 mm	Pebble: 64 to 4 mm	
	Granule: 4 to 2 mm	
	Very coarse sand: 2 to 1 mm	
	Coarse sand: 1 - 1/2 mm	Sables : 2 à 0,2 mm
Sables : 0,5 mm	Medium sand: 1/2 - 1/4 mm	
	Fine sand: 1/4 - 1/8 mm	Poudres: 200 à 1 µm
Poussières, boues	Very fine sand: 1/8 - 1/16 mm	Précolloïdes :
< 0,05 mm	Silt: 1/16 to 1/256 mm	1µm à 0,1 µm
	Clay: below	Colloïdes

Two tables owed to Stéphane Hénin, **Cours de physique du sol, tome 1**, ORSTOM-Editest.

Diameters of the elements in mm	Terminology normal	of Atterberg subdivisions	Sedimentologists terminology
> 20	cailloux		rudites
2 to 20	graviers		
0.2 to 2	sable grossier		arenites
		50 µm to 200 µm	
0.02 to 0.2	sable fin	coarse silt	pelites
		20 µm to 50 µm	
0.002 to 0.02	limon		
< 0.002	argile	coarse clay	
		0.5 to 2 µm	
		fine clay	
		< 0.5 µm	

Properties of the classes of materials.

	ions absorption	water retention	permeability	mechanical properties
clay	strong	strong	weak	coherent in dry
				pasty in wet
silt	weak		weak	incoherent in dry
				pasty in wet
fine sand	null	weak	strong	incoherent, rough
coarse sand	null	null	strong	very rough

10.7.3. Back to the optics of X-Rays in crystallography. To be more accurate: the width of each photon, not absorbed by the crystal but reflected by it, and absorbed farther – historically by a photographic film, and nowadays by a photosensitive sensor of the goniometer – only depends on the geometry and the length of the path, and of the wavelength; it is proportional to the square root of the product [wavelength x path length].

Curious:
- So the link with the remaining of this book, is the width of the beams of X-rays, and mainly of each X-photon or each electron, or each neutron, where it meets this crystal which does not absorb it but reflects it? Have I well summed up?

Open-Eyes:
- Well done. Your summing up is perfect. Never the X-rays cease to be (individual) electromagnetic waves when diffracting on crystals.

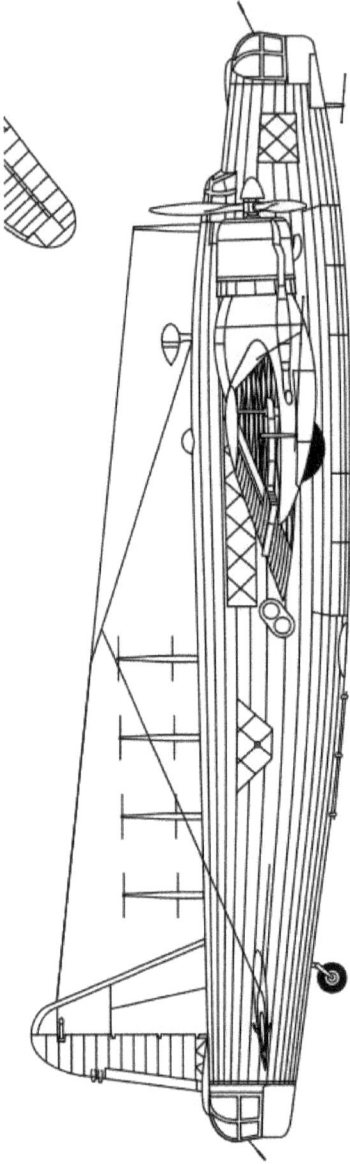

Wellington GR Mk VIII

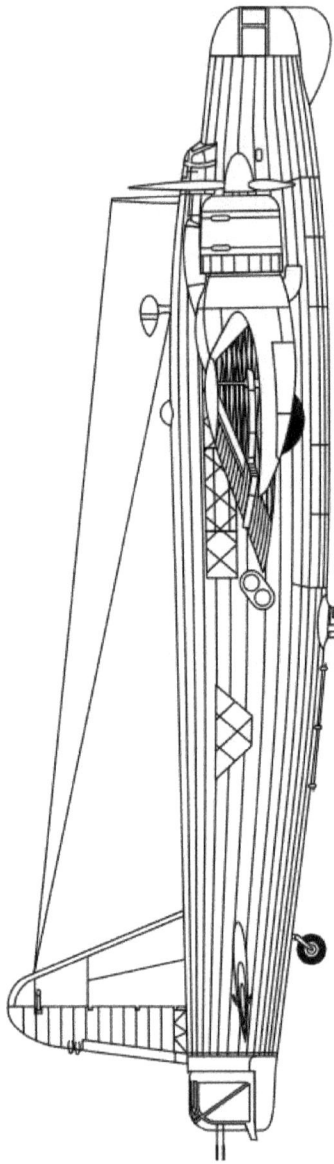

Wellington GR Mk XIV

10.8. Compton scattering of an X photon by a free electron: the Zitterbewegung is indispensable

The Compton scattering of an X or γ photon by an electron, does it prove that the photon is corpuscular? Or that the electron is undulatory?

The paper from Arthur Holy Compton, finished in December 1922, came out in May 1923 in the Physical Review: **A Quantum Theory of the Scattering of X-Rays by Light Elements**. Compton used the line K_α of the molybdenum, indeed a doublet, of mean wavelength 0.070926 nm. Compton concluded that he had ruined the theory of the *big electron*, of a size comparable to the wavelength of the incident photon, and instead he had proved – six years after Einstein had already proved it, in 1916 – that the photon had a definite momentum, and *therefore*, the photon was a small corpuscle. Anyway, nobody ever saw a theory of the Compton cross section under this ideation of these small corpuscles. Still nowadays, the teaching and the popularization repeat that the Compton scattering *proves the corpuscular nature of the light*. I will show that such a proof is invalid, and I had a surprise: the de Broglie wavelength does not hit the target, but the Dirac-Schrödinger wavelength, twice shorter, does.
Link:
http://deontologic.org/quantic/index.php?
title=Calcul_diffusion_Compton_et_Zitterbewegung

10.8.1. Relativist calculation in the frame of the laboratory.
Relativist calculation according to Walter Greiner, **Quantum Mechanics, an Introduction**, page 3:

Conservation of energy: $h\nu = h\nu' + m_0c^2 \left(\frac{1}{\sqrt{1-\frac{v^2}{c^2}}} - 1 \right)$

Conservation of the momentum on the axis of the incident photon:

$\frac{h\nu}{c} = \frac{h\nu'}{c}\cos\vartheta + m_0c \left(\frac{1}{\sqrt{1-\frac{v^2}{c^2}}} - 1 \right)\cos\varphi$

and along the perpendicular axis where the momentum is null:

$0 = \frac{h\nu'}{c}\sin\vartheta - m_0c \left(\frac{1}{\sqrt{1-\frac{v^2}{c^2}}} - 1 \right)\sin\varphi$

Resolving these equations, we obtain:
$\lambda - \lambda' = \frac{h}{m_0}\sin^2\frac{\theta}{2}$
End of the borrow.

Now we remake it, but in the frame of the center of inertia, even considering that it is impossible to experiment in this unpredictable frame. One cannot choose the angle of incidence, nor which electron. We are tight to the frame of the laboratory.

10.8.2. In the frame of the center of inertia. Figure 10.35.

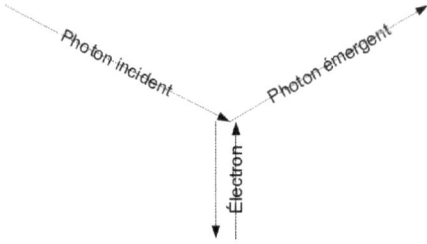

There the calculation is much more simple: the photon does not change energy nor frequency, only direction. We decide it arrives from the left, descending at angle $\alpha = \frac{\theta}{2}$ and continues ascending by the same angle. The electron does not change its speed, only the course of its run, from ascending to descending. We neglect its bonding energy to the metal.

Momentum on z'z transmitted from the electron to the photon:

$-\frac{h\nu}{c}.2\sin\alpha$ (- sign: descending, if z'z is an ascending axis).

Balanced by the momentum transmitted from the photon to the electron:

$2m_e.v$ (non-relativistic form, opposed to the relativistic:

$$m_e c^2 \left(\frac{1}{\sqrt{1-\frac{v^2}{c^2}}} - 1 \right)).$$

Hence the speed of incoming and fleeing electron (non-relativistic writing):

$v = \frac{h\nu}{m_e c}.\sin\alpha$

We deduct the phase velocity: $V = \frac{c^2}{v} = \frac{m_e c^3}{h\nu.\sin\alpha}$

Now we know the intrinsic frequency of the electron: $T_e = \frac{h}{m_e c^2}$

Therefore its Broglian wavelength: $\lambda_e = V.T_e = \frac{V}{\nu_e} = \frac{m_e c^3}{h\nu.\sin\alpha}.\frac{h}{m_e c^2} = \frac{c}{\nu\sin\alpha}$

We notice that this wavelength does not depend on the mass of the electron and would be the same for any charged (or not) particle subject to Compton scattering. It depends not on the Planck's constant either; it depends only on the angle of deviation of the photon, and its period or wavelength before and after the scattering.

Driven by what we already know how to construct in refraction and reflection on a diopter, we now calculate the emission done by this photon mirror, the electron. The horizontal part on x'x is invariant. It's wavelength: $\frac{\lambda}{\cos\alpha}$

The wavelength of the penetrating and emerging part of the photon: $\frac{\lambda}{\sin\alpha} = \frac{c}{\nu\sin\alpha}$

These two wavelengths, of the bouncing electron and the scattered photon, are equal: the total momentum of the system is invariant.

Now we have to choose between the two assertions:

"*The Compton scattering proves that the photon is corpuscular.*",

"*The Compton scattering proves that the electron is undulatory.*".

Now this emission of ascending partial photon and the absorption of descending partial photon are due to the acceleration of the electron along the z-axis.

Up to now, the calculation gave us no idea of the magnitudes of spatial extensions of the X photon and of the electron. But, having used the K_α ray of the molybdenum in metals radiocrystallography, we know that its wavelength is comparable to the equidistances in a metallic crystal, that the conduction electrons in a metal are loosely bound, and moreover, weakly localized, extending each on several tens of interatomic distances. Added to the geometrical demands of the diffraction on interatomic planes, these facts compel us to conclude that the photon and the electron are both wide and deep of some tens interatomic distances during all their Compton interaction.

10.8.3. Numerical application, with the mean K_α ray of the molybdenum: We take a case of strong deviation of the photon, twice 30°, that is $\sin \alpha = \frac{1}{2}$

The mean wavelength of the incident ray is 0.070926 nm. Hence the projection on the direction of propagation of the electron:

$\lambda_{Broglie} = 0.070926$ nm x $2 = 0.141852$ nm.

Hence the speed of the electron:

$\mathbf{v} = \frac{\lambda_{Compton}}{\lambda_{Broglie}}.c = \frac{2.42631}{141.852}.299{,}792{,}458$ m/s $= 5.1278 . 10^6$ m/s.

It is a non-relativistic speed: 1.7 % of $\mathbf{c}$. It would be even less relativistic at lower deviations.

In the frame of the incoming electron, assimilated to the frame of the laboratory, one should retrieve the experimental formulations of Arthur H. Compton. However, with these causes of error:

1. A conduction electron is not at rest, but at Fermi level and Fermi speed in the metal.

2. And we have neglected its binding energy, in metallic binding.

The first mentioned is the main cause of error and enlarging of the Compton scattering, the Fermi Level, plus the fact that the K_α X-ray is a doublet.

The principle objection is that we have just stated the exchange of wave vectors, without establishing the physics of the interaction

Vertical component of the gamma incoming wave vector = electronic wave vector outcoming.

Vertical component of the gamma outcoming wave vector = electronic wave vector incoming.

So the physics of the interaction remains unknown at this step of the calculation. The failure is guaranteed if we extrapolate to microphysics the usual macrophysical scheme, with a massive object that slows, then restarts in the opposite direction, with finite acceleration all during the interaction. In 1926 (Schrödinger 1926) Erwin Schrödinger had shown the path, considering emission of a photon as the beat of atomic (or molecular) electronic wave between its final and initial state. Here again, it is a matter of beating between the initial upping state and final downing state of the electron. During the beating, an intermediate state contains a stationary Broglian wave, to dispatch among four Dirac components, whose two are antichrone, with negative frequencies. Here is obvious another constraint, but we do not know whether it has been experimentally proved (nor whether it can be experimentally tested): the electrical polarization is necessarily in the plane of the drawing. Confirmation however by the plane polarization of the synchrotron radiation on the plane of the turn: globally the mechanism is the same, but the calculation must be relativist, with relativist transformations of the fields.

The Bragg condition in radiocrystallography:
Set **d** as the inter-reticular distance, $\alpha = \vartheta/2$ is the angle of the incident ray on the reticular plane, or half of the total deviation, λ the wavelength of the incident ray, and **n** an integer, order of the reflection: **2d.sinα = n λ**.
Figure 10.36

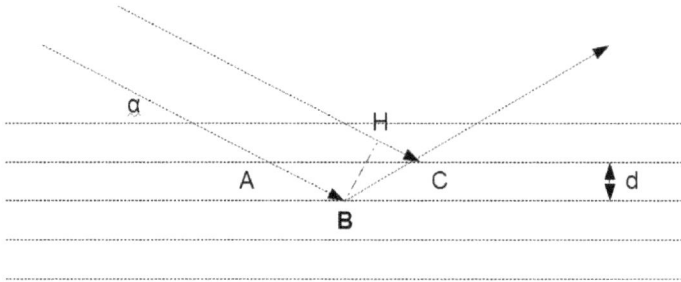

Demonstration: incoming with α on the reticular planes AB etc., the monochromatic wave reflected by the next plane has a path difference equal to BC - HC. The first reflexion only exists if BC - HC is exactly one wavelength. In the isosceles triangle ABC, d = AB sin α = BC sin α. Though in the rectangular triangle BCH, CH = BC cos (2α). The path difference between the two waves is BC - CH = BC (1-cos 2 α) = 2 BC . sin (2 α) = 2 d sin α = n.λ.γ

10.8.4. Bragg condition and Zitterbewegung. Nevertheless, the Broglian wavelength, as calculated above, which was the only known and used by Erwin Schrödinger in 1927 (Schrödinger 1927) only gives us the reflection of second order: $\lambda_e = \frac{\lambda_\gamma}{\sin \alpha}$.
It is a weak reflection, though one should observe the main reflection at order 1, which is never observed (and which would violate the laws of conservation of energy-momentum)... It remains only one outlet: to consider the stationary electromagnetic wave, with doubled temporal and spatial frequencies, the **Zitterbewegung**, the trembling motion discovered by Erwin Schödinger in 1930 (Schrödinger 1930, Dirac 1930, 1958) according to the Dirac equation, which yields the right Bragg reticular equidistance, exactly **d**!
$$\mathbf{d} = \frac{\lambda_e}{2} = \frac{T_e}{2\nu_e} = \frac{h}{2m_e\nu_e} = \frac{\lambda_\gamma}{2 \sin \alpha}$$
Figure 10.37

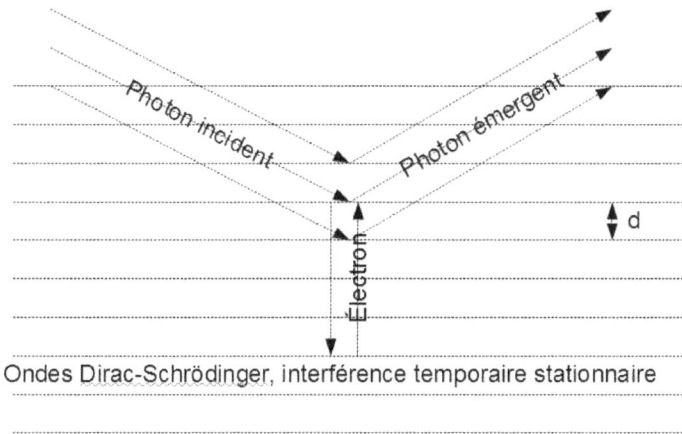

Ondes Dirac-Schrödinger, interférence temporaire stationnaire

Quod Erat Demonstrandum!

10.8.5. Conclusions. Only the spatial frequency of the Zitterbewegung, the trembling motion discovered by Erwin Schrödinger in 1930, stationary during the reflection of the electron on the photon, satisfies the Bragg condition for a first order reflex and gives exactly the Compton scattering of the incoming photon.

We aimed to exhibit the physical and undulatory mechanism that lies under the Compton scattering. Goal accomplished: **Only the equidistance of the temporarily stationary Dirac-Schrödinger waves satisfies the Bragg condition, for the first-order diffraction**. The fact that Erwin Schrödinger did not publicly correct himself in 1930 his error of 1927 gives the measure of the meticulous demoralization obtained on him by Bohr and Heisenberg (Segrè 1984, Selleri 1986). Not the slightest allusion to that in his Nobel lecture in 1933, where the two last pages deny all the work described in the previous pages. Only P.A.M. Dirac mentioned this discovery in his Nobel lecture, not Schrödinger: he had been thoroughly demoralized by Niels Bohr.

Other consequences:
Knowing the spatial extent of a conduction electron at the Fermi level, the Compton interaction electron-photon only imposes to this apex a modest constraint of pinching of the Fermat taperings, about some nanometers, when the reactional conditions on the previous emitter and following absorber, as well for the electron and the X photon can eventually be more pinched; the details depend on the precise physics of these emitters and absorbers.

The electron returns under physical laws, individually. Well, it may sound a dullness; however, under the standard mythology (Greiner 1935, Charpak, Omnès 2004, Hawking, Mlodinow 2010) each quanton individually escaped to any physical laws. Only in great numbers, it rejoined statistically some physical laws, by the mysterious mean of a physical mechanism remaining totally mysterious, though postulated nevertheless.

It is no more necessary to postulate some exorbitant physical mechanism, never observed and never theorized, that could perform the magical transmutation of an electron or a photon into some "*small corpuscle*", or worse: "*punctual corpuscle*".

10.8.6. Bibliography and references
.

Compton 1923: Arthur H. Compton. *A quantum theory of the scattering of X-rays by light elements*. The Physical Review. May 1923, vol 21, n°5. P.A.M. Dirac. *The Principles of Quantum Mechanics*. Oxford University Press, ed 1958. § 69.

Greiner 1981: W. Greiner. *Relativistic Quantum Mechanics; Wave Equations*. Springer Verlag, ed. 1997.

Schrödinger 1926: *An Undulatory Theory of the Mechanics of Atoms and Molecules*. The Physical Review, 28, (1926), 1049-1070

Schrödinger 1927: E. Schrödinger. *Über den Comptoneffect*. Annalen der Physik. IV. Folge, 62.

http://www.apocalyptism.ru/Compton-Schrodinger.htm

Addresses given by Lev Lvovitch Regelson.

Compton effect: Schrödinger's treatment

in: The Science Forum - Scientific Discussion and Debate.

http://www.thescienceforum.com/viewtopic.php?p=235655 , link changed to:

http://www.thescienceforum.com/physics/18025-compton-effect-schroedingers-treatment.html

Schrödinger 1930: *Über die kräftefreie Bewegung in der relativistischen Quantenmechanik*. Sitzungsberichte der Preußischen Akademie der Wissenschaften. Physikalisch-mathematische Klasse, (1930), 418-428

Schrödinger 1933: Nobel Lecture, December 12, 1933. *The Fundamental Idea of Wave Mechanics*.

http://www.nobelprize.org/nobel_prizes/physics/laureates/1933/schrodinger-lecture.pdf

Paul A.M. Dirac - Nobel Lecture, December 12, 1933 Theory of Electrons and Positrons

https://www.nobelprize.org/nobel_prizes/physics/laureates/1933/dirac-lecture.pdf

Segrè 1984: Emilio Segrè. *Les physiciens modernes et leurs découvertes*. Fayard, Paris 1984.

Selleri 1986: Franco Selleri. *Le grand débat de la théorie quantique*. Flammarion, Paris 1986.

Anti-references, a bunch of what not to do (collected at: http://citoyens.deontolog.org/index.php/topic,887.0.html):

Georges Charpak, Roland Omnès. *Soyez savants devenez prophètes*. Ed. Odile Jacob, Paris, juillet 2004. Pages 87 - (92 ?).

Walter Greiner 1935. *Quantum Mechanics, Special Chapters*, pp 365-367. Springer Verlag, ed. 1998.

Stephen Hawking, Leonard Mlodinow 2010. *The Grand Design*. Bentham Books, 2010.

10.8.7. Width and length of the reactional apex.

.

Professor Castle-Holder:
- I object! You have forgotten your initial aim. You were so pleased to show evidence of the Dirac-Schrödinger spatial period that you forgot to compute the diameter of the reactional apex: electron + gamma ⇒ electron + gamma. As the copenhaguist *Great Ancestors* – including Compton – had no idea of the underlying physical mechanism – and you have demonstrated it is undulatory, what they refused *a priori* – they postulated that all that was smaller than small, corpuscular, event punctual, in the femtometer range. You seem to have the proof of the contrary, but now, you have to unroll it completely, this proof.

Open-Eyes:
- You are right. We have spent time on the metallic state, and on the collisions of electrons on phonons, and we deducted the plausible dimensions of the conduction electrons. Mean free path: about 200 Å. A phonon is always sampled on many atoms; it cannot become "*small*". So a conduction electron also cannot become "*small*"; each one is several interatomic distances wide, even several tens of interatomic distances. How many? Cell parameter of the copper is 362 pm, hence an interatomic distance about 256 pm. Set 25 interatomic distances, that is 6.4 nm or 64 Å, about the third of the mean free path at room temperature. Compton used the doublet K_α of the molybdenum, of mean wavelength 0.070926 nm. We conclude that the Compton reaction extended on $\frac{6,400 nm}{0.070926 nm} = 90,243$ wavelengths of the Kα of the molybdenum. Could we say that it is "*small*"? *Corpuscular?*

Curious:
- Do you speak of the width or length of the reaction?

Open-Eyes:
- Neither precisely, or maybe both, as they are angles between the two trajectories, of the electron, and of the photon. Just say that 64 Å is the mean length of the conduction electron, and let some fuzziness, about 3 to 10 nm, and is an approximate value of the diameter of the reaction. We cannot make precise more; we do not have the means of a finer investigation. Up to now, it is far beyond our means.

10.9. The medium of propagation, the changes in polarization

Transactions are between three partners, we said. Here we study the third one: the medium of propagation.

Open-Eyes:
- I take again the speech to evoke a vulnerability of our first modelization of the Fermat spindles for the photons, in August 1997 and May 1998, confronting it to the properties of the polarizers in plane-polarized light.

To make the calculation simple, then I assumed a perfect alignment of polarization of the emitter and the absorber, figured as dipolar antennas. Well, but when we have experiments in polarized light, with Nicols or polaroids partly crossed? When we are the first from 1927 thinking to subject EACH photon to physical laws, when all the literature only considers statistics on great numbers of photons. The Vulgate tells us that EACH photon is exempted from all physical laws, BUT in great numbers ends in verifying the statistics predicted by the formal calculation. What was needed is to understand the fine physics of the anisotropic crystals, and its effects on polarization. "*To transmit a photon*" does not have the same meaning for Richard Feynman – though inventor of the Feynman diagrams – than the one we need to go forward. More humble and experimental, the Russian manuals are invaluable. "*To transmit*"? With the original polarization, or the one of the last polarizer? In 1963, Feynman did not bother of that.

Solved. The polarizer imposes its plane of polarization (but a small error), so is able in near field to **twist the plane of polarization**.
The theoretical vulnerability is over. Sure, the twisting of the polarization has a cost in impedance.

Professor Castle-Holder:
- Impedance? Did you formalize this new notion? It is new in this context.

Open-Eyes:
- To be brief, I will say that a solution of the time-dependent and three-dimensional Schrödinger equation – what they call a wave-function –, or a traditional and re-normalized solution of the Maxwell equations (traditional = with only initial conditions and no final conditions) is the quotient of a transmittance (the inverse of an impedance) by a function of concurrency between potential absorbers. At the scale of the laboratory, this function of concurrency is in r^2, at least in the first approximation. At astronomical scale, or in the mist, one should add a volumic term. Under these conditions, the transmittance or impedance are independent of the distance, the vacuum is a perfect iterative impedance, unless someday someone demonstrates the "*tired light*". The transmittance has some new and nontraditional properties: it depends on the polarizations and spins, and their alignment between emitter and absorber; nobody in the 19th century nor in 1926 thought to that.

Though in free field and without magnetic field nothing can twist a plane of polarization of the light, things go differently in near field, in the matter, and especially in anisotropic crystals, and in chiral molecules.

Between a first polarizer oriented in 2 h – 8 h, and a second one, oriented in 3h – 9 h, the most economical hypothesis is that they share the twisting task, so that in far field, the photon is polarized in 2 h 30' – 8 h 30', at least for those (75 %) which are transmitted to a farther absorber. It is not experimentable, by principle, but at least it is economical and free from contradictions.

Concerning the 25 % of the photons that will be absorbed by the second polarizer, one must consider a twist to 1 h – 7 h when issued from the first polarizer, which will be absorbed in 0 h – 6 h by the second polarizer. Each one carries an about 30° twist. Each of these twists costs in impedance, hence the sharing through a cosine law.

This new state of the theory relieves an explicit vulnerability of the old theory of transaction. What the polarizers taught us is that an anisotropic crystal can twist a polarization plane the same way a solution of chiral molecules does (circular birefringences, with different phase speeds). Even with gas molecules, the possibilities of helicoidal recoil could compensate a slight misalignment of the polarizations.

We still have to understand how, in the quantic domain, a photon emitted under a dipolar electric transition can be absorbed under a magnetic transition and vice-versa. It is a conversion, implying that the emitting and receiving atoms have a different recoil, differing by a **h** quantum of looping or angular moment. It is also usually performed in the (macrophysical) quarter-wave filters. So now we must carry you in a detour in polarized-light optics, especially in the anisotropic crystals. In Appendix C, we have included a refresher lesson on the structure of a plane wave, and comparison with a circularly polarized wave, in order to explain to you how to convert between these two pure cases. See under: **Appendix C. Radiation and propagation of electromagnetic waves. Polarization**.

Curious:
- Up to now, your exposé on waves is deterministic. But the others always speak about probabilities. Have you new solutions to answer the controversy "*God does not play dice*"? Do you explain the obviously random character of the quantic events?

10.10. Which randomness and where? The Fourier transform, fraudulently relabeled

Here the randomness and each particle find their right place, and the quantic particles rejoin the physical laws.

Open-Eyes:
- We must distinguish two different facts, which are usually presented to the students as **only one** principle of *uncertainty*.

The first one is concealing the properties of the Fourier Transform, relabeled into *Heisenberg's principle of uncertainty*. The vocabulary also is egocentric and deceptive. In fact, it is indefinition: if a photon is very well defined in frequency, it is very long, so its *position* is very poorly defined. Conversely, if it short, concentrated in time and space, then its frequency is poorly defined. The product of these indefinitions is defined by the properties of the Fourier Transform and its reverse.

Curious:
- You won't escape without explaining this Fourier Transform.

Open-Eyes:
- Ready to spend enough time for that? At the gymnasium, you had operated with an alternate tension and an oscilloscope? Preferably a low-frequency generator, giving a sinusoidal signal, or with a step-down transformer, where the secondary circuit is well insulated from the primary. You have seen a perpetual sinusoid, following the tension upon time. If it undulates so for a long time to a long time, when is the center of this wave packet? Where you like it: any date works provided it is not near the handling of the switch. Similarly, any abscissa in the course of a continuous laser beam is as bad and fuzzy as the others to pretend to be the center of the wave packet. On the contrary, a photon has a beginning and an end; it transmits only a finite momentum and a finite energy. If you want it short, to have a precisely definite beginning and end, then its spectrum spreads: it is no more well definite in frequency. To define a frequency, you need some time, a duration; then it is the frequency in a frame which defines the momentum and the energy of the photon in this frame. Mathematically, in the most simple case, when the signal is a Gaussian, its transform is again a Gaussian, and the product of their widths is the Planck quantum. Therefore, it is not a *cruel uncertainty* as in the cuckold story for schoolboys, but an **inevitably fuzzy definition** for any kind of undulatory packet. To translate this fact into "*uncertainty*" requires a corpuscular postulate: an unsustainable hypothesis, but then admitted in this tribe.

Professor Castle-Holder:
- You exaggerate! You have not yet given any maths. I close your gap. Let $f(x)$ a complex or real function of real variable x. Its **Fourier transform**

(or **spectrum** in physicist language) is the complex or real function of a real variable ν :

$\hat{f}(\nu) = \int f(x) . e^{-2i\pi\nu x} dx$

Symbolically written: $\hat{f} = \mathscr{F}(f)$.

Theorem: Any integrable function f(x) has a Fourier transform $\hat{f}(x)$.
It is continuous, limited, and tends to zero when ν tends to infinite.
The reverse transform: $f(x) = \int \hat{f}(x) . e^{2i\pi\nu x} d\nu$ symbolically written: $f(x) = F(\hat{f})$.

Open-Eyes:
- Thanks! Some of our students have to handle several F.T. in their professional lives; they use spectrum analyzers. A square periodical signal has a spectrum in harmonic sines and cosines, and these harmonics are all even or uneven, depending on the choice of origin. The F.T. of a square pulse is cardinal sine:

$\prod(x) \overset{TF}{\to} \frac{\sin(\pi\nu)}{\pi\nu}$

The transform of a Gaussian is a Gaussian, and the product of their widths is constant:

$e^{-px^2} \overset{TF}{\to} e^{-\pi\nu^2}$

Some more properties:

Even $\overset{TF}{\to}$ uneven

Uneven $\overset{TF}{\to}$ even

Real $\overset{TF}{\to}$ Hermitian (the function is also the transposed of its conjugate).

Pure imaginary $\overset{TF}{\to}$ anti-hermitian (the function is also the opposite of the transposed of its conjugate).

Properties of the Fourier transform:
It is linear as the integration is:

$\mathscr{F}[\lambda.f(x) + \mu.g(x)] = \lambda.\hat{f}(\nu) + \mu.\hat{g}(\nu)$

I skip the transposition and the conjugation, to point the changing of the scale:

$\mathscr{F}[f(a.x)] = \frac{1}{|a|}\hat{f}\left(\frac{\nu}{a}\right)$

If you spread the pre-image, the spectral transform is more concentrated.

I skip the translation, the modulation, the derivations, the convolution, and the Parseval-Plancherel relation: please report to an undergraduate lesson.

A consequence of these unavoidable properties of the Fourier transform: As all the measures you will take for a good definition of the spatial position of this wave packet that the copenhaguists call "*particle*" and think "*corpuscle*", end in deteriorating the definition of its frequency, and consequently of its momentum, you could tell yourself that nature conspires to hold you in a *cruel uncertainty* about the behavior of this *corpuscle*. It was the folkloric part of the hegemonic fairy tale.

Curious:
- Let us sum up: you throw the terminology "*Heisenberg's Principle of un-certainty*", and only it, arguing it was already published a century before by Joseph Fourier.

However, you do not throw the relation that gives a floor value to the product of the indefinitions on positions and momentum: it is merely a consequence of the Fourier transform applied to a wave packet, for each photon, and by extension to each other particle, the fermions included.

Open-Eyes:
- Right! Not "*indeterminations*" which would be anthropocentric again, in-definitions is the right word. The product of the spreadings in frequency and position of any quantic "*particle*" is not at will; it has **h** as floor value.

The second indeterminism is fundamental. It was an axiom in copenhaguist semantics; now it is an unavoidable consequence for us transactional physicists.

Professor Castle-Holder:
- No! Not so quickly! Our reader will not escape now without doing some basic exercises! The theorem from Emmy Noether states three couples of variables linked by the Planck's constant:
The couple position-momentum,
the couple angle-angular moment,
the couple duration-energy.

The principle of the game: I give you the lifetime of the initial state (a metastable state), the lifetime of the final state (often infinite), deduct the length of the emitted wave pack, the photon.

Curious:
- Objection: mathematically, I cannot do that.

Open-Eyes:
- And the second objection, alas: your statement forgot the properties of the absorbers, and of the intermediate space and optical devices, where the photons could boson together. Whichever worthy is your problem, it aims far beyond the scope of this handbook.

10.11. The Dirac-de-Broglie ground noise, the impossibility to delimit a quantic system

Our adversaries the anti-transactionists believe that believing is enough to ob-tain that a quantic system could be delimited and isolated from the remaining of the world. So is one of their main logical faults.

10.11.1. Taking the properties of our artifacts for the properties of what they are pretended to describe. To take the properties of our artifacts for the properties of what they are pretended to describe, is alas a

classical trap in sciences – and it is worse in non-sciences – and the misadventures they prefer to shove under the rug are many. Not speaking of the look of the rug, blistered.

The language: the first tricky artifact. Some examples are known for long. The antique Greeks loved to reason, but those whom writings are now known despised to experiment as they despised the craftsmen: they were aristocrats. So they bequeathed a pretty set of blunders, due to their idolatry of their Greek language. Aristotle, for example, projected the categories of the language into physics: *the heavy bodies fall as it is their nature.* The result, the Aristotle mechanics is not good at all; it is a projection of adjectives. Zenon d'Elee pushed the logic of the greek language into paradoxes: Achilles will never catch the turtle because one can describe the movement into an infinity of Greek phrases.

The trapping artifacts of the sheet of paper and the line by the typewriter. About fifty years ago, Joseph Davidovits perceived that our lectures and handbooks in macromolecular chemistry are fooled by the properties of the typewriter: we type a straight line, in thin lines, and the printer also works by horizontal lines. Silently, the common believing deducted that the macromolecules coming from a linear synthesis by adding monomers at the end of the chain in a solvent should remain unrolled, only little bending from a straight line. Even now, all the handbook of macromolecular chemistry exhibit drawings of **unrolled** noodles. Now Davidovits was an enough good chemist to perceive that the unrolled macromolecule in the solvent, asks more energy than a folded molecule, next than a curled up molecule during its growing; after that, each micelle may be strained when passing through a spinneret to obtain a weaveable thread, it will remain micellar, will never unroll. Only the contacts between the micelles take charge of the shear strain. Only his model predicted many macroscopic properties; the most notable in his view was the melting entropy, but also the viscosity. Afterward, Davidovits had the mishap to exaggerate the geometric regularity of its curled up micelle: in his turn, he was hypnotized by the artifact of the flat sheet of paper where we draw and write, and use to communicate with to the others scientists. He modelized the micelles as plane rackets. It was a mistake. Paul John Flory did the inverse exaggeration: for him, the macromolecules were purely *statistical*, so broadly spherical, without distinction between the long axis and short axes. My own experience in the laboratory as a scientist of the solid-state mechanics validated more the model with a definite long axis: we had traction a ribbon test piece of *Nylatron*, a polyamide with MoS_2 as filler. Similarly, the sailors buy pre-strained halyards (more expensive) where the long axes of the micelles are already aligned with the traction on the rope. On the contrary, the alpinists use ropes with an annealed polyamide (the long axes are at random) which can absorb the energy of a falling alpinist, but only once: when the long axes are aligned by the strain resulting from the energy of the shock, the rope must not be used again in alpinism.

Flory had the Nobel prize for his micelles. Davidovits was already doing other things, working within the industry.

I have had on my shelf a report from the Laboratoire Central des Ponts et Chaussées which dealt on a micellar model for the bitumen, globally spherical, with success. The genesis of the bitumen does not require an oblong geometry, on the contrary to the polyamides we daily use.

Professor Marmot:
- You drown the fish, now! Polymers and bitumen are only kitchen plays for low peasants! They are not maths. It proves that this *scholardventurer* is just a mere *crank*!

10.11.2. In mechanics, the economy of variables and equations is sound. .

Open-Eyes:
- In astronomy, even the movements of three gravitating bodies are incalculable. There are very good reasons for the astronomers and the engineers to be fond of simplifications and economy of means. Without simplicity or simplifications, working would be impossible. In macroscopic mechanics, at our scale, it is sound to think we can isolate a system. The artillery with the resistance of the air and with irregular and poorly known winds, it is already complicated, but if we can suppress the air, all is simplified and becomes easier to calculate. So is the mechanics in the cosmos; sure it differs from our daily experience on Earth, where there are frictions everywhere. A gyroscopic compass, it is expensive and difficult to build, it needs an internal vacuum to minimize the fluid frictions, and great cares against the solid frictions. In the end, it works well, giving a far better precision and stability than what gives a magnetic compass (still there, just in case).

Yes, by thought and even experimentally we can delimit and isolate a mechanical system, provided it is at our scale, but the danger was to blindly extrapolate to microphysics, postulating without proofs that it would work as well. Alas, this fool trap worked well.

10.12. Sure and admitted frontiers? Some caveat from other crafts

What a quantum system is composed of? In the dominating chapel, the copenhaguist physicist does not doubt that his list of quantic objects he puts in equations, is a complete and sure list. Now we will argue against this hasty and incorrect certitude. In this prior sub-chapter, I will come back to the pathologies induced by a failing psychic delimitation of the persons.

Professor Castle-Holder:
- Wow! Fasten your seatbelts! You do not suspect what waits for you!

Open-Eyes:

- When we teach the first rudiments of mechanics, that is the statics, to fifteen years old pupils, one of their difficulties is striking: many have a hell of time to delimit a mechanical system, to state its frontiers, to list the entering and the outgoing actions. If this step is not got, the remaining of the building is founded on mud, and soon all fall down.

This pathology is exacerbated in the inventors of perpetual motion devices, or of any kind of *"over-unitary"* machines: they are never clear on the frontiers of their *messygimmikclutter*, on the inputs and the outputs. When you discuss with them, they are not long before exhibiting lots of psychotic symptoms; their technical delirium is the projection from their own psychic malformations and acquired infirmities. You will be perplexed by the tragic example of "AIXO-GEN MOTORS":

http://deonto-ethics.org/impostures/index.php?board=33.0

The notorious controversy in the years thirty between Albert Einstein and Niels Bohr puts in evidence the contrast between an Einstein, secure in his frontiers, and a Bohr, still invaded by the mother irrationalities:

At my left, the champion Einstein, who shouts: *"My father, he is rational and legalistic. He does not play dice!"*.

At my right, the champion Bohr, who shouts: *"My mother was never rational nor predictable. We must limit our curiosity to the questions she likes, and do not attire me a pair of slaps."* And never ask where do the babies come from!

At my left, Einstein speaks once more: *"My father, he is subtle but not mischievous"*.

You have recognized the controversy that opposed for the Solvay Congress in 1927, Albert Einstein to the Göttingen-København chapel, initiated by Max Born and Werner Heisenberg, followed by Niels Bohr. The climax of the controversy was the paper signed by Einstein, Podolsky, and Rosen in 1935, still known as the "EPR paradox".

Alas, I could give more tragic examples, such as the damages obtained on my son as he was invaded by the growing paranoia of his mother, more and more despotic and invading. In contrast, at less than three years, he had been able to set limits to his mother: *"You, you want me to be well-behaved, but I do not like to behave so much!"*. Under the law of corruption, the fruits did not hold the promises of the flowers.

10.12.1. Delimiting a quantic system? If it is a delusion, it is high time to know that. It has largely proved, by all the variants of *"delayed choice"* experiments (see in Appendix F) that the dream that the emitter of two correlated photons could be isolated from the two absorbers, is a fallacious dream. The emitter and the two absorbers remain tight until the last absorption is complete, so the no isolation from the future exist at the individual scale. What we will demonstrate now is that a lateral isolation, and a delimitation of a system do not exist either. A poetic image of this permanent intrusion of the external world was done by the staging by the Fura del Baus team, in February 2007 at the Garnier Opera, **A Kékszakállú Herceg Vára** (the Blue-Beard's castle), with Willard White playing the duke Kékszakállú and Béatrice Uria-Monzon playing Judit: the impossible bed of their impossible love, harassed by all the hands and furtive bodies of the previous wives of Kékszakállú. https://www.youtube.com/watch?NR=1&v=wDaIe-vWmp8& feature=endscreen at 4' 10".

In a note to the Académie des Sciences in September 1923, confirmed in 1924 by his thesis, Louis de Broglie established his theorem of Harmony of phases, where he proved that the celerity of phase is $\frac{c^2}{v}$ where $\mathbf{v}$ is the group velocity, identical to the usual speed in macrophysics. Hence follows that in the frame of the electron, where its speed is obviously null, its phase velocity is infinite in all its spatial extension; the electron is everywhere in phase with itself. Furthermore, for an "*observer*", to observe the Lorentzian transform of the intrinsic frequency of the electron $\frac{mc^2}{h}$, it is necessary that this wave, that is the electron itself, has an appreciable extension both finite, and fuzzy. De Broglie was not able to draw lots of important consequences as he resisted to conclude that the wave he discovered **was** the electron: he remained in the corpuscularist illusion, and only gave to the wave the role of piloting the mythical corpuscle.

The consequence of this non-negligible spatial extension and this locally infinite phase velocity is that at least in condensed matter, and maybe in all circumstances, every quantum particle, every fermion is lapped by the Broglian waves of all its neighbors, though we cannot set a list of who is neighbor, and who is not. So is the Broglian noise. In 1928, P.A.M. Dirac revolutionized all that by proving that the electron wave has four components (instead of only one), and two of them are retrochronous. So became questionable the extrapolating to microphysics the irreversibility of the macro-time, although valid and proved in macroscopical physics. It is only a popularization here, so we will not teach here the algebra of the 2-tensors on a space of dimension 4.

In 1930 and following, Schrödinger proved that according to the Dirac equation, and for any electromagnetic interactions, one has to consider a second intrinsic frequency $2\frac{mc^2}{h}$, and that the spatial Dirac-Schrödinger wavelength of the electron brings back the quantitative laws of the Compton scattering under the Bragg law, the foundation of the radiocrystallography.

In 1941, John Archibald Wheeler and Richard Feynman exploited this success of Dirac and Schrödinger, by a theory of the absorber; they computed that all the mass of the electron came from the electromagnetic mass, arising from its interactions with all the others electric charges in the Universe. As it could shout "*Help me! The Legion! They accelerate me!*"... But this leaves intact the mystery of the origin of the mass of the muon and the tauon, the two heavy electrons.
Link: http://authors.library.caltech.edu/11095/1/WHErmp45.pdf

Now can we shield the Dirac-de-Broglie ground noise? Nothing at all! Just as nobody can shield the gravity. With all this lapping of Broglian waves, impossible to monitor by any instrumentation, it is impossible to predict when nor which emitter-medium-absorber transaction will occur. Never the frontiers of a real quantic system will be at hand: irremediably, they are far, fuzzy, and fluctuating. The de-excitation of an atom or its nucleus may only be statistically predicted, on great numbers. Only the great numbers can statistically flatten the fluctuations of the Broglian ground noise. Only the great numbers can show that many absorbers are seen as equivalent, from the point of view of the emitter which sees their admittances, excepted if they are in resonance with the emitted frequency, which yields far higher admittances. Now, *the curse of the astronomers*, the thermodynamic laws imply that the emitters are far less numerous and far more detectable than the absorbers. It excuses the deny of the absorbers by the hegemonic chapel of the anti-transactionists. Not so good an excuse, however, as we know for Fraunhofer the dark lines, or absorption lines in the solar spectrum, and their interpretation by Kirchhoff in the 19th century; it was formalized in 1916 by the Einstein coefficients.

This debate was already held on Usenet in December 2003, May 2004, January 2008, June 2008... The names of the participants are on the wiki:
http://quantic.deonto-ethique.eu/index.php?title=
"Probabilité_de_présence"_qu%27ils_disaient...

We sum up: the end of the illusion, you cannot delimit nor isolate a system at the microphysical or quantic scale. The wild complexity of the environment will always foil all our experimenter tricks.

The hopes to manipulate the environment enough to prove this statement are thin: the frequencies implied in the Broglian noise are far above our instrumental means. Moreover, **the theorem of the requisite variety**, from William Ross Ashby, proves we will never have experimental access to the details of the Broglian ground noise. Forever, we are very far from the account.

Only one direct measure of the electromagnetic frequency of the electron is performed, at the ALS of Saclay:

10.12.2. Experimental observation compatible with the particle internal clock, by M. Gouanère, M. Spighel, N. Cue, M.J.Gaillard, R.Genre, R.Kirsch, J.C.Poizat, J. Remillieux, P. Catillon, L. Roussel.
. http://aflb.ensmp.fr/AFLB-331/aflb331m625.pdf
http://aflb.ensmp.fr/AFLB-301/aflb301m416.pdf

Of course, when one recalls these facts, he is submerged under insults and despise by the believers.
An example:
https://www.researchgate.net/post/Is_a_subquantal_structure_

possible_which_is_compatible_with_relativity_and_free_will
The physicists are territorial animals like the others, rats and dishonest like the others as soon as they feel that their domination on the territory is defied.

I had published this popularization: **Coluche nous avait expliqué pourquoi l'expérience de Gouanère & al. ne sera jamais refaite**
at the address:
http://www.agoravox.fr/culture-loisirs/culture/article/
coluche-nous-avait-explique-154321
(absent-mindedness: I wrote SLAC instead of ALS).
The experiment by the team bossed by Michel Gouanère at the ALS (Linear Accelerator in Saclay) in 2004, published in 2005 will never be made again before decades as its result bothers too many people, and compels to reform the whole teaching of the quantum mechanics, as it is hegemonic for 1927. *"No one may tell the truth on the television because too many people watch it".*

On the experimental conditions
On the administrative level, this experiment was half-clandestine, done during a time the accelerator was in maintenance, at very reduced power. It gave the direct proof of the second intrinsic frequency Dirac-Schrödinger $\upsilon_{DS} = \frac{2mc^2}{h}$ of the electron. It is the double of the Broglian one
$\upsilon = \frac{mc^2}{h}$.
The intrinsic Broglian frequency is valid for any massive particle, including the electron; Louis de Broglie deducted it in 1923, by joining the Planck formula of the quantum of action ($\mathbf{E = h.\upsilon}$) established for the light in December 1900, and the formula from Einstein (1905): $\mathbf{E = m.c^2}$.
Notations:
m is the mass of the particle.
c is the speed of the light.
h is the Planck quantum of action per cycle (or of angular moment), 6.6260755 . 10^{-34} joule.second/cycle.
υ (pronounce: "nu") is the intrinsic frequency of the particle, established by de Broglie.
E is the energy of the particle, rated in a frame, here the frame of the particle if it has a mass.
When in a move, it results in a spatial frequency, or its inverse a wavelength, whose experimental evidences are many, especially in all the experiments of diffraction of electrons or neutrons, for 1925.
For an electron, the intrinsic frequency is 1.23559 . 10^{20} Hz (cycles per second). The second intrinsic frequency of these perpetual oscillators, shown by Schrödinger in 1930 on a solution of the equation of Dirac (1928) in free field, is valid for the fermions: electrons, neutrons, protons. The Broglian period intervenes in all the interferences of a particle with itself, for instance in the Aharonov-Bohm shift of interference fringes. The Dirac-Schrödinger frequency intervenes in the electromagnetic interaction of the fermion with other things. It is the one proved by the experience at the ALS by the Gouanère team; it is

also in evidence in the Compton scattering of an X photon.

The ten experimenters found the absorption resonance (expected for a move in "rosette") at k = 81.1 MeV/c; it is ultra-relativistic. Hence, seen by us, the phase speed and the group speed seem very near to **c**. It is really far from the values we are accustomed to, in electron diffraction. See for instance at http://citoyens.deontolog.org/index.php/topic,1570.0.html

Indeed, for electron diffraction, intervenes the phase wave, largely supraluminic under 100 to 400 V of accelerating potential difference.

But at the ALS, they moved the electron clock through the silicon crystal, so what intervenes is the group speed. And 384 pm of interatomic distance is a large pace. The only available adjustment variable is the relativistic slowing of the clock. So one needs a true electron accelerator.

Let us convert the units, and do as if the electron was alone, in the vacuum.

1 MeV/c = 534.4288314 . 10^{-24} kg.m/s

162.2 MeV/c = 86.68435646 . 10^{-21} kg.m/s

Divide by the rest mass of the electron: 95.159358 . 10^9 m/s = 317.41741 **c**

Hence the rapidity **u**, expressed in unit **c**:

u = Argsh(317.4174) = 6.453367289

cosh(u) = 317.4190

Apparent slowing of the internal Dirac-Schrödinger clock ($\frac{2mc^2}{h}$ at rest): $\frac{1}{\cosh(u)}$

Hence its apparent period in the frame of the laboratory:

$\frac{h}{mc^2}$ .cosh(u) = 4.04665 . 10^{-21} s/cycle * 317.4190 = 1.28448 . 10^{-18} s/cycle.

When covered at the speed **c.tanh(u)** = 0.9999950374 c = 299,790,970 m/s, it yields a spatial period of 385.1 pm. However, this calculation was done under the assumption of an electron alone in the vacuum, while in the crystal, the electron is a dressed charge, so with a different effective mass. It implies that the internal clock is affected by the crystalline environment.

One could dream a less expensive and quicker method to measure this variation of effective mass. . . And I wonder whether the thermal dilatation of the crystal under the electron bombing was calibrated.

What is still to perfect when confirming the Gouanère & Al. Experiment?

The authors (unrolling them: M. GOUANÈRE, M. SPIGHEL, N. CUE, M.J. GAILLARD, R. GENRE, R. KIRSCH, J.C. POIZAT, J. REMILLIEUX, P. CATILLON, L. ROUSSEL) worry about the slight discrepancy: resonance found at 81.1 MeV/c, instead of 84.874 MeV/c expected, that is a 0.28 % gap. A so small variation in an effective mass will not disturb the solid-state physics, but they would prefer that the experiment would be repeated, in better conditions.

On balance of the slight complexification by the transactional physics, we obtain a gigantic lightering of the calculations when we take account of the intrinsic frequency $\frac{mc^2}{h}$ so a wavelength in flight, which severely constrains the width of Fermat spindle in propagation. Out of the Fermat spindle, the contribution is null. Of course, interferential configurations manage several branches of Fermat spindles, differing by an integer number of wavelengths; this for an electron, photon, neutron, etc. up to an entire molecule.

10.13. A set of two dynamical equations of Schrödinger

Yakir Aharonov, Peter Bergmann, and Joel Lebowitz did it in 1964: no intrinsic frequency of the fermion (electron here), the wavelength *"flippantly"* treated, the stiffness of the trajectory thrown over the windmills, they only drew statistical conclusions. These guys were drilled to never go through with their ideas.

10.14. Quantic vacuum fluctuations

Though many believers love to throw at our heads the Broglian fluctuations that must be non-existing because they are not taught... other believers, graduate believers work on the fluctuation of the vacuum, the Casimir effect included. The Casimir force is proved with a 1 % precision, and it is fundamentally the same thing, but viewed only from the point of view of the vacuum, excluding the potential emitters and the absorbers which can carry out transactions at random of some *rogue wave*.

10.15. "Antiparticles", and their part in the transactional microphysics

It was evoked by John Archibald Wheeler, at least orally, and reported years later by Richard Feynman: the fact that all electrons are the same, especially in their electric charge, may be interpreted so: all the electrons are only one electron, bouncing back and forth in the past and in the future. But what is the effect of the fields on these antiparticles?
Principle of equivalence: Considering the ordinary particles or the antiparticles which move backward in time, they are only two complementary representations of the same physics.

So, to a photon deviated by the gravity of the mass of a star or a galaxy along its journey from a star or a quasar, corresponds an antiphoton with negative energy, which is deviated the same way along the same trajectory in the same thin Fermat spindle. The same identity of bendings of the trajectories by the electrostatic plates of a cathodic oscilloscope, or the engraver of microchips; same bendings of trajectories under magnetic lenses for a television set or an electronic microscope. Negative mass indeed, but inversed charge, and inversed arrow of time.

Professor Marmot:
- You are stuck now! When it runs the inverse path, your anti-electron charged + should be repelled by the plate charged -. You violate the laws of electrostatics.

Open-Eyes:
- You forget that the flow of the micro-time is inverted too. The invariance of

the laws of electromagnetism is conserved. Curious: So you postulate a new kind of antiparticles: negative mass, negative energy, inversed charge, inversed temporal flow of time?

Open-Eyes:
- But it does not denote new entities. It is just the second half of the description of the same, already known particles. While the positron experimentally discovered in 1932 was a new entity.

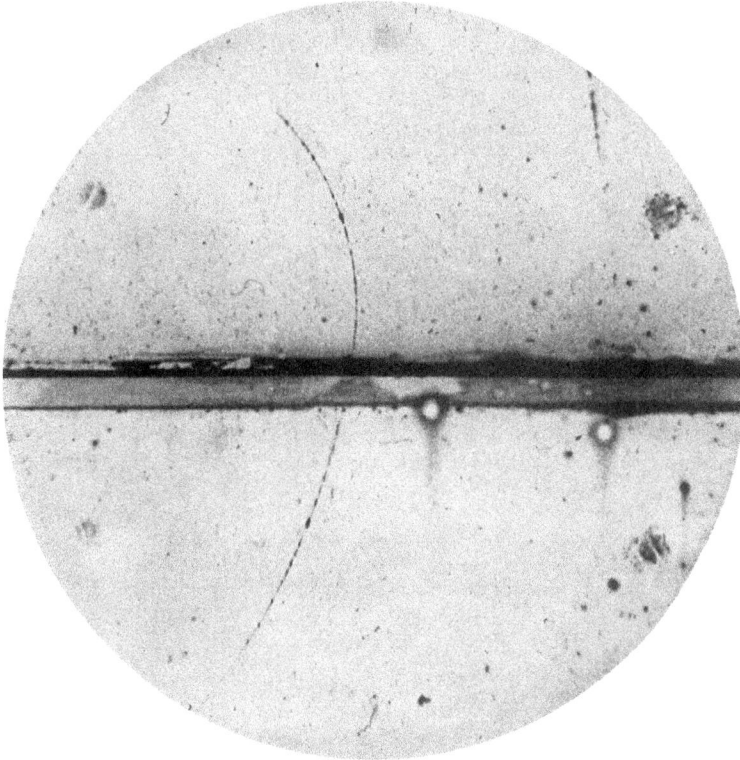

Figure 10.38.
Anderson, Carl D. (1933). "*The Positive Electron*". Physical Review 43 (6): 491–494. DOI:10.1103/PhysRev.43.491

Professor Castle-Holder:
- Here is the historic photo, taken in a fog chamber by the team of Carl David Anderson in August 1932.
Among the results of a collision by a cosmic ray, we see a trace whose parameters are incompatible with a proton – bending in a magnetic field, length of the trace – but prove an electron, of positive charge. This positron has a positive mass, a positive energy; its behavior is like any ordinary positive charge in the ordinary world; for it, the arrow of micro-time along its path is the same as our macro-time. The running direction is proved by the crossing of a thin lead plate: the positron lost energy and momentum by crossing it, so the bending

of its path is tighter.

Open-Eyes:
- Let us return to the alpha radioactivity. The conditions of synthesis of a nucleus 232 thorium are pretty rare: an implosion of supernova, we will not see one again here on Earth. The emission of an alpha, then the damping of this alpha in the matter show all the characteristics of irreversibility at our scale. It is more than improbable to synthesize a 232 thorium with an alpha and a 228 polonium, an endothermic process. And it is impossible to accelerate an alpha by giving it the ionization energy of the atoms it has shaken up. Thermalization and irreversibility are there at each step.

10.16. The dyes: the spatial extension of the resonating electrons

Curious:
- In the beginning, you had announced developments on the dyes and their electronic chemistry. Your promise is not yet held, and you have not yet explained why you need them.

Professor Castle-Holder:
- The dyes are a particular case of the spectral absorptions; particular in that the absorbed band is in the range of the visible for human eyes. Our retinas are wired to accentuate the contrasts by subtracting some signals of the specialized cones. What is routed to the Lateral Geniculate Nuclei (LGN) of the thalamus and hence to the visual cortex V1 has already been preprocessed by the retinal wiring. The most ancient color way, present in all the mammals, the koniocellular way, sends to the thalamus the signal [blue - yellow], that is for us [blue - (green + red)]. More recent, less than forty millions years, and only on the apes of the Old World, our parvocellular way sends the signal [green − red]. If in our retinas we had very different opsins, as have the diurnal birds and most fishes, who have four diurnal opsins, the colors we would perceive would be very different. Here are two well-known dyes: The methylene blue, which is a big flat cation:

Figure 10.39

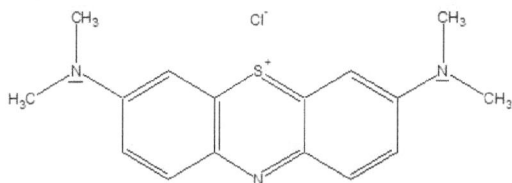

Figure 10.40, the alizarine yellow (this name is controversed):

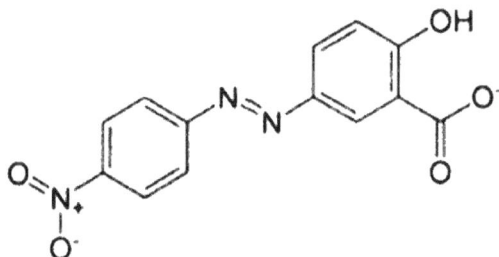

These absorptions in the visible range are due to a proper mode of electron oscillation in the molecule. An electron is enough delocalized to accumulate alternatively at one welcoming end and another end of the molecule, and it oscillates between these ends. It can oscillate as the intermediate path contains conjugated double bonds, whose properties make a conductor, at least for this loosely bound electron, or more precisely, whose bonding is widely shared.

Open-Eyes:
- "*Conjugated double bonds*". This vocabulary, specific to the chemists, was fixed in a far past, about since Kekule (1829-1896) and its solution for the molecule of benzene, next Thiele (1865-1918). Indeed it implies that these bonding electrons are not confined to one basic covalent bond but are more shared. So for the benzene, three delocalized electrons constitute the π bonding, shared by the six carbon atoms. These conjugated bondings influence some physical properties. Thus the molar refraction is abnormally high, proving a strong coupling to electromagnetic radiations. Consequently, the measure of the index of refraction can deduce the presence of conjugated double bonds in molecules.

Another special chemist vocabulary for these conjugated bondings is "*chromophore*" (Theory of the chromophores by Witt, 1876). The longer is the conjugated chain, the lower is the absorption in resonance, from ultraviolet (the benzene) to the red.

Besides the double bond C=C, the remarkable chromophores are the group carbonyl C=O, the group azo N=N, and the group nitro -NO_2. The ends of the electron oscillation are said *auxochromes*. The typical auxochromes: - NH_2, -OH, -$N(CH_3)_2$. The acid groups -SO_3H and -COOH are weakly auxochromes but as they are polar, and give to the dyes the solubility in water and affinity to silk and wool fibers.

Curious:
- And what is the interest in quantum physics?

Open-Eyes:
- The long diameter of such molecules, often from a nitrogen to another, is nearly a floor value of the spatial extension for this delocalized electron. What may be awkward is to guess this diameter. We may be fooled by the artifact of the plane paper.

Classically, one represents the second bonding of the ethene, the π bonding, as *around* the strong σ bonding, above and under the molecule plane. Each bonding is composed of two electrons, of opposed spins. I did not myself solve the Schrödinger equation for the ethene; we know that this molecule is plane, which implies that the π bonding has not a rotational symmetry, rather a plane one. I remain unsatisfied with what is given in the handbooks of organic chemistry; I suspect that a complete resolution should show that the σ and π bondings are hybridized. About the conjugated bondings, alternate of single and double bondings according to the graphic habits of the chemists, I lack even more of a solution. We were taught that there could be an oscillation – said "*resonance*" – between the possible configurations, and I still ignore whether it the better description. If it is, one could find spectrographic proofs. The main resonance of absorption by the benzene is in the ultraviolet at 185 nm, but I do not have information on the proper modes, neither for this main resonance nor for the two weaker at 204 nm and the triplet at 260 nm.

Professor Castle-Holder:
- Let's immediately exploit the information. Wavelength: $\lambda = 185$ nm, hence the frequency $v = c/\lambda = 1.620$ PHz. Energy of the photon captured at this frequency: $E = h\,v = 1.074$ attojoules $= 6.7$ eV.

Now the aromatic binding energy of the benzene is rated at 150kJ/mol, that is 249 zJ per molecule. Rated by comparison with the original Kekulé formula, where the double bonds were fixed, without oscillation.

Open-Eyes:
- Stop! We ignore the proper mode excited by the photon, and we ignore the polarization. The property of the ground mode at $\mathbf{h/2}$ is valid for a true dye, which absorbs in plane polarization, but is invalid for a rotation mode of the π electrons around the ring. So we must wait for complete information before concluding.

Figure 10.41.
The double bond of the ethene.
Figure from Organic Chemistry, by K. Peter C. Vollhardt. W.H. Freeman and Company.

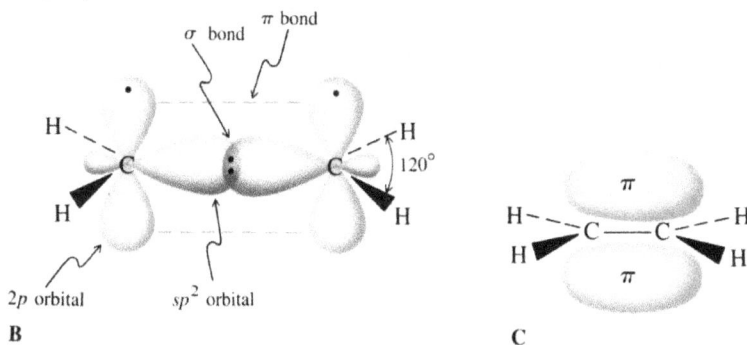

A simplified model of the behavior of the electron in a resonant molecule, either in the visible range to give a visible color, either and more common in the ultraviolet, is a box with two compartments: sometimes the electron is in the A compartment, sometimes in the B one, and the Hamiltonian (the energy) is calculable. But such an oscillator cannot step down under a half-quantum of action. What happens when this dye absorbs a photon at the resonant frequency? The oscillation rises in intensity, without increasing the numbers of involved electrons: at one end of the oscillation, when there is 1.414 electron in compartment A, there is -0.414 in B, and so on exchanging. Again one more photon? 1.732 electron in compartment A, -0.732 in B. These results from the formalism are incompatible with any corpuscular ideation.

Professor Castle-Holder:
- I remind that for a true dye without fluorescence, the de-energizing must be by non-optical ways, without emission of photons, but mechanically by shocks or emission of phonons, a thermal de-energizing: the body or the gas heats. The practice of the dye lasers proves that a high energizing of the proper modes of these molecules breaks them in a noticeable proportion. The breaking of a hyper-energized molecule becomes probable. Most of the organic dyes, inks, and paints fade when exposed to sun light.

Open-Eyes:
- This is the end of the chapter exposing the theory of the transactional microphysics. This sub-chapter on the dyes confirmed the defeat of all corpuscularist ideations, for the photons as much as for the electrons. When an electron oscillates in a dye molecule, it oscillates as a wave. The convergence of a photon onto the resonating molecule or atom which will absorb it was already established. This revision on the dyes reminds that it is a fact encountered many times in the day of everybody.

CHAPTER 11

The photo-electric effect revisited. Laws of Lenard (1900)

Professor Castle-Holder:
- You will not escape so easily! You have thrown away the interpretation of the photoelectric effect as given in 1905 by Albert Einstein, without reviewing the physics of the photoelectric effect, as discovered in 1867 by Heinrich Hertz and investigated by Phillip Lenard (1900).

Summing up: Light can extract electrons from a metal, but there is a threshold in frequency, under which it is inefficient. For the zinc, the metal of the historic discovery, the threshold is in the ultraviolet. The electrons are emitted only when the frequency of the light is high enough and is above a **threshold frequency**.
1. This threshold frequency depends on the material and is directly linked to the energy of binding of the electrons which may be emitted. It varies according to the crystallographic face of an illuminated monocrystal.
2. The number of the extracted electrons by the light, which induces the intensity of the current is proportional to the intensity of the illumination.
3. The speed of the extracted electrons does not depend on the intensity of the light source.
4. The kinetic energy of the extracted electrons is proportional to the difference [frequency of the light – threshold frequency].
5. The photoelectric emission comes very shortly after the illumination, less than 10^{-9} s. It is a quasi-instantaneous phenomenon.

Figure 11.1.

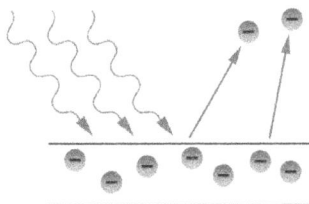

References to the founding works:

Heinrich Hertz: *Ueber den Einfluss des ultravioletten Lichtes auf die electrische Entladung.* Annalen der Physik 267 (8),
S. 983-1000, 1887.doi: 10.1002/andp.18872670827
http://onlinelibrary.wiley.com/doi/10.1002/andp.18872670827/abstract

P. Lenard: *Erzeugung von Kathodenstrahlen durch ultraviolettes Licht.* In: Annalen der Physik. 307, Nr. 6, 1900, S. 359–375.
doi:10.1002/andp.19003070611.
http://onlinelibrary.wiley.com/doi/10.1002/andp.19003070611/
abstract;jsessionid=E4385DBD654732F7AE8429461F56284F.f01t01

A. Einstein: *Ueber einen die Erzeugung und Verwandlung des Lichtes betref-fenden heuristischen Gesichtspunkt.* In: Annalen der Physik. 322, Nr. 6, 1905, S. 132–148. doi:10.1002/andp.19053220607.

Open-Eyes:
- Gladly. It will immerse us in the solid-state physics, that at the beginning you wished to reject. In 1905, Einstein admitted the postulate of the independent electrons in the metal, and in that time, it was admitted the electron to be a small sphere. Poincaré calculated its flattening at relativistic speeds. Now, we know more. We know they are organized in plasmons: electron gas waves, elastically bond to the array of metallic ions; their coupling to the photons are called polaritons. So many *quasi-particles*, which have a good quantic mathematization, without any possibility of a corpuscular reference. In 1905, Einstein did not question about a size of the absorber, as he presumed the photon was already pre-concentrated into a small corpuscle. In transactional physics, we know that a reflecting photon does not have any reason to be concentrated by the surface of a metallic mirror. As long as the mirror is plane (we set aside the concave or convex mirrors of a telescope) the metallic reflection does not change the general geometry of a Fermat spindle: a photon concentrates only onto the absorbing reaction. The diameter of the absorption reaction determines the final diameter of the photon, and these two diameters are intrinsically fuzzy. Surely, along the light path, a converging or diverging optics changes the angle of convergence of a Fermat spindle, for each photon. While an absorbed photon by a metal always comes concentrated at the diameter of the waiting absorption reaction. Whether this reaction will emit an electron or not.
So are the implications of the transaction. Our new task is to identify these absorption reactions, and to investigate their fine physics.

Either by the photoelectric effect or by thermionic emission as in the cathodes of vacuum tubes, the output work of an electron depends on the metal or alloy. Here is a table of mean results (variations about 5 % according to the method of measurement and the crystalline face, when not specified). Table extracted from the Ashcroft & Mermin, *Solid State Physics*:

Work functions of typical metals

Metal	W (eV)	Metal	W (eV)	Metal	W (eV)
Li	2.38	Ca	2.80	In	3.8
Na	2.35	Sr	2.35	Ga	3.92
K	2.22	Ba	2.49	Tl	3.7
Rb	2.16	Nb	3.99	Sn	4.38
Cs	1.81	Fe	4.31	Pb	4.0
Cu	4.4	Mn	3.83	Bi	4.4
Ag	4.3	Zn	4.24	Sb	4.08
Au	4.3	Cd	4.1	W	4.5
Be	3.92	Hg	4.52	Ta	4.1
Mg	3.64	Al	4.25	Mo	4.3

Source: V.S. Fomenko. Handbook of Thermionic Properties. G.V. Samsonov, ed., Plenum Press Data Division, New York, 1966. (Author's distillation of many different experimental determinations).

Another source (H. J. Reich): Thorium on tungsten, 2.63 eV.

The difference of electron affinity in different metals is also exploited in the thermocouples, best mean to measure the temperatures. The electron output work out of metals depends on the crystalline state (good crystal of bad crystal), and of the index of the illuminated face: 4.26 eV for polycrystalline silver, but 4.74 eV on the face (1 1 1) of a monocrystal, 4.64 eV on a face (1 0 1), and 4.52 eV on a face (110)[1]. So at least for silver, the most emissive zones in a polycrystalline metal are the grain boundaries: zones of the highest energy; we have seen above in metallurgy, that they are largely eliminated by annealing. And it is not thick, a grain boundary: two to three atoms. Now our main problem is solved: the convergence of the incident photon onto the (photoelectrically-efficient) absorber, the small zone transiently apt to emit an electron. Moreover, the (1 1 1) planes in a face-centered cubic metal are the densest planes, and (1 1 0) is the less dense of the three where the output work was measured.

The thermionic emission was a crucial point for the makers of electronic tubes. The usable technologies were not many:

• The cathodes in pure tungsten, for the big triodes for the emission industry in high tension, more than 3,500 V. While the cathodes in tantalum were hardly used.

• The cathodes in tungsten with thorium, through a process of complex (anodic) thermionic activation, yielding a mono-atomic layer of thorium on tungsten.

• And the most common when the book from Herbert J. Reich (*Principles of Electron Tubes*) was translated into French at the Editions Radio (*Techniques et applications des tubes électroniques*), in 1951: the indirect-heated cathodes, where a tungsten wire heats a tube made of a sheet of nickel, covered with barium and strontium oxides. Here again, the activation process is complex, conversion of carbonates to oxides, the partial anodic reduction of the oxides.

---

[1]Dweydari, A. W., Mee, C. H. B. (1975). "Work function measurements on (100) and (110) surfaces of silver". Physica Status Solidi (a) 27: 223.

According to the temperature of heating (1,000 K to 1,300 K), we obtain an electron affinity in the range of 0.5 to 1.5 V. The lifespan of these oxide cathodes was a few thousand hours; like all the activated cathodes, they were hypersensitive to any inlet of gas.

Professor Castle-Holder:
- You have not solved the problem of the confinement of this photoelectric absorber, so it releases this electron, and not another one, under the energy injected by this photon, and not another one.

Open-Eyes:
- The electrons are indistinguishable; the photons are indistinguishable.
First, we will explain why under the threshold energy, nothing photoelectrical occurs: it is a metal, in the metallic state. So any photon slow enough in its periodicity meets a mirror of conduction electrons. But a small minority absorbed by another mechanism (to study separately), they are reflected, do not penetrate the metal. We have seen above that the most emissive cathodes are in the worst metal, full of plane defects, the grain boundaries, and with punctual defects, impurities, and lacunae. I do not see studies on the effect of the surface traces of the linear defects, the dislocations. Any defect on the surface or emerging at the surface hinders the development of the polariton, hinders the perfection of the mirror.
What is the effect of the absorptions of photons that do not extract electrons? They increase the temperature of the metal, first by increasing the energy of the plasmons (elastic waves of the electron), which in turn transfer energy to the phonons.
We lack a serious set of indispensable data: the spectral absorbances for at least some metals, and the active mechanisms. A summing up in 1966:

François Cabannes: *Facteurs de réflexion et d'émission des métaux.* HAL-Inria
https://hal.inria.fr/file/index/docid/206511/filename/
ajp-jphys_1967_28_2_235_0.pdf.
It summed up the shortage of reliable data in this domain.

Thereon, I conjecture until the existence of a more thorough study, that the Lenard-Einstein absorber of a photon that will give an electron emission, is a small defect in surface or near the surface, which hinders the full development of the polariton necessary to reflect the photon. It is more critical when this polariton must be quick, excited by a high-frequency photon. This defect may be transient, by collision or concentration of phonons. Up to now, nothing indicates that this defect radically changes its nature when the photon is quicker than the threshold of the photoelectrical emission; we have only proof of a change in its efficiency.

Professor Castle-Holder:

- But you do not cover the Lenard laws of the photoelectrical emission.

Open-Eyes:
- Indeed, I do! The threshold of extraction still depends on the metal and its surface state, more or less crystalline, and the temperature, facts already known for 118 years. The problem I had to solve was the size of the defect which can concentrate the photon to eject **one** electron. In 1905, Einstein did not question about a size of the absorber, as he presumed the photon was already pre-concentrated into a small corpuscle. The practice of the vacuum tubes has proved how much the output work depends on the temperature of this cathode; it pushes us to scrutinize the transient defects by transient concentrations of phonons and plasmons. It remains to verify whether this transience is not too short for the length of coherence of the incident photon.

We still have to solve the question of the retrochronous signal from the absorber to the emitter, allowing the transaction to achieve, and provoking the convergence of the photon onto the defect, surely far smaller than could be the reflected photon *en route* towards another absorber. Concerning the molecule of carbon monoxide, it was enough to look at the frequency of the irreducible vibration of zero, at **h/2**.

Professor Marmot:
- Well! You are stuck again!

Open-Eyes:
- Sorry, I am not! It is evident; you have just to read the long paragraphs H. J. Reich wrote on the space-charge around the thermo-emissive cathodes, § 2.8 to 2.13 in my edition. The cathodic metal continuously emits and retakes electrons, at least at short distances. Now the combination [reflection + photoelectric effect] is not a resonating phenomenon, but a low-pass filter; so we cannot expect a clear signal from an absorption spectrum. Spectral data about the physical-chemistry of surfaces seem lacking, and for obtaining spectral data from the low-pass filter [reflection + photoelectric effect], we need spectrally-constrained sources.

Here we must conclude that absorbed photons at a too low frequency to frankly extract electrons are nevertheless able to increase and dilate the space-charge, and the apparatus that Lenard used in the end 19th century was not able to investigate this. Above the threshold, the space-charge is torn: some poorly bonded electrons, near the impact of the photon, are freed. There are developments in spectrometry in UV photo-emission, to investigate the adsorbed atoms.

Professor Castle-Holder:
- For our Curious reader, who has no experience in physical-chemistry of surfaces, I must add that atoms of oxygen or nitrogen which remain at the surface of a solid are said **ad**sorbed. While the photon, though not so much penetrating, is said **ab**sorbed as it is eaten: it was the end of its run.

I stop you on one point: the space charge is a notion developed in electronics and physics of the vacuum tubes, with heated cathodes, so thermionic emission. No valid extrapolation to room temperature, unless bombing by radiation at a frequency higher than the Lenard-Einstein threshold; the vacuum photodiodes were used in time.

Open-Eyes:
- Mmh. . . Nevertheless, in electrochemistry, we are taught a physical-chemistry of the double layers, which looks very like. Moreover, the electrical contacts work very well by extension of the electron at some distance from the metallic crystal: in a switch, as you have dozens at home, the space charge works well, without any authorization from the teachers of MQ. The thermocouples also work well: each conduction electron extends beyond the welding joint.

Moreover, the space charge intervenes also in the gas tubes like the thyratrons used in the practice works of the students on the Ramsauer-Townsend transparency effect, above in the sub-chapter 8.3. Please report to the handbooks of the years 1940-1980, on the vacuum tubes and the gas tubes, as the Kaganov already quoted. The space charge intervenes around the cathode and the grid.

Professor Castle-Holder:
- Reluctantly, as I had preferred to conclude to a draw, I admit that here also, you have taken a slight advantage: nothing could be proved against you, while you could point some defects in the standard theorization. According to you, we do not know enough, and maybe we were presumptuous, sleeping on the laurels of the *Great Ancestors*.

CHAPTER 12

Territorial animals like the others. The concurrency in fairy tales

Engineer Marmot (shortened in the english version):
- *... The QM is not a theory but a model. ...*
A quantic state is a proper function of the equation of Schrödinger, and an observable is an operator acting on a closed space of proper states. They are mathematical applications of the model. Period. The books by Messiah, Cohen-Tannoudji and al., Landau and Lifshitz treat all these aspects and its ties with the measure with a great rigor (and great modesty). But it is deep: when I was a student (in the years 60-70), I had spent three weeks to read the twenty pages of the Landau on the notion of the measure.
While the formulation of the model is ultra-simple, either for Schrödinger (H x Psi = E x Psi) or Dirac (H = alpha x P + beta x M), the calculations are heavy, even with a simple case.
For example, for collisions of protons on a nucleus (protons of 1 GeV, Saturn II accelerator, 1976), the cross-section is given (Born approximation of distorted waves) by an integral of overlapping of the incoming and outgoing distorted waves with a potential. The results give an incredible accord of the theory and the experience! But the price to pay was heavy, in toil. It was my thesis: five years of calculation and development of the code.
http://www.agoravox.fr/commentaire4224498
...
http://www.agoravox.fr/commentaire4224527

Curious:
- And you let him say you such violent things, Mr. the *scholardventurer*?

Open-Eyes:
- The *professor Marmot* is an assembly of real quotations from several people, so do not be surprised that someones contradict some others. In this one, whose severity remains respectable, he has just admitted "*The QM is not a theory*", while most of its thurifers pretend the contrary. It is just a mathematical phenomenology, no more, more or less cumbersome. Concerning the Landau and Lifshitz, I think ill of the famous sadism of Lev Davidovitch Landau, who took care of being the most obscure and the less usable as possible, especially in the volumes 3 and 4, here we are concerned with. While, still with the Mir editions, I think very well of the **Physique atomique** by Chpolski, or the two

volumes of **Optique** by Sivoukhine, or the **Fondements de la physique des cristaux**, by Sirotine and Chaskolaskaïa. Concerning the *"twenty pages of the Landau on the notion of the measure"*, they are the paragraphs 1 and 7, pages 9 to 14, 32 to 36, ten pages in total, but beautified by the heroic memory. From the beginning, I explain that if the classical QM is so obscure and awkward, it is because it is corpusculist in its core ideas, therefore contradictory, with no heads and no tails; there is no *subtle depth to understand*, but a hodge-podge to get rid of; we have to rebuild from scratch on sane bases. Though the mathematical phenomenology and the formalism are correct, the semantics with which they envelop the maths is scrap, strictly for the birds.

Professor Marmot:
- *"scrap, strictly for the birds"*! Who do you think you are?

Open-Eyes:
- This engineer Marmot has not obliged the illustrious deceased Landau and Lifshitz, as I will give them a tour on the merry-go-round. At the end of page 12, beginning of 13, 3rd edition, 1975: *"Let measure at intervals Δt the **successive** coordinates of **an** electron. Generally, the results will not materialize a regular curve. On the contrary, the most precise the measurements are, the most the results are chaotic, with bounces, as the notion of trajectory is invalid for an electron. A more or less continuous trajectory is obtained only if we measure the coordinates of the electron with low precision, for instance the droplets of fog in a Wilson chamber. But, if conserving the precision of the measurements, we reduce the intervals Δt, neighboring measures will give neighbor results, of course. However, the results of a series of successive measures, though in a small portion of space, will be dispersed in chaos, with no alignment on a regular curve. Particularly, when making Δl approach zero, the results of the measures will not at all align in a straight line."*.
End of quote.
The authors are prisoners of the confusion between the fate of an individual electron and the properties of a crowd of electrons, whose we never master the initial conditions nor the final ones. Each electron from the crowd has chaotic initial and final conditions, under the sway of the Broglian ground noise, that we will never rule.
When they teach their chaotic trajectory of a mad dog, it is a shameful bluff; never any such experiment, under such corpuscularist ideation, will be made. They just bet that never any student would dare to ask proofs. The conservation of the momentum applied to a relativist particle issuing from the accelerator and/or a collision forbids the erratic behavior postulated by Landau and Lifshitz; the law of the physical optics, from Fresnel (1819) forbids it too.

Professor Castle-Holder:
- Have you some proof?

Open-Eyes:

- Say! I was at the ends of the nights at the C.E.A. at Saclay, scanning the views taken at the big bubble-room Gargamelle. I certify to have never seen any *zigzagodromy*, but remarkably fine and stiff trajectories. In fact, the computer in charge of processing my points did not admit any zigzagodromy, only beautiful stiff arcs.

But what may be true in the words from the two illustrious men, quoted above? Maybe, if we could scrutinize the beginning of each bubble in the liquid hydrogen, each ion starting in one side, and each ripped-off electron starting the other side, each initiating a beginning of a bubble, slightly out of the trajectory of the ultra-relativist particle? But we do not have means to take a snapshot of the primers of a bubble by ionization. Indeed, these highest authorities have written bluff and blunders, grounded on never-validated corpuscularist presuppositions.

And worst! The second is the duration of 9,192,631,770 periods of the radiation corresponding to the transition between the two hyperfine levels of the fundamental state of the cesium 133 atom. But, say the *authorized authorities*, this stability and liability come from the madness of mad electrons, patrolling and *zigzagodroming* in all directions... Huh!

Curious:
- Bang! You have fired down two highest authorities of the scientific edition: Landau and Lifshitz!
Zigzagodromy? Some sailor slang?

Open-Eyes:
- A pen-sailor, indeed, a colorful pen fumbling-yachtsman: Jacques Perret. In oceanic navigation, we mainly distinguish the orthodromic route, the shortest, from the loxodromic one, at constant heading – seeming straight on a Mercator projection.

Professor Castle-Holder:
- Maybe Landau and Lifshitz were under the influence of the kinetic theory of gas, where the trajectory of a molecule bumping from shock to shock is erratic.

Open-Eyes:
- Affirmative: this is a plausible explanation.
Franck Laloë re-presents again, on and on the « *many worlds* » delirium from Everett, with tender indulgence. It is not innocent: it is a ruse of war: the folly of Everett is so weird, that it cannot make shadow to the fairy tales of the copenhaguists. On the contrary, it serves to implicitly malign all who are not themselves.
In the English version, I skip another involuntary comic, Antoine Moreau, as he is not worth the work of the translator.
Links:

<http://www.e-scio.net/mecaq/imaginer.php3> and
<http://www.e-scio.net/mecaq/libre.php3>

CHAPTER 13

Shoven causality. *Mektoub?*

13.1. Example of the Göttingen-København dead alley for 1927

The original message was in English, no need to translate.

Backward in time influence in the microscopic domain. A reality, or a wrongly posed problem?

Consider a pair of photons, A and B, prepared in the polarization singlet

(1) $|\psi> \,= 2\text{-}\frac{1}{2} \,(|x>A \;|x>B + |y>A \;|y>B).$

Assume that the photon A is tested in the lab of the experimenter Alice, and the photon B in the lab of the experimenter Bob. Assume also that the labs are in movement with respect to one another, see

H. Zbinden, J. Brendel, W. Tittel and N. Gisin, "Experimental Test of Relativistic Quantum State Collapse with Moving Reference Frames", arXiv:quant-ph/0002031v3 .

Assume however, that according to the clock of a third lab, of the observer Charlie, Alice and Bob perform their experiments simultaneously.

Consider a trial in which Alice and Bob did their test according to the same directions, {x, y}, and both obtained the result x. Let's see how do they interpret this result:

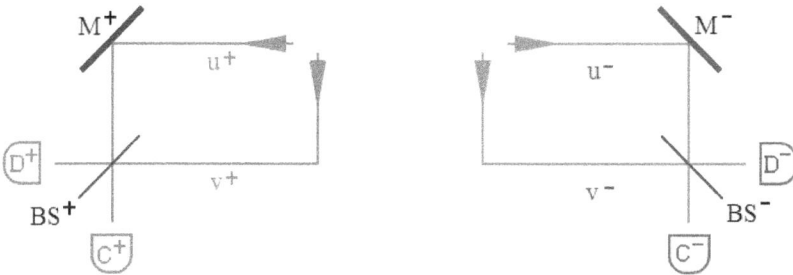

1) According to the time axis of Alice's lab, her test occurred first and she obtained the result x. No doubt, since by the time of her experiment Bob didn't yet do his text, the result obtained by Alice was independent on Bob's choice of axes and result. However, the wave-function (1) shows that if Alice obtained x, Bob's result for the same choice of axes should also be x. So, Bob's result YES depends on Alice's result.

2) According to the time axis of Bob's lab, his test occurred first, so, the result he obtained should be independent of what Alice will do in the future, i.e. which axes will she choose and which result will she get. **Then why cannot Bob obtain a result y?**
Putting the problem in another way, the wave-function (1) shows that the two particles have equal "rights". It's not a tableau of "leader" and "led", i.e., no particle is defined as producing its result independently with the other particle producing its result dependently.
However, if the two results are mutually dependent, that means that the result obtained by Alice depends on the future choice of Bob about the system of axes, and the result he will obtain. And also vice-versa.
All the physics we learned until now taught us that a present even can depend only on the past history, never on a future event. Should we believe that in the microscopic domain it goes otherwise?

End of quotation.

What are the standardized mistakes this Israelian researcher subscribes to, and block her in the same blind alley for years?
Of course, it is a **wrongly posed problem**, as it is traditional in the anti-transactional tradition.
Her first reasoning flaw is anti-relativist: If it depends on the position of the observer that the detection by A and B are simultaneous, or A before B or B before A, therefore none is in the light cone of the other, each one is in the elsewhere of the other; they are spatially separate, therefore causally separate in macroscopic physics.

The author of the question believes to the Newtonian macro-time for each of the laboratories. Now, these macro-times do not have any pertinence to describe the physics of the transactions.
The macro-time is only a statistic emergence; it has the property to flow in the same direction as the entropy does – also a statistic emergence. All along the 19th and 20th centuries, they took the habit to consider the macro-time as a fundamental parameter, but it is not yet grounded. And it is never causal in microphysics.

The Dirac equation for the electron (1928) gives two orthochronous components and two retrochronous ones. Hence results that for any more complex and heavier fermion, there are always as many orthochronous components as retrochronous ones. Hence, the Dirac-de-Broglie ground noise where the successful transaction emerges from is always bidirectional in all the micro-times.

One uncorrelated photon is a successful transaction between three partners: an absorber, an emitter, and space and optical devices between them. Two correlated photons are a successful transaction between five partners: two absorbers, an emitter, and the spaces and optical devices between them. Never the macro-times of Bob and Alice intervene in the physical laws of the transactions.

When struggling with other contributors (but avoiding any exchange with me, as the transactional physicist lives in a quite another physics than the anti-transactional), the Israelian interjects « *collapse, measurement...* ». However, the « *collapse* » she clings with only exists in the Göttingen-København newspeak but never in the real world; and the traditional « *measurement...* » does not belong to the microphysics, but to the local folklore of the local tribe that the student must revere to obtain his/her exams.

13.2. The mishaps of the "causality" in microphysics

Several authors panic on the mishaps that the quantum physics generally speaking, and notably the transactional physics inflict to the previous notions of causality; Ruth Kastner and Louis Marchildon: **Causal Loops and Collapse in the Transactional Interpretation of Quantum Mechanics**, in 2006.

Bernard d'Espagnat pushed the matter farther: According to him, the whole Universe should be entangled. Therefore, "*Mektoub*" (fatality, in Arab)? *All is written*?
Let see first who were the pioneers.
P.A.M. Dirac in 1938. **Classical theory of radiating electrons**.
http://rspa.royalsocietypublishing.org/content/167/929/148.full-text.pdf or
http://imotiro.org/repositorio/howto/artigoshistoricosordemcronologica
/1938%20-%20Dirac%20-%20Classical%20theory%20
of%20radiation%20electron.pdf
J. A. Wheeler and R. Feynman in 1941. **Interaction with the absorber as the mechanism of radiation**:
http://authors.library.caltech.edu/11095/1/WHErmp45.pdf
and in 1949: **Classical electrodynamics in terms of direct interparticle action**.
http://link.aps.org/pdf/10.1103/RevModPhys.21.425
In university, traditionally we were taught to calculate retarded potential, never advanced potentials, because of "*Causality! What else?*". We had also calculated the dispersion of a spherical wave emitted at one point, next we were warned in a half-sentence that the inverse, the convergence of a wave could not exist: "*Causality! What else?*". However, Dirac in 1938, next Wheeler and Feynman in 1941 began to disobey the taboo, with notable success, but did not dare to carry on the same way. In his famous series of lectures in Caltech, issued in 1964, Feynman incidentally told it, and all the transactional physicists confess having been intrigued by this paragraph where Feynman says they obtained by their absorber model that all the mass of the electron (only the light electron, not the heavy muon nor the heavier tauon) comes from the electromagnetic interactions with all the other electric charges of the universe, by retarded **and advanced waves**.

Well, with all these elements of retro-causality from the absorber and the intermediate medium with have put in evidence, up till which past can we roll up backward in time? Not very far indeed, except some photons and some neutrinos which may be thirteen milliards of years old. The *stones-eater* who quickly destroys all the coherences remains the Dirac-de-Broglie ground noise. Not very efficient to muddle in the synchronous transfer of a quantum particle, it remains dominant during every stable state between two transactions. We may hope to give it at least some qualitative laws, but have very small hopes to verify the laws of the handshake experimentally.

So we feel better: No predestination, no *mektoub*, no divination, no telepathy. Gladly, as the everyday experience proves that each robin who visits the campus at lunchtime surprises us, that our babies surprise us, that at wartime the enemy surprises us, and that the avalanche surprises us...

Indeed, the experience often shows that the personal preferences about causality are largely projective and transferential, hence passionate attitudes. Sooner, we have reviewed the Bohr and Einstein attitudes. If you see a researcher claiming that she wants a theory which is compatible with her "*free will*" and "*god*", do not hesitate to detour some miles to avoid her.

Curious:
- You have not yet answered in a positive way to the question: **If time is not the universal and ubiquitous time of the god of Isaac Newton, then what it is, for you relativist scientist?**

Open-Eyes:
- If you report to the questionings in the 19th century, in the time of Michael Faraday, and before the Maxwell equations were established in the scholarly community, they considered *foams of vortex* turning around the conducting wire, and the worse bad luck, it was for the electric field and the electric current that they imagined the most complicated solutions; they were carried haywire by the facility of obtaining magnetic spectra as Faraday did, so imagined that the magnetic field should have the most simple mathematic expression: the vectorial structure that Hamilton had just invented. To research and conjecture is taking the chance to err more than once, to be betrayed by bad luck. As the bad luck accumulated during the territorial struggles, only in 1894 Pierre Curie proved that only the electric field has the symmetries of a vectorial field. But Pierre Curie died before knowing the adequate mathematical tool for the magnetic field: the works by Hermann Grassmann and Gregorio Ricci-Curbastro did not percolate up to him.

During the same end of the 19th century, it was proven against violent opposition, that the temperature of a gas is not a variable at our hand, but is a statistical emergence of the individual kinetic energies of the molecules of the gas. It took a half-century more to generalize to condensed states with phonons, plus the conduction electrons in the metals.

In 1971, when he accepted to stay in his domain of competence, Roger Penrose demonstrated that the orientations of our familiar macroscopic space might be obtained as a statistical emergence from a network of spinors. It remained to obtain a similar result for the distances. For that, Penrose initiated a research program generalizing from spinors to twistors. It has yielded nothing. The design brief must have been inadequate. Must redo!

In conclusion, concerning the macroscopic time too (the relativistic one, but locally considered, in the laboratory where you are), it is a statistical emergence of all the interactions, from anything to anything, and it flows in the same direction as the entropy. Again a kind of *foam* of particular micro-times, all incompatible among them. The complete demonstration is not for soon, as many theoretical prerequisites must be set before.

The mathematics of the handshake in the frame of the Broglian lapping are still to be done: their mission is to convert from tridimensional in symmetric and unlimited time into unidimensional, whose results are obvious in our macro-time.
Not a small challenge!

Professor Marmot:
- *"to convert from tridimensional in symmetric and unlimited time into unidimensional, whose results are obvious in our macro-time"*. It means nothing! What a preposterous gobbledygook!

Open-Eyes:
- Flummoxed, huh! Each word counts. And what would be your counter-proposal? What would be your mathematics of the handshake in the frame of the Broglian lapping?

Curious:
- But if we have this quantic time-symmetry with past-future, why not in the macroscopic level?

Open-Eyes:
- The time-symmetry exists only for the micro-times of some transactions, and for the fermions. It does not imply anything about the macro-times, which are only statistical emergences.
Some Cramer's followers like L. Marchildon and R. Kastner are running in round in their *"causal loops"*.

Three challenges:
* Please explain what could be the time-symmetry of the trace of a particle in a fog or bubble chamber. How will you manage to use the de-ionization of a

molecule of ionized di-hydrogen molecule to accelerate the particle toward its emitter?

** How will you manage to shoot an alpha, right on a nucleus of radium 228, to obtain a nucleus of thorium 232?

How will you manage to break a helium nucleus into four hydrogen atoms, that is the reverse of the reactions occurring in the center of our star, the Sun?

Do you really expect to defeat the statistical thermodynamics?

*** (*Still more awkward!*) Shoot together an antineutrino and an electron into a nucleus to produce the reverse of a beta disintegration.

Professor Castle-Holder:

- If the experiment succeeds, we will strike a big cymbal clash like at the circus! No optical device exists for the neutrinos. With bent monocrystals one can do some monochromatization and focalization with the neutrons, but nothing yet with the neutrinos.

CHAPTER 14

"Measures", Schrödinger cat, "Principle of uncertainty", "entangled states". . .

When we open any popularization paper or book, impossible not to read some magical words which found the parlance of the sect, such as *"measure, observer, Schrödinger cat, many worlds, to prepare in a state, entangled states, uncertainty..."*. Why we never had use of them?

Professor Castle-Holder:
- I will borrow your parlance, and I distinguish three possible principles for a **measure**:
1. An absorber triggers a changing in the state of the apparatus, to carry a signal to the recorder, or formerly a human operator (who counts the clicks of a Geiger counter, for instance). So is the majority case.
2. In a minority of cases, this is the change of state of the emitter which is probed by the apparatus.
3. And a very small minority: the crossed-through medium can do a *"weak measurement"*, only slightly perturbing the monitored process; eventually, the perturbation may be discretely compensated.

Open-Eyes:
- For instance, with his sardonic apologue of the *dead-alive* cat, Erwin Schrödinger respectfully made fun of the triumphant egocentrists, the copenhaguists. Eighty-three years later, they have not yet determined how much he was laughing at their faces. According to the Göttingen-København pack, and mostly according to Eugen Wigner, the apparatus was waiting until an august copenhaguist deign to lean his august attention, and became aware, and only then, the apparatus could "know" whether the atom had disintegrated in the prescribed time or not. It was a complete fool idea, but this idea was hegemonic in 1935: the traumatism of the World War 1914-1918 and its futile butcheries still ruled the minds and their collective pathologies. In October 2006, the popularization magazine "Sciences et Avenir" published a special issue on the Schrödinger cat. In 78 pages, no one of the highest academic authorities could get clear in it. It was a festival of egocentrism and anthropocentrism in the Bohr manner: " *And I measure, and my measure changes, etc. and the Wigner's psychism, and the information and I destroy the information, and I, and Me and Myself! Etc.*"

Curious:

- The "*Uncertainty principle*", it was already treated in the § 10.10 " **Which randomness and where? The Fourier transform, fraudulently relabeled.**" Your re-reading holds the road if you are right to throw to the bin the corpuscular notions.

"*Observer*", according to you, he does not play any role in microphysics, so you throw him in the garbage can.

"*To prepare in a state*"? Please develop.

Professor Castle-Holder:

- It is a kind of "*the logistics will follow*". Here: It is up to the experimenter to cope with concretizing the initial (even final too?) conditions the theorist has mathematically written.

Open-Eyes:

- And it seems to me that we have already shown that what the theorist imagins to be some initial condition is rather far from the reality: no way to shield from the Dirac-de-Broglie ground noise.

Regarding "*entangled states*", for which they shout their admiration, it is just the consequence of the undulatory formalism and the transactional reality it describes; they are surprised because, in the back of the formalism, they still thought corpuscles – which is contradictory with the formalism. The fact is an emitter which emits two complementary "*particles*" to two absorbers. As the long as no decoherence intervened on both absorbers, it remains a transaction with three main partners: an emitter, two absorbers. And the physical laws of the real transaction are not concerned with the chronological order in which we perceive each absorption.

Professor Castle-Holder:

- **Decoherence**: a new word for our Mr. Curious. In the course of the following other reactions on the absorber side, very soon is lost the pseudo "*information*" ruling the only transfer considered by the theorist: frequency, phase, polarization, momentum, etc.

Open-Eyes:

- A very near notion is **thermalization**. Alas, it is not practiced by the same persons. Thermalization considers the repartition and dilution of the received energy, for instance brought by a neutron or a moderately heavy nucleus, ramming a target nucleus. The received photons as they are considered in atomic physics are always thermalized (but the bosonic effects, particularly in a laser cavity): the duration of an absorption is always short compared to the duration of the metastable final state. The ulterior de-energization by re-emission of a photon has no more undulatory correlation with the previously received photon, "*once, many, many Broglian periods before*" (Broglian periods of this atom). Surely, in an energized laser medium, where nearly all the atoms of the gas resonate in phase, according to the frequency dedicated by the energized transition and by the cavity, there is no more thermalization for the electrons of the concerned orbital; though there is still a thermalization for the deeper electrons.

Concerning « *information* », sure it is a fashionable word, especially by Stephen Hawking; alas, it is only an egocentric and childish fantasy when they pretend to apply it to physics. No law of nature deals of some "*conservation of the information*", not even to a definition of "*information*". Only an opportunist animal, perhaps a predator, can give a meaning to "*information*": it is what he can exploit for his profit. It is not a notion which may be transferred from a species to another, not even from one individual to another one: we are too diverse. The roe deer is a very olfactive animal (and the wolf also is), arms race obliges for at least 80 million years in the family. An olfactive information he uses, how could we use it? We are diurnals animals very visual, and quasi-anosmic in comparison. And if I write in blue the word "red", is it an *information* for the roe deer?

Curious:
- But what about the famous Schrödinger's cat?

Open-Eyes:
- OK, we will do again the exercise already given at the end of chapter 2, we will calculate the global Broglian frequency of a cat weighing 3 kg.
3 kg . 1.35639 . 10^{50} kg^{-1} . s^{-1} = 4.069 . 10^{50} Hz
Admit you translate it at 1 cm/s, hence its wavelength:
$$\frac{662.6076.10^{-36}kg.m^2/s}{3kg*0.01m/s} = 2.2087 . 10^{-32} \text{ m.}$$
Hin hin! And you planned to make of "*wave function*" of that? Its respiration, the heartbeat, the movements of its vibrissae, eyes, and ears, are tens of magnitudes bigger than this kind of "*wavelength*".

These Göttingen-København guys did not have the eyes lined up with the holes, and never realized the monstrous gap of scales between the microphysics and their mythical "*observer*".

Transactional (Quantum) Microphysics, Principles and Applications

CHAPTER 15

Didactical advantages? Does the transactional physics do better than the copenhaguist one?

15.1. Previous disasters in mechanics and electromagnetism

Open-Eyes:

- In *Surely you are joking, Professor Feynman*, Richard Feynman tells how he was severe in Brazil about the "scientific" teaching which turned into scholasticism - even frauds - when no more connected with the practical and industrial world, and having no more scope than producing science professors. The teaching of sciences lacks scientific methods for piloting itself, and lacks the qualiticians discipline; it is a toy for political brawls between clans and factions, where the contests of wills and tricks for conserving or taking territories and privileges take over any concern on quality, neither didactic, neither scientific. Afterward, obeying the eldest in the higher grade, the strength of habit and tradition are enough to maintain the blunders of past years from generation to generation; excusable blunders in the 19th century, already unforgivable for long.

And if by chance, a student or a young teacher realizes the blunders he/she was told, and exposes the remedies? The scholastic feudalities promptly eliminate this inconvenient finder. "*Move along, there's nothing to see here*"!

We have collected in the Annex G, the most common collections of silliness in mechanics. Here we will concentrate on the disasters in the teaching of electromagnetism, dominated by the "*vector product*", imposed by Oliver Heaviside, now reigning for 1888, which nobody has enough guts to correct.

The major blunder in the Bachelor Degree in 2012 in Pondichéry!

Or: these inspectors who do not know the symmetries of the fields of electromagnetism.

The subject was at

http ://www.ac-polynesie.pf/spip/IMG/zip/BCG_2013_S_PHYSIQUE_INDE. zip 13PHYCOIN1.pdf, exercice 3, pages 14 and 15, but is is deleted now.

Look at the question of Physics, on an apparent modification of gravity on a weight pendulum, supposed to be provided by a magnetic field in the center of a Helmholtz double coil.

Un pendule dans un champ magnétique Pour vérifier l'influence de l'intensité de la pesanteur sur la période d'un pendule simple, il est difficile d'envisager de se déplacer sur une autre planète. En revanche, il est relativement simple de placer un pendule, constitué d'un fil et d'une bille en acier, à l'intérieur d'un dispositif créant un champ magnétique uniforme dans une zone suffisamment large pour englober la totalité de la trajectoire de la bille du pendule pendant ses oscillations. Ce dispositif peut être constitué par des bobines de Helmholtz.

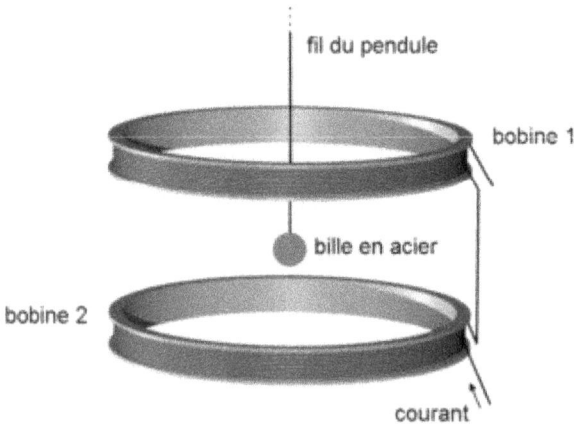

Bobines de Helmholtz

Bobines de Helmholtz

Lorsque l'axe des bobines est vertical, le passage du courant électrique crée un champ magnétique uniforme vertical dans la zone cylindrique située entre les deux bobines. Une bille en acier située dans cette zone est soumise à une force magnétique verticale.

2.1. Expliquer pourquoi ce dispositif expérimental permet de simuler une variation de l'intensité de la pesanteur.

2.2. Comment doit être orientée la force magnétique exercée sur la bille pour simuler un accroissement de la pesanteur ? Justifier.

2.3. Comment peut-on simuler un affaiblissement de l'intensité de la pesanteur ?

2.4. Si le dispositif a été correctement installé pour simuler un accroissement de la pesanteur, comment cela se traduit-il sur l'évolution de la période du pendule ? Justifier.

2.5. Le système utilisé ne permet pas de simuler une forte variation de la pesanteur mais il permet cependant de constater une variation de la période, à condition de choisir un protocole optimisant la précision de la mesure.

2.5.1. Proposer une méthode expérimentale pour obtenir une mesure la plus précise possible de la période.

2.5.2. Dans le cas d'un pendule de longueur 0,50 m, on mesure une période de 1,5 s lorsque les bobines sont parcourues par un courant électrique.

2.5.2.1. Le dispositif simule-t-il un accroissement ou une diminution de la pesanteur ? Expliquer.

2.5.2.2. Déterminer la valeur de l'intensité de la pesanteur apparente.

Translation:

In order to verify the influence of gravity on the period of a simple weight pendulum, it is difficult to move onto another planet. However, it is easy to put a pendulum, made of thread and a steel ball, in a device creating a uniform magnetic field on a large enough zone where the pendulum oscillates. This device may be the Helmholtz double coil.

When the axis of the coils is vertical, the current in the coils produces a uniform magnetic field in the cylindrical volume between the two coils. **A steel ball in this zone is subjected to a vertical magnetic force.**

2.1. Explain why this experimental device permits to simulate a variation of gravity.

2.2. How must be oriented the magnetic force on the ball to simulate an increase of the gravity? Justify.

2.3. How to simulate a decrease of gravity?

2.4. If the device has been installed for simulating the increase of gravity, how it modifies the period of the pendulum? Justify.

2.5. This device cannot simulate a strong variation of gravity, but it allows to state a variation of the period, provided a precise protocol.

2.5.1. Please propose an experimental method to obtain the most accurate measurement of the period.

2.5.2. With a pendulum of length 0.50 m, one measures 1.5 s as the period when an electric current runs in the coils.

2.5.2.1. Does the device simulate an increase or a decrease of the gravity? Explain.

2.5.2.2. Calculate the intensity of the apparent gravity.

End of quotation.

A colleague in Belfort asked: "*Just for curiosity, who can explain how the device of Helmholtz coils in the second part of the exercise n° 2 can create a vertical force on the ball?*"

Thanks for his alerting: this problem statement in Pondichéry is as false as possible. It is one of the consequences of the battle at the beginning of the 20th century, won by the Heaviside followers, against the warnings by James Clerk Maxwell in 1873 and Pierre Curie en 1894, representing by a vector - here vertical - the magnetic field, which is a being of rotation with nothing vectorial, is a gyror, here horizontal. Obviously, the teacher who wrote the problem statement never tried the experiment, nor could do the calculation he asks the pupils, and guesses with his/her wet finger. He demands a believing to the pupils. The inspector who validated the problem statement also did not understand the matter.

The "*vertical magnetic force*" that the writer has postulated, for a ferromagnetic test body, only exists where there is a gradient of magnetic field. Meanwhile, the main experimental interest of the Helmholtz's double coil is a constant field in the middle, so no gradient. Thus no "*magnetic force*" for a ferromagnetic test body. Moreover, if he/she had tried the experiment, the damping by Foucault's currents would be obvious, if the ball is conducting and of a non-negligible radius.

So they are, the feudalities which reign on the teaching of sciences.

15.2. The disaster in the teaching of Quantum Mechanics

Open-Eyes:
- Let us begin with a reference: ***Insights into teaching quantum mechanics in secondary and lower undergraduate education***, (21 pages) issued in PHYSICAL REVIEW PHYSICS EDUCATION RESEARCH 13, 010109 (2017), by K. Krijtenburg-Lewerissa, H. J. Pol, A. Brinkman, and W. R. van Joolingen, all four Dutch persons.
Link:
http://journals.aps.org/prper/pdf/10.1103/PhysRevPhysEducRes.
13.010109
They acknowledge that to teach according to the hegemonic tradition, gives painful results. « *the introduction of probability, uncertainty, and superposition, which are essential for understanding quantum mechanics, is highly nontrivial* ».
They point as guilty that all that is "*counter-intuitive*", and point the "*classical world*" as the big *in the wrong*. They are not able to detect that if it is so difficult to implant into students not yet selected on their obedience and tolerance to absurdity, maybe the reason could be that the doctrine is silly and ill-conceived.

We recall the ten postulates practiced in transactional microphysics:
1. **The absorbers and their properties exist. No "*corpuscular aspects*" exist**.

2. **Planck and phase postulate.** The unit of phase or angle comes back in the dimensional monomial.

3. **De Broglie-Dirac-Schrödinger postulate**. If a particle has a mass, the intrinsic frequencies mc^2/h and $2mc^2/h$ play each one their role. The Broglian mc^2/h for each interference of a quantic particle with itself. The Dirac-Schrödinger $2mc^2/h$ for all electromagnetic interactions, for instance, the Compton scattering.

4. **Fermat-Fresnel Postulate**. For each **individual wave**, all the real journeys come in phase to the absorber, eventually at an integer number of periods (it is then an interference). Hence the geometry of the Fermat's spindle between emitter and absorber - several spindles in case of interference on the travel.

5. **Every photon has an absorber**. A photon is a successful transaction between three partners: an emitter, an absorber, and space or optical devices between them. This transaction transfers by electromagnetic means, a quantum of looping **h**, and an energy-momentum whose value depends on the respective frames of the emitter and the absorber.

6. The properties of a **crowd** of individual waves flow from the properties of **the individual waves**, and not the inverse.

7. **Macro-time $\neq$ micro-times**. The god of Isaac Newton, in charge of all seeing simultaneously, does not exist. The time of Isaac Newton, a supposed universal and ubiquitous parameter, does not exist either. **We distinguish the macro-time** of macro-systems such as the laboratory, **from the micro-times** where dwell all the gropings of Broglian waves from which emerge the successful transactions. The macro-time is a local and statistical emergence, and it flows the same way as the entropy, a statistical emergence too. It has no causal properties in microphysics.

8. **Kirchhoff's Principle of retrosymmetry**.
In 1859, Gustav Kirchhoff proved that the Fraunhofer dark spectral lines from a cold gas or vapor correspond to bright spectral lines of the same elements in a hotter gas. So the spectral emission of a photon is exactly the same physical phenomenon as the absorption. Generalization: the retrosymmetry applies to the low energies of all the atomic physics, the molecular spectroscopy, and all the solid state physics.

9. **No, it is impossible to isolate a quantic system** as we isolate its equations at the blackboard: No mean exists to shield the Dirac-de-Broglie noise. It is impossible to predict which transaction will emerge from this lapping, nor when. The implied de Broglie frequencies are inaccessible from our human scale, and the theorem of the requisite variety, from William Ross Ashby, is here to ruin all our fantasies of panoptical omniscience. Moreover, the innumerable involved micro-times are bi-directional: orthochronous and retrochronous.

10. Then the moral principle: we refrain from censuring the experimental results that embarrass the doctrine in power.

Hiding so many experimental facts to the students is wrong, and it violates the scientific deontology.
Sure, so many experimental results embarrass the copenhaguists:
All the spectral absorptions, all the interferences as the anti-reflective coatings, quarter-wave plates, interferential colors, the Goos-Hänchen effect in plane polarization, Imbert-Fedorov in circular polarization, all proofs of the non-negligible width of each photon. A very long list!
They hid from you the transparency effect Ramsauer-Townsend, which is strictly undulatory. But if the electron is strictly undulatory, how will they continue to impress the gullible public, by their mystic "*wave-corpuscle dualism*"?
Many other everyday experimental results are incompatible with the corpuscular ideation of the Göttingen-Københavnists.

The economy of postulates and concepts is on our side. No more need to erect the properties of the Fourier transform as a new postulate: they are merely inherited. The magical concepts of *"superposition of (corpuscular) states"*, *"intrication* (of supposed theoretical and corpuscular states), *measurement, psychism and consciousness of the observer"*, all that is dropped: **Your Majesty, I did not need that hypothesis!**

Curious:
- *"But why the emperor is naked on the main street?"*

Professor Castle-Holder:
- And what about the « *kosher switch* » imagined by Elitzur and Vaidman, of which Roger Penrose was so fond? Page 279 of
https://altexploit.files.wordpress.com/2017/07/roger-penrose-shadows-of-the-mind_-a-search-for-the-missing-science-of-consciousness-oxford-university-press-1994.pdf

Open-Eyes:
- For those who make fun of laughing, two links:
http://www.jforum.fr/linterrupteur-cacher-qui-fait-peter-les-plombs-des-juifs-u-s-video.html
http://www.chiourim.com/polemique-autour-de-l-interrupteur-de-shabbat9261-html
It remains a swindle, perpetrated by presenting with earnest faces an ill-conceived trick.

Professor Marmot:
- Now we have enough to hang this grouchy *scholardventurer*: guilty of anti-semitism!

Open-Eyes:
- Like for the other gods, a whole clergy earned his prestige and power by ill-presenting things, in front of a flabbergasted public, perplexed by so much theological obscurity. With the transactional microphysics, the daylight dispels the darkness; and no more a cat can receive a *"wave function"*, even in dreams.

Curious:
- You had pretended that while your semantics wholly differs from what is taught everywhere, you do not differ in mathematics. I wish to see the two versions, side by side, to judge with facts.

Open-Eyes:
- Suggestion: to compare the pages on the ammonia maser, or the hydrogen maser for the cosmic line at 22 cm. Alas, already plenty of presuppositions have been introduced as soon as the page 3, being as many tricks of a hypnotist using absurdities to sedate the vigilance; among them a fallacious geometry about the angular moments and the spins. These hidden flaws have resemblances to the hypnotist tricks practiced as a professional by Joël Sternheimer to swindle the french Académie des Sciences, via the naivety of André Lichnerowicz :

http://jacques.lavau.deonto-ethique.eu/Theorie_fondee_sur_l_hypnose.html

I give up. Maybe in a later edition?

15.3. Structural reasons for the non-quality

Curious:
- Bad management of quality? Are they some structural reasons?

Open-Eyes:
- Already written in April 2013 in the forum of the *Union des Professeurs de Physique et Chimie* (UdPPC); a now closed and destroyed forum. These structural defects are communitarianist and feudal.

In my lifetime, I have seen significant improvements in the teaching of Chemistry. In the IUPAC, International Union of Pure and Applied Chemistry, the customers are powerful and organized, so that they can demand quality.

In geology also, especially the geology of the sedimentary basins, the customers are sized enough to be heard. A micro-paleontologist has more career opportunities in the oil industry than in the University. My *Manuel de sédimentologie* comes from the Editions Technip, settled in 1956 by the *Institut Français du Pétrole*, directly linked to the custom industry, and it is excellent.

The industrial concurrence in chemistry may have perverse effects, for instance in high polymers; it is of fair game to let the academics and the competitors wander in macromolecular models lacking any realism, with unrolled spaghettis; and hiding from them that we reinterpret them with a micellar model, far more realist and effective.

Nothing such in the IUPAP (International Union of Pure and Applied Physics), nor in mathematics: the customers are small and atomized, they remain a minor power in front of the mighty producers. The quality control is powerless when the monopolist supplier imposes his monopolistic whims. Be a century late when teaching maths, and nobody will notice. I have proved that the teaching of the QM, held for 1927 by a Göttingen-København tribe, who conquered the power by sheer violence, is ninety-one years late in front of the paper by Dirac, **The quantum theory of the electron**, February 1928; there they behave like a hen who has found a fork. Soon they will be eighty-nine years late on the paper by Schrödinger: ***Über die kräftefreie Bewegung in der relativistischen Quantenmechanik***, July 1930; and eighty-five years late on the Nobel lecture by Dirac, 1833. The reactions from the tribe are only territorial: their monopoly over all, and shooting at the finder who is to be destroyed before any public hears about him!

Professor Castle-Holder:
- Please elaborate a more general and interprofessional statement or tool.

Open-Eyes:

- A requisite is that the use of this tool is not reserved to few specialists, but can establish an interprofessional Esperanto. The best draft I know today is the *Analyse Modulaire des Systèmes*, by Jacques Mélèse (no translation into english). J. Mélèse transposed from industrial cybernetics to management of administrations and firms. It proved to be effective, mainly because it begins with a survey of the needs and requirements of the different services: *Here is what I need, to do my part of the work.* I pass the speech to an unsatisfied objector:

Objector Leypanou:
- *"Encourage your users or pupils to reformulate what they think to have understood"*: but the vast majority of the maths as they are taught fly at light-years from any practical purpose. Worst: many are an only intellectual gym, and will never have applications. So speaking of specification notices has no pertinence.

Open-Eyes:
- Such a detachment and disdain from the needs and from the technologies is unacceptable; it is a denying of public service.
Up to now, I still use the theorem of the inscribed angles in a circle, while the Poncelet's arrays of circles only served me in CAPES written trials, for an electrostatic problem in dimension 2. Geometrical inversion also was never used in professional life. Apollonius's theorem of the median in a triangle still serves in all optical interferences. Conic sections are of constant use. The Bragg's relation, founding all crystallography is based on basic trigonometry we had learned when fourteen; while its application to crystallography requires Euclidian tensorial geometry, which is still not taught, though the use of the metric tensor in mechanics and electromagnetics is a critical need. Not for the birds, I have recalled the superlative blunder in the scientific Bachelor degree in Pondichéry en 2012: only the relativistic transformation of the electrostatic field gives the Laplace, Biot and Savart magnetic forces. But these future teachers never study it.
Feudalities, and secular routines... Alas, the communities are communitarianist - even those who pretend to be "scientific" and who mock each other. Again a deny of public service.

CHAPTER 16

Conclusion, by the Curious

Curious:
- It is up to me to conclude: I am the final customer; we will see whether I have well caught what is to know.

This "*Open-Eyes*" pioneer has thrown to the garbage bin the whole of the waves-corpuscles duality, has thrown away the corpuscles and the corpuscular aspects, conserving only the waves.

Moreover, he innovates by distinguishing the individual waves, such as each electron, from the collectives and the beams of waves such as an electron beam, or many other collectives of waves.

Instead of the corpuscular theory, he put in evidence the absorbers of the waves and gave them the same causal importance than to the emitters. It is revolutionary, as the transactional microphysics implies a retrochronous causality, just as strong as the ordinary orthochronous time, but only at the individual microphysical scale, in micro-times. No extrapolation may be drawn at the macrophysical scale, nor any parapsychology. Never any supraluminal signal will ever come to our service.

Up to now, nothing very new, compared to what John Cramer did in 1986. But two unexpected innovations: he rehabilitated the geometrical optics, though practiced in the 17th - 18th centuries, of which he says it is irreplaceable regarding the economy of computer time; he stated that an astigmatic eye does not receive less light, so a photon still converges onto the cis-retinal in an opsin.

However, Mr. "*Open-Eyes*" took quite a new benefit from the theorem of the harmony of phases, discovered in 1923 by Louis-Victor de Broglie: he deducted that the electron wave is implicitly expanded in space, and at least in the simplified mathematization of 1924, everywhere in phase with itself. Therefore, the macroscopical relativity is invalid inside a particle such as an electron. Hence, he deducts that everywhere, a quantum particle is immersed in the lapping of Broglian waves from all the other particles, and that this Broglian ground noise is universal, as much impossible to shield from, as the gravity is. According to Mr. "*Open-Eyes*", this Broglian ground noise is enough to chart all the random and statistical aspects of the Quantum Mechanics, which had irritated so many physicists in the 1930 years, among them the most famous was Albert Einstein. According to Mr. "*Open-Eyes*", it is impossible to delimit *ad libitum* a quantic system, and to isolate it perfectly. The remaining of the world still self-invites on the bench of the experimenter.

Above, I have temporarily over-simplified by staying on the first theory from Louis de Broglie and the initial theory from Erwin Schrödinger; we have to take account of the radical update performed by Paul Adrien Maurice Dirac, in 1928. Dirac demonstrated that the electron is a wave with four components; two of them are with negative energy, negative frequency, and negative flow of time (micro-time). Hence results an electromagnetic oscillation at frequency double of the Broglian one, said **Zitterbewegung** or *Schrödinger trembling motion*, which plays the major role in all the interactions of the electron with something else, say in the Compton scattering of a photon by a nearly free electron, and also in the materialization of a gamma into a pair electron-positron, and in the annihilation of a positron when it meets an electron. So we must correct this *"everywhere in phase with itself"* by taking account of the advanced waves and retarded waves in the micro-times. The Dirac-de-Broglie ground noise also takes a much more complicated form, simultaneously with advanced waves and retarded waves. It escapes to any direct experiment, though some theoretical formulation still may be possible, may be done.

Moreover, Mr. *"Open-Eyes"* roughly sketched the geometry of a *Fermat spindle*, its width and the angle of the tangent cone, by using the 17th century Fermat's Principle, and by using his training with the laser beams; he demonstrated that this geometry has important applications as well in astronomy as on the bench, even in instrumental optics.

However, Mr. *"Open-Eyes"* left opened a conceptual problem: though we know, or presume to know that a helium atom, or heavier, a molecule have complex **spatial** structures, however, all these were proved apt to propagate following several distinct paths, just as if they were reduced to their Broglian waves, or to a more complex generalization for the composite particles such as a proton, atoms, or molecules. As such, it is not a difficulty specific to the transactional theory – it has not invented these experimental facts, already established – but a hard challenge for all. There is an astonishing inversion: the most a molecule is big and heavy, the smallest is its Broglian wavelength, the highest is its intrinsic frequency. On the contrary, the neutrinos which cross through us could have long wavelengths.

Maybe the professor Castle-Holder could draw his own conclusion? Do you think that the simplifications they brought in semantics are interesting?

Professor Castle-Holder:
- Above all, they made a non-local theory, and they are extremists on this point. Extreme and even shocking. Just what Albert Einstein hated, and he (with Podolsky and Rosen) challenged in 1935 with their EPR paper.
Another reproach: in contrast to the innovators-publicists to whom Science

News compose roaring and hooking titles with no real innovation inside, this transactional-man only once had the diplomatic politeness to mention the magical couple of words *"wave function"*; and worse: it was to tear the concept and split it into impedance function and concurrency function. In my opinion, it is an audacity that nobody in the trade will pardon him. While John Cramer had the diplomatic politeness to write *"wave function collapse"*, so he faked to remain in the tribe... This *scholardventurer* does not respect anything...

Open-Eyes:
- You know, such sacrileges, I have committed more! For example, I never sacrificed to the rhetoric *"classical/quantic/classical/quantic..."*, which tricks to make you induce *"We are the modern forever, and the Others are just retired cavalry colonels, dimwitted"*... Moreover, as a historian, I have opened some of their closets, and extracted skeletons hidden there, somehow in the manner of Mrs. Blue-Beard.

Professor Castle-Holder:
- One reproach more: obviously, their theory is still unfinished. The historic transactionist John G. Cramer only considered two-partners transactions, only the absorber, and the emitter; it was enough to chart *delayed-choice* experiment, in the style of Marlan Scully, but failed on the black body problem, which Max Planck scrutinized and solved in 1900. It also forbade him to think the *"weak measurements"*, on which many physicists are busy today. Now, this more recent transactionist brings more robust theoretical elements: he takes the intermediate medium in the transaction, so he can treat the case of the interferential astronomy on a long base, or of the laser medium. But it is an unfinished theory.

Open-Eyes:
- One may wonder why the transactional microphysics was not already elaborated about 1932: they already had enough facts at hand. Some intellectual infirmities made the disaster still in power. Not any of the Solvay Congress participants had professional training in heuristics, the art of finding, of going through a file until finding the faults and listing the clandestine (wrong) premises. Only Louis de Broglie had a professional practice in radioelectric transmissions (as a soldier), but Niels Bohr could stifle him to never learn anything from him. Erwin Schrödinger also was methodically demoralized by Bohr; an eloquent proof is to compare the Nobel lectures (1933) of Dirac and Schrödinger: only Dirac mentions the last results from Schrödinger, the application of the Zitterbewegung to the Compton scattering. The shameful secret, the skeletons in the closet, is that in 1927, there were winners and defeated, and that the two minority competitors, defeated, were eliminated from the scientific stage; the brawl was territorial, it was won by the most territorial and combative animals from the most combative pack. They made you believe that the dispute was scientific, and the griots glossed it so, flattering the winners who rewrote the history to their profit.

So yes, the transactional theory may be the less local of all, it was inevitable.

Curious:
- In the beginning, you said: *"If you dared to have your standard output supervised by a qualitician and by a didactician, your pride would be unpleasantly surprised"*. Please develop now!

Open-Eyes:
- Hmm... Dangerous for the thematic unicity of the book! In a sense, the reproach I tend to formulate is an anachronism: winners or defeated, the protagonists of the Solvay Congress had not in that time any methodologic tools, we now have around us. Now we just have to pluck, learn and transpose from other crafts where these tools were elaborated. In 1927, nobody had them.

Quality is free, provided it is integrated at the beginning of the design. Concerning the teaching of sciences, I see some elements of quality, which are alas neglected:
1. Always have a reality-test which waits for you at the next turn, and that you cannot escape. Without severe experimental sanctions, you do not more teach science, but only how to become a *"science"* teacher in next turn, from generation to generation, from off-ground *scholaste* to off-ground *scholaste*.
2. Stop fantasizing you are *the modern* forever. Think that someones will have to correct your most triumphant blunders, and please have the courtesy to facilitate them the task. Please do not more code with the soldering iron. Please do not more use strong brazing to weld together your calculation with the deliria you inherited from the predecessors. Make corrigible, instead of playing the prophet or the evangelist. A known tune: *"And after me, there will be no more other prophets!"*.
3. Please modularize, code and rank the heritages from class to class. Plan the interfacings so a module may be corrected without the obligation to correct also all the remaining.
4. Please prepare the crossed, interdisciplinary fecundations, by querying by neighbor crafts their own reality-testings[1]. Reality-testings, we will never have too many. Against the *negative gravity* of the *phlogistic*, which seemed a no-problem to his colleagues in chemistry, Lavoisier opposed the reality check from the astronomers: no negative gravity exists.
5. Do not hesitate to ask neighbors crafts, providers or customers, what do they need, and what use do they do. Visit your customers in the workshop.
6. Between trades and crafts, negotiate by relevance trees[2]. This tool is not reserved for the high-technology weapons industry, even though it was born there. It is worth be broadly diffused, popularized and used.

---

[1]On the door of the automate piloting a cement oven, the engineer had written the Laplace transform of the oven, representing its response to a pulse, with the real figures.

[2]Relevance trees. Gathered in Erich Jantsch: **Technological forecasting in perspective**. OCDE 1967. Pp 218-233

7. Encourage your users or pupils to reformulate what they think to have understood, and verify with them whether their reformulations are better or worse when applied.

8. Be careful about the costs of learning; take care to optimize the didactic run, carefully punctuated with experimental verifications. There is still a lot to improve.

9. Check what they understood, in point of fact.

10. Stop dreaming without proofs nor checkings, that what was transmitted to you was **therefore** *already well understood.* For instance, Anatole Abragam boasted[3] of the reign of the rumor in the teaching of sciences. Now a rumor is anything but scientific, even when it propagates in a scholar population.

So, if I am heard, it will be substantial progress. We do not yet know how to pilot the scientific teaching scientifically: the feedbacks on the terrain efficiency are terribly slow to come back; therefore the course without early retro-action often becomes hazardous, even disastrous, driven by political chumminess or other pathologies.

Professor Castle-Holder:

- It is public that you borrowed the expression "*to invent the inventors*" to Paul Delouvrier. It goes beyond the frame of the science teaching. What would be the right frame of thinking, indeed?

Open-Eyes:

- Excepted at the beginning of Chapter 12, where intervened an authentic doctor-engineer, elsewhere the real persons who compose the composite "*Professor Marmot*" are content to hurl anathemas and malevolence, deny of other people, for the benefit of their narcissistic wounds, ill-managed from the childhood. They represent a very real social problem, and it would be safe that they would be charted, not only by their colleagues, but also by the broad public. The laymen are not scientifically equipped to decide on our debates and arguing, but they can become competent in the moral matter. It is not innate, but it may be learned.

As she intervened in the CNAM (Conservatoire National des Arts et Métiers), in the course of Technological Forecasting and Managing the R&D, Florence Vidal concluded by « *Please respect the minority. You do not know where from will come the new ideas which will save your firm. Nothing guarantees that it will come from those you presently consider as the most beautiful and the nicest.* »

A contrario, if your goal is to "aid" a people to fall into disrepair and to suicide, inspire them the cult of harassing the minority and the scapegoats, the cult of civil war against the learned, the cult of the dictatorship of the emotion, and

---

[3]Anatole Abragam. **De la physique avant toute chose**. Editions Odile Jacob, 1987. Pages 67 – 68: « *A partir du moment où les résultats sont suffisamment établis et suffisamment bien compris pour ne plus soulever de contestations dans la communauté des savants, on écrit des livres pour exposer leurs résultats et plus personne à part les philosophes et les historiens ne lit les mémoires originaux. Certains le regretteront mais c'est ainsi, et selon moi* **c'est très bien ainsi**. »

therefore, the cult of the civil war against the analytical mind. This trick is working very hard, all around us.

Curious:
- Please sum up!

Open-Eyes:
- The presence or absence of ethics of knowledge, of deontology or the efficient interprofessional relationships, of the respect of the social contract, you should become trained to verify these presences and their efficiency.

Professor Castle-Holder:
- Social contract? Here nobody has the slightest idea of what it could be!

Open-Eyes:
- Sure! In our academic feudalities, the implicit internal direction is:
Article 1. The boss is right.
Article 2. The boss is always right.
Article 3. In any other cases, apply the article 1.

So Dan Shechtman, now Nobel laureate in chemistry (Nobel in 2011) for the discovery of the quasi-crystals, was first fired from his laboratory.

In other words, it may be said by the fable of the charming and appetizing little rabbit.
This is the story of a charming little rabbit. A day he was furiously busy in the keyboard of his wifi-connected laptop in grassland of cress, a wolf came out of the bushes and thought he would happily eat a so appetizing rabbit for his five-o-clock lunch. However, he is puzzled by the doggedness of the rabbit on his keyboard, he asks:
- *But what in the heaven are you doing, charming and appetizing little rabbit?*
- *Well, I do researches for my thesis.* Answered the charming and appetizing without even stumbling an ear.
- *A thesis!? Ha ha ha!* Mocked the wolf. *And what I your subject?*
- *The Superiority of the Rabbits over the Wolves.* Answered the furry creature. Laughing till it hurt, and holding his hairy belly with his four legs, the wolf did not believe a word. But the rabbit insists *yes yes*, and proposes the wolf to see a complete demonstration in his small burrow, and luck, it is only at two leaps from here. The wolf thinks he is not so hungry, and can eat this rabbit when he wants, and accepts.
Nevermore the wolf was seen again.

A month later, a leopard notices the same rabbit, still in the middle of his cress glade, extremely busy to do calculations on his glowing Cray X11.
- *What are you doing, charming and appetizing rabbit?*
- *You see: calculations for my thesis*, answers the charming and appetizing

without raising his head.

- *A thesis!? Ha ha ha!* Mocked the leopard. *And what I your subject?*
- *The Superiority of the Rabbits over the Leopards.* Answered the furry creature.
The leopard rolls and suffocates in laughing, and even pees on his beautiful fur.
So laughing, he accepts to follow the rabbit in his small burrow for a demonstration from A to Z, and luck, it is only at two leaps from here.
So the leopard went to the rabbit's hole... and never came out.

A month later, a fox crosses the rabbit, and again the same scenario. The rabbit is busy chatting on IRC, explains that his thesis is on *the Superiority of the Rabbits over the Foxes*. Nearly suffocating from laugh, the fox falls on the cress, follows the rabbit for the demonstration. As well, the burrow is still at two leaps from here.
Two leaps later, in the bottom of the burrow, the fox has the surprise to discover a heap of bones of wolf, a heap of bones of leopard, a heap of other bones, and in the middle of the room, a lion. A superb, well-fed and terrific lion.
- *I introduce you my thesis advisor*, says the rabbit.
And the lion jumps on the fox, and eats him without fuss.

The moral of the story: The subject of your thesis does not matter. All that matters is the power of your thesis advisor.

Social contract:
As taxpayers, the broad public pays our wages and laboratories. They are entitled to expect correct and verified information, and valid and fruitful methods of reasoning, whose limits of validity are explicit. If we teach to the students that the magnetic field $\check{B}$, the angular speed $\check{\omega}$, or an angular moment $\check{J}$, are of *vectorial nature*, though their symmetries and their behavior are the opposite, then we breach the social contract linking us to the taxpayers and our students. If a hundred and twelve years later, the scientific community has not yet rectified the blunder hazarded by Albert Einstein in 1905 "*Light voyages in small grains*", they did not play their roles; there is still there a collective pathologic behavior, which falls within of the field of investigation of the researcher in social sciences.

I perceive as very abnormal the fact that one had to wait until the years eighty for F. Bohren, H. Paul, R. Fischer, and John G. Cramer, years ninety for myself, to have the first bases of the transactional quantic established. Here the social contract is severely flouted. For a researcher in social sciences, it is a valid subject of thesis.

A surprise: the vital epistemologic information came from a biologist, the Nobel laureate Jacques Monod, in his book « **Le hasard et la nécessité** » (*Chance and Necessity: Essay on the Natural Philosophy of Modern Biology*) where he recalls the long struggle, an independence war, the biologists had to sustain against the animist burden the dominating churches imposed; their struggle for setting an impersonal and objective science. Bohr and Wigner pegged their egocentric and anthropocentric animism « *Me, myself and I* » right in the middle of what would be the quantum physics. They had the excuse that the monster butcheries of the world war were still fresh in the minds, and still muddled their minds. In 2018, have we still this excuse?

Scientific maturity does not fall all roasted in your beak: it requires many disillusions, and to have managed the sorrow due to drastic revisions of the blunders to which one had sometimes clung.

Curious:

- Maybe it is the calcium: when you lack calcium, you bite the dogma.

APPENDIX A

A: Occupation of the electron layers and sub-layers with the atomic number

Element and atomic number		Principal **n**, et secondary **l** atomic numbers.																		
	n	1	2		3			4				5					6			7
	l	0	0	1	0	1	2	0	1	2	3	0	1	2	3	4	0	1	2	0
1 H		1																		
2 He		2																		
3 Li		2	1																	
4 Be		2	2																	
5 B		2	2	1																
6 C		2	2	2																
7 N		2	2	3																
8 O		2	2	4																
9 Fe		2	2	5																
10 Ne		2	2	6																
11 Na		2	2	6	1															
12 Mg		2	2	6	2															
13 Al		2	2	6	2	1														
14 Si		2	2	6	2	2														
15 P		2	2	6	2	3														
16 S		2	2	6	2	4														
17 Cl		2	2	6	2	5														
18 A		2	2	6	2	6														
19 K		2	2	6	2	6		1												
20 Ca		2	2	6	2	6		2												
21 Sc		2	2	6	2	6	1	2												
22 Ti		2	2	6	2	6	2	2												
23 V		2	2	6	2	6	3	2												
24 Cr		2	2	6	2	6	5	1												
25 Mn		2	2	6	2	6	5	2												
26 Fe		2	2	6	2	6	6	2												
27 Co		2	2	6	2	6	7	2												
28 Ni		2	2	6	2	6	8	2												

29	Cu	2	2	6	2	6	10	1										
30	Zn	2	2	6	2	6	10	2										
31	Ga	2	2	6	2	6	10	2	1									
32	Ge	2	2	6	2	6	10	2	2									
33	As	2	2	6	2	6	10	2	3									
34	Se	2	2	6	2	6	10	2	4									
35	Br	2	2	6	2	6	10	2	5									
36	Kr	2	2	6	2	6	10	2	6									
37	Rb	2	2	6	2	6	10	2	6		1							
38	Sr	2	2	6	2	6	10	2	6		2							
39	Y	2	2	6	2	6	10	2	6	1	2							
40	Zr	2	2	6	2	6	10	2	6	2	2							
41	Nb	2	2	6	2	6	10	2	6	4	1							
42	Mo	2	2	6	2	6	10	2	6	5	1							
43	Te	2	2	6	2	6	10	2	6	6	1							
44	Ru	2	2	6	2	6	10	2	6	7	1							
45	Rh	2	2	6	2	6	10	2	6	8	1							
46	Pd	2	2	6	2	6	10	2	6	10	—							
47	Ag	2	2	6	2	6	10	2	6	10	1							
48	Cd	2	2	6	2	6	10	2	6	10	2							
49	In	2	2	6	2	6	10	2	6	10	2	1						
50	Sn	2	2	6	2	6	10	2	6	10	2	2						
51	Sb	2	2	6	2	6	10	2	6	10	2	3						
52	Te	2	2	6	2	6	10	2	6	10	2	4						
53	I	2	2	6	2	6	10	2	6	10	2	5						
54	Xe	2	2	6	2	6	10	2	6	10	2	6						
55	Cs	2	2	6	2	6	10	2	6	10	2	6		1				
56	Ba	2	2	6	2	6	10	2	6	10	2	6		2				
57	La	2	2	6	2	6	10	2	6	10	2	6	1	2				
58	Ce	2	2	6	2	6	10	2	6	10	2	2	6	2				
59	Pr	2	2	6	2	6	10	2	6	10	3	2	6	2				
60	Pr	2	2	6	2	6	10	2	6	10	4	2	6	2				
61	Pm	2	2	6	2	6	10	2	6	10	5	2	6	2				
62	Sm	2	2	6	2	6	10	2	6	10	6	2	6	2				
63	Eu	2	2	6	2	6	10	2	6	10	7	2	6	2				

64	Gd	2	2	6	2	6	10	2	6	10	7	2	6	1			2			
65	Tb	2	2	6	2	6	10	2	6	10	9	2	6				2			
66	Dy	2	2	6	2	6	10	2	6	10	10	2	6				2			
67	Ho	2	2	6	2	6	10	2	6	10	11	2	6				2			
68	Er	2	2	6	2	6	10	2	6	10	12	2	6				2			
69	Tm	2	2	6	2	6	10	2	6	10	13	2	6				2			
70	Yb	2	2	6	2	6	10	2	6	10	14	2	6				2			
71	Lu	2	2	6	2	6	10	2	6	10	14	2	6	1			2			
72	Hf	2	2	6	2	6	10	2	6	10	14	2	6	2			2			
73	Ta	2	2	6	2	6	10	2	6	10	14	2	6	3			2			
74	W	2	2	6	2	6	10	2	6	10	14	2	6	4			2			
75	Re	2	2	6	2	6	10	2	6	10	14	2	6	5			2			
76	Os	2	2	6	2	6	10	2	6	10	14	2	6	6			2			
77	Ir	2	2	6	2	6	10	2	6	10	14	2	6	7			2			
78	Pt	2	2	6	2	6	10	2	6	10	14	2	6	8			2			
79	Au	2	2	6	2	6	10	2	6	10	14	2	6	10			1			
80	Hg	2	2	6	2	6	10	2	6	10	14	2	6	10			2			
81	Tl	2	2	6	2	6	10	2	6	10	14	2	6	10			2	1		
82	Pb	2	2	6	2	6	10	2	6	10	14	2	6	10			2	2		
83	Bi	2	2	6	2	6	10	2	6	10	14	2	6	10			2	3		
	n	1	2		3			4				5					6			7
	l	0	0	1	0	1	2	0	1	2	3	0	1	2	3	4	0	1	2	0

Transactional (Quantum) Microphysics, Principles and Applications

Three fields: gravity, electric, magnetic

B.1. Defining a new expression: test body

It is the body you have to put in a field, for observing the force it will result on it, which reveals what is the field, its characteristics, its magnitude, its orientation. For the gravity field, any body is a test body: an apple, a ball, a stone; they fall...

B.2. Reusing our knowledge on the gravity

B.2.1. What causes the gravity? Any body with a mass produces a field of gravity around it: it attracts the other masses. The only easy to perceive and measure fields are those produced by big masses: heavy celestial bodies such as Earth, Moon, other planets, the Sun, or other stars.
Roughly said: *"Gravity field"* means *"There are big masses on this side, not far"*.

B.2.2. Direction and sense. Gravity is always attractive. There are no anti-gravity, no anti-masses, no shields.

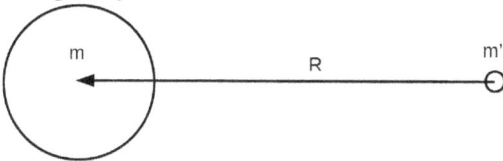

Figure 18.1.
If we define a radius $\vec{R}$ as starting at the test body (where the field is active), with its end at the center of the mass which produces the field, the force on the body test is always in the sense of $\vec{R}$.

B.2.3. Mathematical law. We define R as the length of the vector $\vec{R}$. The inverse of $\vec{R}$ is defined so: $\left(\vec{R}\right)^{-1} = \frac{1}{\vec{R}}$ is the vector having the same direction and sense as $\vec{R}$, and whose norm is the inverse: R^{-1}.
In other words: $\vec{R} \cdot \left(\vec{R}\right)^{-1} = 1$.
Universal constant of gravity: $\eta = 66.726 \cdot 10^{-12}$ m3 /(kg. s^2).
Newton law: The body producing the main field has mass **m**. The test body has mass **m'**. The force induced on the test body is in the direction and sense of $\vec{R}$.

Hence Force of attraction on the test body: $\vec{F} = \frac{\eta.m.m'}{R.\vec{R}}$.

η is a very small constant, so the gravity force is by far the weakest force of all existing forces. Therefore it is the most difficult one to measure accurately, though it is the most formerly known.

Application:
Calculate the gravitational attraction between two masses of 50 kg, when one meter separates their centers of gravity.

Curious:
-

Open-Eyes:
- Note this result. Later, we will compare it to a similar case in electrostatics. We define the field as the quotient of the force by the mass of the test body. Hence the expression of the gravity field: $\vec{\gamma} = \frac{\eta.m.}{R.\vec{R}}$.
Physical dimension: $\vec{\gamma}$ is an acceleration, its unit is the m/s^2.
Application: calculate the gravity acceleration at the North Pole (at the poles, you do not need to take account of the Earth rotation). Mass of the Earth: 5.979 . 10^{21} kg. Radius at the North Pole: 6,356,912 m.

Curious:
-

Open-Eyes:
- In most applications, it is more economical to use the gravity potential:
$G = -\frac{\eta m}{R}$.
G is only defined to within a constant; here null at infinite distance. But the difference of potential between two points is defined unambiguously. The difference of potential energy between two positions of the test body **m'** is the product of **m'** by the difference of potential. Its unit: **m^2/s^2**, like the square of a speed.
Figure 18.2.

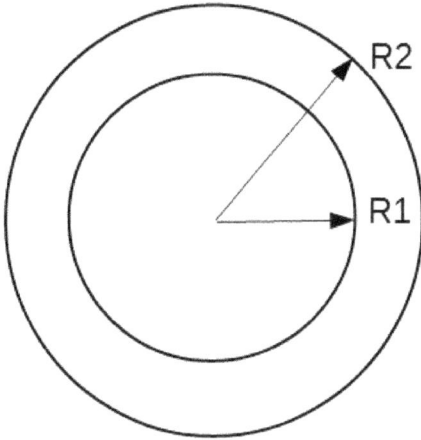

$E_1 - E_2 = m'\,(G_1 - G_2) = \eta.m.m'\left(\frac{1}{R_2} - \frac{1}{R_1}\right) = \eta.m.m'\frac{R_1 - R_2}{R_1.R_2}$.

Application: calculate the potential energy delivered to a 1 (metric) ton satellite, to hoist from the altitude (rated from the center of Earth) of 6,378,550 m (6,378,533 m + altitude of the launchstep of Kourou over sea level) to 42,145,530 m (geosynchronous altitude, still rated from the center of Earth), that is 35,767 km higher.

Earth mass: $5.979 \cdot 10^{21}$ kg.

62 546 kJ - 9 466 kJ = 53 080 kJ.

Curious:

- 11

Open-Eyes:

B.2.4. How to measure a gravity field? The main method is to measure the period of a weight pendulum: this period depends on the equivalent length of the pendulum, of the gravity, and nothing else. Period: $T = 2\pi\sqrt{\frac{l}{g}}$.

Precision?

One must measure several tens of periods. The pendulum must be very little damped, to last enough time. The angle to the vertical must be little, say less than 5°, for keeping the above formula precise enough. The length of the pendulum must remain invariant, though generally, the metals dilate when heated. And we depend on the precision of the clock.

One measures so only the apparent gravity: the gravity acceleration minus the effect of the Earth rotation.

<u>Numerical application</u>: A pendulum, equivalent to a simple pendulum of length 4 m, oscillates at a period of 4.01355 s. What is the apparent gravity in that

place? Please give the result up to five significant digits.

Curious:

-

Open-Eyes:

-

B.3. Electric field

B.3.1. Effects on electric charges. We use electric fields every day in our oscilloscopes, in any FET (Field Effect Transistor) and microchips with built-in FETs, in the electrolyzes, in the fluorescent tubes which light our rooms and halls, etc. When electrolyzing, the cathode attracts the cations Na+; the anode attracts the anions Cl-. The electric field has sorted the ions according to their charge. In our oscilloscopes, a pierced anode (the Wehnelt) accelerates the electrons e-, and lets them continue until a final anode, the fluorescent screen. The deflecting plates drive an electric field the deviates the electrons up or down, right or left, proportionally to the difference of potential between plates (resp. horizontal, vertical).

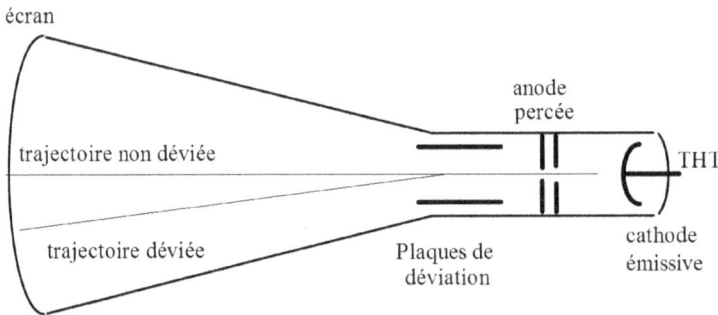

Figure 18.3.

The force acting on a charge **q** is the product of the charge by the electric field. **F = q E**. And this law is vectorial: The unit of electric field is the volt per meter. According to the law above, it could also be said the:

Curious:
- Newton per coulomb.

Open-Eyes:

- Notice: All the bodies fall the same way, toward the center of Earth in our neighborhood. While two kinds of test bodies exist in electrostatics: the charges + and the charges -. This law is easier to draw if the test body is a positive charge:

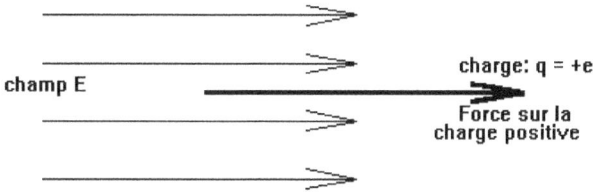

Figure 18.4.

+ e represents the charge of a proton.

Please achieve the legend of the drawing: the most positive potentials are at the left or the right?

So we draw the electric field as a field of arrows, flowing from the + charges to the - charges. Is it the same sense as the conventional sense of the electric current?

Your answer:

Open-Eyes:

- If the test body is a negative charge, here is the drawing:

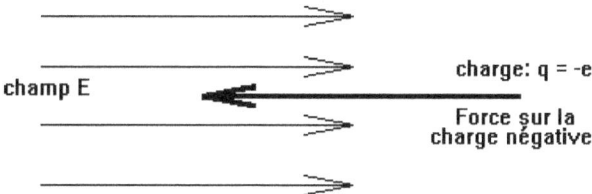

Figure 18.5.

- e represents the charge of an electron.

In other words, the negative electron flees from the other electrons.

B.3.2. How the electric charges produce a field. The field produced in the vacuum by a "punctual" charge:

$$\vec{E} = \frac{q}{4\pi\varepsilon_0 . R} \vec{R}$$

The electric permittivity of the vacuum e_0, equals $8.85412 \cdot 10^{-12}$ farads per meter.

When summed upon all the directions of space: $4\pi\varepsilon_0 = 111.265 \cdot 10^{-12}$ F / m.

Its inverse: $\frac{1}{4\pi\varepsilon_0} = 8.9876 \cdot 10^9$ m / F. Nearly nine milliards of meters per farad.

For media other than vacuum, one replaces the vacuum permittivity by the one of the medium ε. The quotient $\varepsilon_r = \varepsilon / \varepsilon_0$ is the relative permittivity. ε_r depends on the temperature, on the field, on the frequency...

Some relative permittivities, or dielectric « constants » (though not constants):

medium	ε_r	medium	ε_r
air at the normal pressure	1.0005	water	80.5
petrol, oils.	2.1 to 2.2	ethanol	26
polypropylene	2.2	muscovite mica	6.8 to 7.5
polytetrafluoroethylene	2.1	glass for capacitors	4
polyethylene terephthalate	3.2	alumina Al_2O_3	8.4
Cellulose paper	3.4 to 5.5	Tantalum oxide Ta_2O_5	26

Application:
Calculate the electrostatic attraction between two charges, respectively + and
− 100 nanocoulomb (624 milliards of electrons), separated by one meter in a
nonionizable gas.

Curious:
-

Open-Eyes:
- It is 539 times bigger than the gravitational attraction for 50 kg masses. Now,
these 624 milliards of electrons are not more than one in more or in less, in 1.3
millions of milliards (bound electrons). It shows how meticulous the precautions must be in measuring the universal gravity constant h.

Numerical applications: A capacitor has a thickness of dielectric (alumina) of
0.1 μm. It is charged under 70 V, and has a capacity (quotient $\frac{Q}{V}$) of 0.01 F
(farad). The alumina has a relative permittivity $\varepsilon_r = 8.4$. Compared to the
same problem in the vacuum, at equal charges and distance, the energy, the
difference of potential, the field and the force are all divided by ε_r .
What is the electric charge accumulated on the + electrode?
What is the electric charge accumulated on the − electrode?
What is the mean electric field in the gap?
The total force of attraction between the plates, that compresses the dielectric?

Curious:
-

Open-Eyes:

B.3.3. Electric potential. The most usually in the calculations, it is more economical to use the electric potential, in volts. For example, $U = \frac{q}{4\pi\varepsilon_0.R}$ at the distance **R** of a point charge **q**.

Note: If the charge would really be "*punctual*", the potential and the field would be infinite at that point. It never occurs.

U is only defined to within a constant depending on an arbitrary constant; here null volts at infinite distance. But the difference of potential between two points is defined unambiguously. The difference of potential energy between two positions of the test body **q'**, is always the product of **q'** by the difference of potential.

$$E_1 - E_2 = q' \, (U_1 - U_2) = \frac{q.q'}{4\pi\varepsilon_0}\left(\frac{1}{R_1} - \frac{1}{R_2}\right) = \frac{q.q'}{4\pi\varepsilon_0}.\frac{R_2-R_1}{R_1.R_2}.$$

In a region where the electric field $\overrightarrow{E}$ is constant, let orient the x-axis parallel to the vector $\overrightarrow{E}$. Note E its norm. Then the electric potential in this zone is
U(x) = - **E.x** + **U(0)**.
Why the minus sign?

Curious:
-

Open-Eyes:
Numerical application:
Let's again take the example of the small spheres, each 1 cm of radius, at a distance 1 m, one with 10^{-7} coulomb in excess, the other in lack. What is their difference in potential? It is enough, as the radii are small in front of the distance, to superimpose the potentials of each one of the charged spheres, as if alone. $U_1 - U_2 = \frac{2q'}{4\pi\varepsilon_0}\left(\frac{1}{0.01m} - \frac{1}{1m}\right) = \frac{2.10^{-7}C}{4\pi\varepsilon_0}.\frac{1-0.01}{0.01m} =$
Curious:
-

Open-Eyes: 178 kV.

B.3.4. How to measure the electric field?
B.3.4.1. *Measuring an unbalance of charges: the electrometer.* It is made of two very thin sheets of gold, hanging together. If the device is not electrically neutral, the sheets repel each other. One measures their spacing.

B.3.4.2. *The antennas.* Any piece of well-isolated wire, connected to an entry of a high-impedance operational amplifier (like the CA-3140) is a surprising sensor of electric potential, rated from the remaining of the measuring apparatus. The antennas are much more impressive as sensors of **variable** electric fields, hence their universal use in transmissions: radio, television, radar, radio relay links, etc. With phase detector screwdrivers, your body serves as an antenna; it is one of the capacitor-armature while the remaining of the room is the other armature. The capacity is in the range of 100 to 200 pF.

B.3.4.3. *The biological sensors of electrical field.* Every fish, and mostly those who live in cloudy water, are able to probe their near environment by emitting differences of potential, and sensing how they are modified. It allows catfishes to detect their preys. A shark can find a fish hidden under sand, just by its electric field.

B.3.5. Now sum up by a table, what we know on the two fields, and their test bodies:

Type of field	Test body of the field?	What produces the field?	geometry	associated potential
Gravity field	every material body	every material body	Vector: m / s²	Scalar: m² / s²
Electric field	Every electric charge	Electric charges	Vector. Unit: V / m.	Scalar: volt, ou J / C.

Calculate the gravitational field of the Earth, where the Moon is, at the mean distance of 384,403 km (in fact varying between 363,299 km and 405,506 km).

Curious:
-

Open-Eyes:
- The mass of the Moon is $73.54 \cdot 10^{21}$ kg. Deduct the gravitational force applied to the Moon by the Earth.

Curious:
-

B.4. Magnetic field

B.4.1. Legal definition of the ampere. A current of 1 ampere, running in two straight and filiform conducting wires, at a distance 1 meter on a great length, produces between them in the vacuum, a force of $2 \cdot 10^{-7}$ newton per meter. This force is attractive if the two currents run in the same sense, repulsive if they run on the contrary sense.

Figure 18.6.

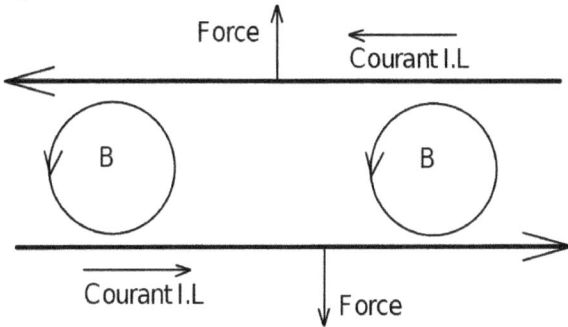

B.4.2. Production of the magnetic field. A magnetic field is nothing else than a small variation of an electric field, but seen through a mirage of the space-time. It means that an electric charge has a moving speed, and it translates that like a roller conveyor, or a roller bearing. For a beginning, we will only study the fields in vacuum or air. Later we will see the much more complex case of the magnets: permanent magnets and magnetizable materials. Figure 18.7.

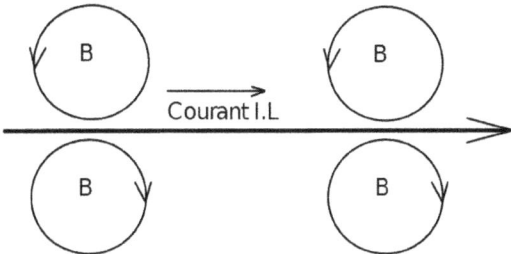

This magnetic field $\check{B}$ translates the fact: *I see a current running in front of me*. The product **i.l**, intensity of current, by the length of an element of circuit, is a vector: $\overrightarrow{i.l}$. A magnetic field is a **turning being, and it makes turning a speed**. This being is in a plane, the plane of rotation. Unit: **the tesla** or joule.second by radian, by square meter and by coulomb.

B.4.3. Effects of the magnetic field.

B.4.3.1. *No effect on an immobile charge.*

B.4.3.2. *Effect on the mobile charges.* The force deviating a current, or a moving charge, is vectorial. And the magnetic field is the quotient of two perpendicular vectors. So it cannot at all be a vector itself. It had been called a *bivector* in the States, but we call it a **gyror**.

Figure 18.8.

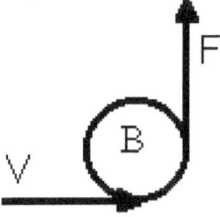

Relation between the speed vector of the electron, the gyror magnetic field, and the resulting force on the electron:

The effect of the magnetic field is to make turn - deviate from its trajectory - an electric charge **already moving**, or in other words, or apply a lateral deviating force on a current.

Effect of a magnetic field on a running electron: it curves its trajectory into a circle arc. Now take the case of a running electron in a uniform field:

We study the simple case, where the initial speed is already in the plane containing the magnétic field: Experimental mounting: a glass ampoule filled with rarefied dihydrogen, in which is injected an electron beam, which may be deviated with a magnetic field. You notice in which sense turns the beam of electrons: in the same sense than the current in the coils.

Figure 18.9.
Here the equiplane of the magnetic field contains the vector speed of the electron, so the trajectory is a circle (otherwise, it is a helix). The force applied to the electron remains perpendicular to the speed, so the norm of the speed does not vary.

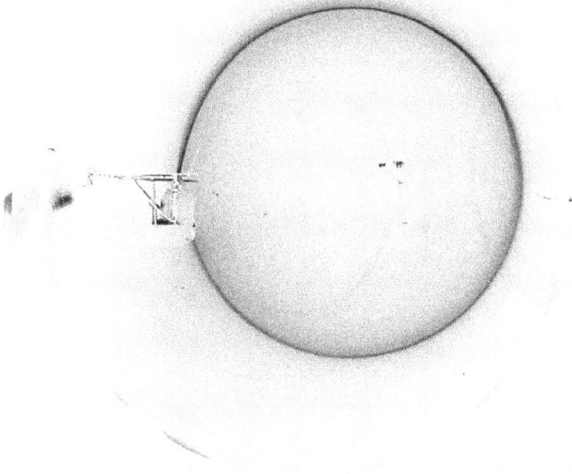

Figure 18.10.

Courant dans les spires d'inducteur.

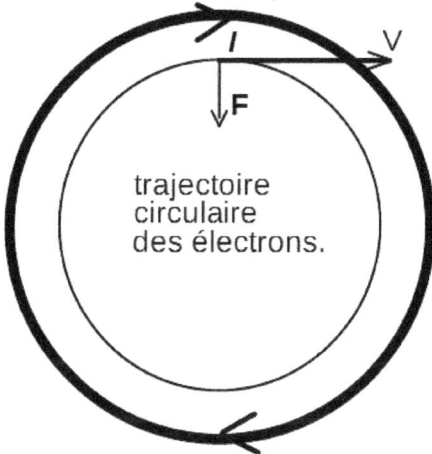

trajectoire
circulaire
des électrons.

Figure 18.11.

We see a circle in a circle, and turning the same! Of course, there is a trick! I have taken electrons, whose charge is negative. If I had taken a current, its curve would have circled on the contrary sense

And what about the general case? The trajectory is hélicoidal, with the circular section of the helix coplanar with the equiplane of the magnetic field. The field $\check{B}$ acts only in the inner projection of the speed onto the stable plane of $\check{B}$. The outer projection of the speed is unaltered and gives the axis of the helicoidal trajectory.

Figure 18.12.

Courant dans les bobines extérieures, qui créent le champ

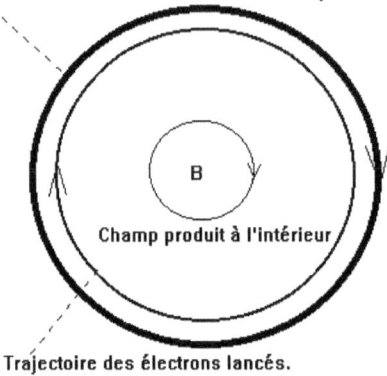

Champ produit à l'intérieur

Trajectoire des électrons lancés.

B.4.4. Mathematical expression of the law, effect of the magnetic field on the running electron.

B.4.4.1. *For a mobile charge, a "particle":* $\overrightarrow{F} = -q.B.\overset{\smile}{\overrightarrow{v}}$. In the general case, only intervenes the inner projection of $\overrightarrow{v}$ on the stable plane of $\overset{\smile}{B}$
So $B.\overrightarrow{v}$ is qualifiable as an inner product.
No commutativity exists: the gyror $\overset{\smile}{B}$ acts on the vector $\overrightarrow{v}$. The inverse writing is nonsense.
The analytical expression of $\overset{\smile}{B}$, antisymmetrical operator of rank 2, reflects it is the composition of a projection on the stable "proper" plane, and a quarter-turn rotation in this stable plane.

Figure 18.13.

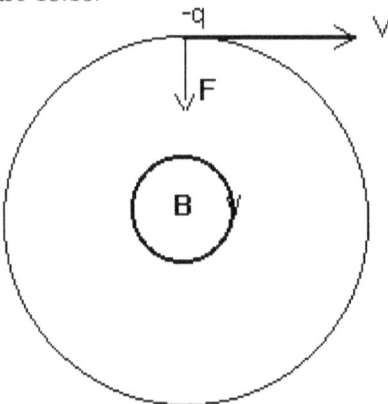

Trajectoire d'un électron dans un champ
magnétique B, coplanaire à sa vitesse V

The trajectory of an electron in a magnetic field $\overset{\smile}{B}$ coplanar to its speed $\overrightarrow{v}$.

B.4.4.2. *For a current of intensity i in a conducting wire,* differential law:
$\mathrm{d}\overrightarrow{F} = -\overset{\smile}{B}.\left(\overrightarrow{i.dl}\right)$ (Ampère-Laplace law).

Meaning of the minus sign: two currents of opposite sense repel each other, attract if in the same sense. The field $\check{B}$ is due to the other current elements.

B.4.5. Mathematical expression of the law: production of the magnetic field by a current. Orientation and symmetry:

Figure 18.14.

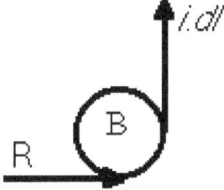

The complete expression demands an external product of two vectors: Contribution of each element of current $i.\vec{dl}$ (or q.$\vec{v}$):

$$d\check{B} = \frac{\mu_0}{4\pi} \cdot \frac{(\vec{R})^{-1}}{|R|} \wedge (i.\vec{dl})$$

. The element of current q.$\vec{v}$, and the vector $\vec{R}$ from the point M to the element of current determine the plane of $d\check{B}$ (contribution to the magnetic field).

Figure 18.15.

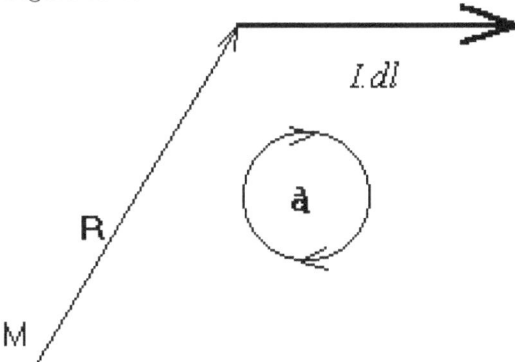

The field produced by a turn, or more, say in the inside of a solenoid, A circle in a circle, and turn alike!
In a long solenoid, without iron: $\mathbf{B = \mu_0\ n_l\ i}$
With **i** intensity of the current, by turn,
n_l : number of turns per meter length,
μ_0 : magnetic permeability of vacuum.
$\mu 0 = 4\pi \cdot 10^{-7}$ H/m $= 12.56637 \cdot 10^{-7}$ H/m.

Figure 18.16.

Courant dans une spire,
ou dans la spire moyenne d'un solénoïde

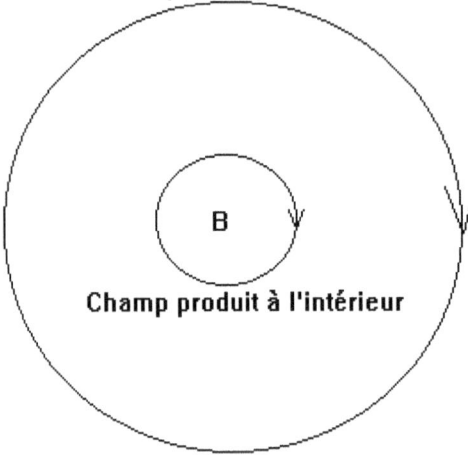

Notice: $\mu_0 * e_0 * c^2 = 1$, where **c** is the light speed in vacuum.

B.4.6. Magnetic potential (vector). $d\vec{A} = \frac{\mu_0}{4\pi} \cdot \frac{i.\vec{dl}}{|R|}$ The magnetic potential copies the near currents, with attenuation by the distance. Its *rotational* is the gyratorial field $\check{B}$.

Unit: the tesla.meter, T.m.

Expressing the vector Laplacian (spatial second derivative) of $\vec{A}$ as a function of the density of current $\vec{i}$: $\vec{\Delta}\vec{A} + \mu_0 \vec{i} = 0$

If you sum the element $d\vec{A}$ on an unlimited straight wire, the result is a logarithm of the inverse of the distance to the axis of the wire. But a logarithm is infinite at both the zero and the infinite. So we must take the origin at a finite distance from the axis, say at the surface of the wire. The symmetry of the problem is cylindrical of revolution.

Calculation for a limited wire, limited from **-L** to **+L** on the **z**-axis. The point of observation is on the plane **xy** or **rϑ**, at the distance **r**. The frame **Oxyz** is orthonormal. Figure 18.17.

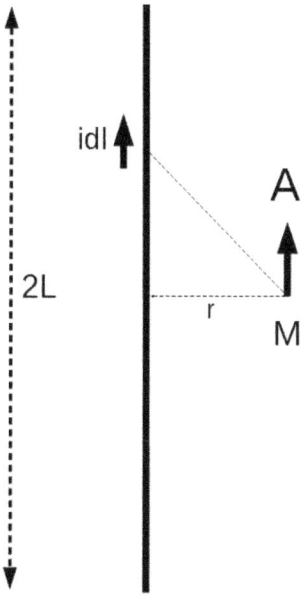

We write $\overrightarrow{u_z}$ the unitary vector along z.

$$\vec{A_L}(r) = \frac{\mu_0 \overrightarrow{I.u_z}}{4\pi} \int_{-L}^{L} \frac{dl}{\sqrt{l^2+r^2}} = \frac{\mu_0 \overrightarrow{I.u_z}}{2\pi} ln \frac{L+\sqrt{L^2+r^2}}{r}$$

Now we approximate that the half-length **L** is much larger than the distance **r**.

Then $L + \sqrt{L^2 + r^2}$ tends toward 2L and $\vec{A_L}(r) \approx \frac{\mu_0 \overrightarrow{I.u_z}}{2\pi} ln \frac{2L}{r}$

We take the reference of potential at the radius **R**.

$\vec{A_L}(r) - \vec{A_L}(R) = \frac{\mu_0 \overrightarrow{I.u_z}}{2\pi} ln(\frac{R}{r})$ and L does not more intervene. We have obtained

$\vec{A}(r) - \vec{A}(R) = \frac{\mu_0 \overrightarrow{I.u_z}}{2\pi} ln(\frac{R}{r})$

The field $\check{B}$ is the rotational of $\vec{A}$ and it lays on the sagittal plane containing the axis of the wire, and the vector $\overrightarrow{R}$ linking the point of observation M to the wire conducting the current.

$\check{B}(r) = \frac{\mu_0}{2\pi} \cdot \left(\overrightarrow{R}\right)^{-1} \wedge \left(\overrightarrow{i.u_z}\right)$

As already drawn above, the orientation of the gyratorial field $\check{B}$ is the one of balls or rollers in a balls or rollers bearing.

And for **a thin and long solenoid**?

The current density is tangential, and is written **J**, or vectorially $\overrightarrow{J}$. **a** is the radius of the cylinder. The simplified drawing is in the plane perpendicular to the axis of the solenoid; this plane contains the currents $\overrightarrow{J}$, the field $\vec{A}$ and the field $\check{B}$.

Figure 18.18.

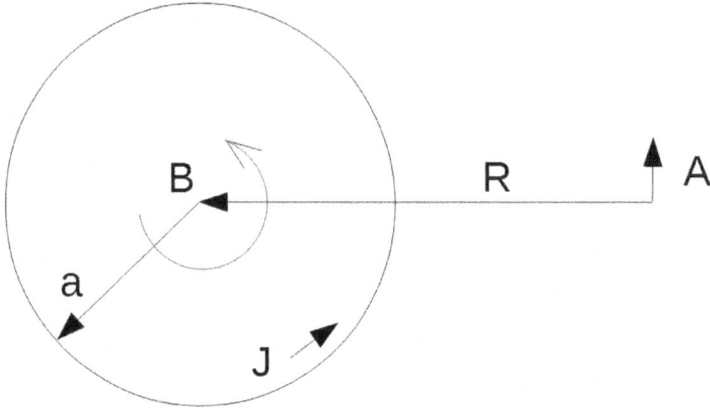

In the solenoid, the field $\check{B}$ is constant, and its modulus is $\mu_0 \mathbf{J}$, it turns in the same turning as the current does in the helicoidal wire. Its integral $\vec{A}$ copies the nearest vector $\vec{J}$, and is null in the axis of the solenoid.

In the solenoid, $\vec{A}$ behaves like the field of speeds in a rotating solid with constant angular speed, and is proportional to the distance from the axis: $\mu_0 \mathbf{r} \ \mathbf{J}$ ar the distance $\mathbf{r}$. Its rotational $\check{B}$ is therefore constant.

Out of the solenoid, $\vec{A}$ behaves like a vortex; it decreases in $\mathbf{1 \ / \ R}$ with the distance $\mathbf{R}$. Its modulus is $\mathbf{A} = \mu_0 \mathbf{J} \ \frac{a^2}{R}$

It is null an infinite distance in this plane. It is irrotational, and the field $\check{B}$ is null - provided that the solenoid is long enough to seem infinite.

B.4.7. Motors and generators.

B.4.7.1. *The principle of the electric motors.* Figure 18.18.

Courant dans les spires d'inducteur.

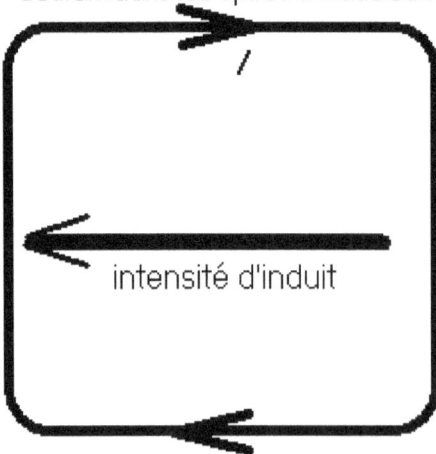

I

intensité d'induit

Knowing the sense of the current in the armature element in a DC motor, predict the sense of rotation. Draw it with an arrow, up or down. The drawing is in the plane of the air gap, between the pole piece and the rotor or armature. You see the sense of the current in the pole piece coil. The central arrow represents the intensity in an armature element. It does matter whether the coil is behind or before the page.

Solution :

Courant dans les spires d'inducteur.

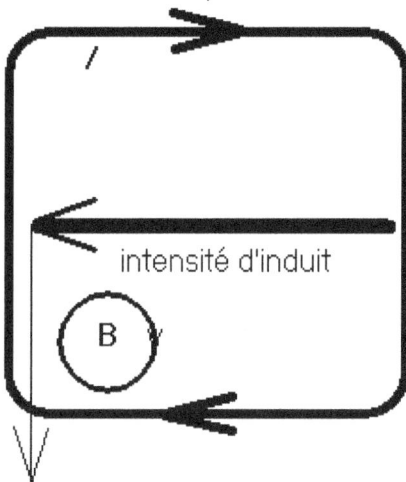

intensité d'induit

B

Force électromagnétique
s'exerçant sur l'élément d'induit.

B.4.7.2. *The motors are reversible: generators.* Figure 18.19.

Courant dans les spires d'inducteur.

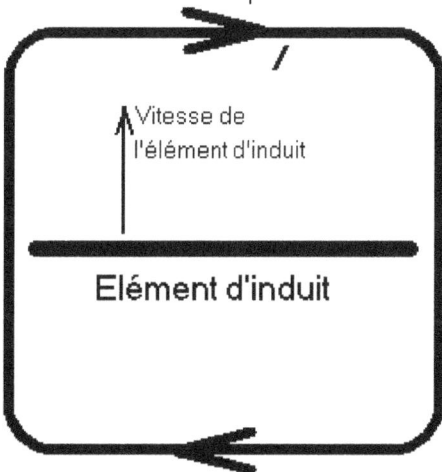

Vitesse de
l'élément d'induit

Elément d'induit

Take the case of an armature element, moving before a pole piece. We can draw the sense of the current in the turns of coils around the pole piece. The drawing is completely held in the plane of the turn, parallel to the plane determined by the speed of the armature element, moving before the pole piece and the armature element itself. Your mission is to forecast the sense of the e.f.m. (*electromotive force*) in the rotor. The + is at right? Or left?

As soon as a DC motor turns, it begins behaving as a generator: it produces an electromotive force (e.m.f.), coming in opposition to the one received from the feeding generator, and it reduces its consumption. Beyond some speed, this motor becomes the main generator. About the alternators, the problem is more difficult: you must take the frequency and the phase into consideration.

Solution:

Courant dans les spires d'inducteur.

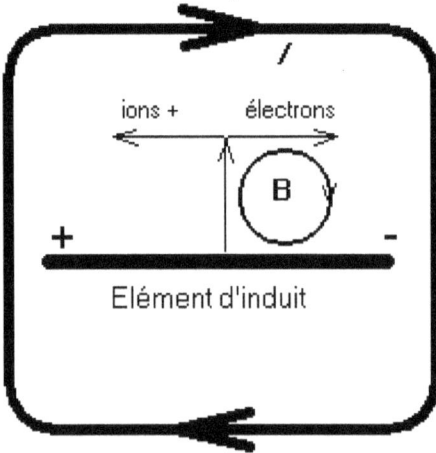

ions +   électrons

B

+   -

Elément d'induit

No motor could have acceptable size and yield without the providential reinforcement and conducting of the magnetic field, by the magnetizable materials: iron for the polar pieces, iron for the rotor, iron for the casing.

B.4.8. The magnetic phenomena in the matter.

The electron. Each electron is by itself like a current turn. In a manner of speaking, it *spins* on itself. Therefore it has a magnetic moment, and the norm of this moment is invariant.

This explains the covalent bond in chemistry: in each atom or molecule the electrons associate in pairs. Each pair is formed of electrons with opposite magnetic moments; so each pair has a null total magnetic moment. If it is more economical in energy, so it is stable.

A difficult to understand paradox: at our scale, we are accustomed to can orientate a spin or a magnet in all directions of space. However, the elementary particles, including the electrons, have a very peculiar way to "*spin*": they have only two ways of spinning, to be with or against their neighboring.

The only approaching image in our scale are the two states of twisting of a belt: Attach one end of a belt. If you twist by one turn the other end, nothing can wipe out this twist. But if you twist it by two turns, it is easy to erase the twisting, by leaving fixed the ends, but bending the belt. Two turns are equivalent to null turn.

The ferrimagnetic and ferromagnetic materials. Some metals (iron, nickel, cerium), some alloys, some oxides (ferrites, like the magnetite Fe_3O_4) are magnets. In other words, instead of canceling each other, the magnetic moments of some electrons in neighbor atoms in some crystalline structures cooperate to align their spins in the same magnetic orientation. Hence strong magnetic fields in these materials, for instance iron. Nobody knows why, but we know how to benefit from this phenomenon.

However at our macroscopic scale, we are convinced that soft iron (with less than 0.2 % carbon) is not magnetized, just because our macroscopic scale is not competent: in fact, iron under 774°C, is completely magnetized. But it is organized in small domains, which close their fields each other. These **Weiss domains** have sizes about 0.1 mm or less.

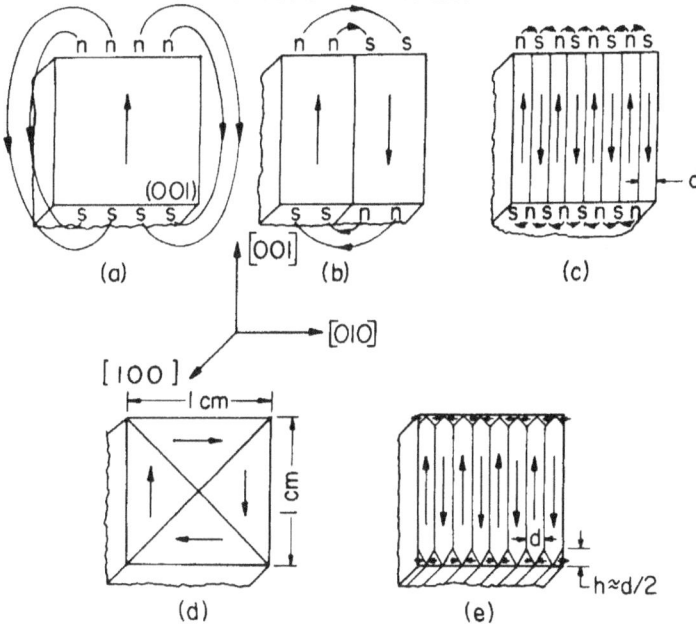

13. Possible domain structures in a square single crystal slab of iron
    having the top and bottom faces in the (001) plane.

Under the action of an external magnetic field, in the soft iron, the domains oriented in the cooperative direction grow, at the expense of the opposed domains. So iron reinforces any external magnetic field. When all the domains are oriented the same, iron is said saturated. It occurs around 2 teslas. The existence of the ferromagnetic materials made possible all the electrotechnics: the electric motors, the transformers, the relays, the electro-magnets, the alternators, etc.

Figure 18.21.

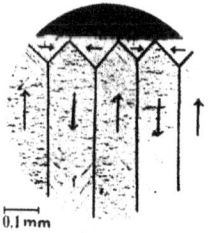

Fig. 3.14. Closure and lamellar domains observed on the (001) surface o
silicon–iron (Chikazumi [1964]).

Figure 18.20.

Some alloys may be elaborated, so the walls of their Weiss domains are blocked: they are permanent magnets. Nowadays, one uses mostly ferrites for making permanent magnets. For instance, in the loudspeakers, in the small DC motors, in the alternators for bicycles. Earth itself is a natural magnet, but we cannot explain how. And this magnet flips sometimes... Sometimes after 50,000 years, sometimes more than 600,000 years.

Let us sum up in a table what we know on these three kinds of fields, and their test body.

Type of field	Test body of the field?	What produces the field?	geometry	associated potential	Mathematical form for...
Gravitational field	any materiel body	any materiel body	vector: m /s²	scalar: m² /s²	for punctual mass **m**
Electric field	Any electric charge	electric charges	Vector. Unit: V / m.	scalar: volt, ou J / C.	for punctual charge **q**
Magnetic field	Any moving electric charge	Every moving electric charge, magnets	gyror: Unit: weber, ou V. s . m⁻²	vector: V. s / m (Volt/vitesse)	for element of current $\overrightarrow{q. v}$
Electromagnetic wave	any electric charge, atoms	every change of position or speed of an electric charge	gyror in 4-dimension: time + space	vector in 4-dimension: time + space	
Nuclear field. Very short distance field (but strong): a nucleus.	protons, neutrons	protons, neutrons			

One notices that the mathematical expression of the potential always has a degree less complexity. For the gravitational and the electric field, the potential is a mere scalar, and it varies in $\frac{1}{R}$.

The magnetic potential looks very alike the currents that produce it: it expresses the *"glue"* which binds the electric charges together. It has the sense and direction of the electric currents, but with attenuation in the distance in $\frac{1}{R}$.

In any case of electromagnetic field (electric + magnetic), the force is on the straight line which links the two charges. Therefore, the magnetic field is just a modification of the electric field, due to the relative movement of the charges. We have given the laws of production of the fields only in differential forms for just one charge or just one element of current: it was not to frighten those who are terrified by the sum symbol $\int$. It remains evident that for any real problem, one has to sum over all the charges and the complete circuits.

Transactional (Quantum) Microphysics, Principles and
Applications

Radiation and propagation of the electromagnetic waves. Polarization.

C.1. Evidence by radio waves.

For a radio wave (or UHF as transmitting television), to observe the polarization is immediate: just orient the antenna (conducting wire of conducting stick), to see the reception enhance to the maximum, or almost extinct. On each place, only one orientation is optimal. However, this experiment requires a simple and neat field. In the courtyard of a building, you only receive multiple reflexions of a broadcasting, and no orientation is clearly good nor clearly null.

One may experiment with two cheap talkies-walkies. If their antennas are parallel, and perpendicular to the propagation, the reception is good. Cross antennas, or parallel to the propagation yield quasi-null reception.

Figure 19.1.

Draw here the first experiment: the emitter hold its antenna vertical. How should the receptor be oriented?

Second experiment: the emitter hold its antenna horizontal. In which conditions the receiver will well receive?

Figure 19.2.

Curious:

-

Open-Eyes:

C.2. The line direction of the emitting antenna is the one of the electric field.

The electric field was explained above.

C.3. The plane of propagation of a plane polarized e.m.wave.

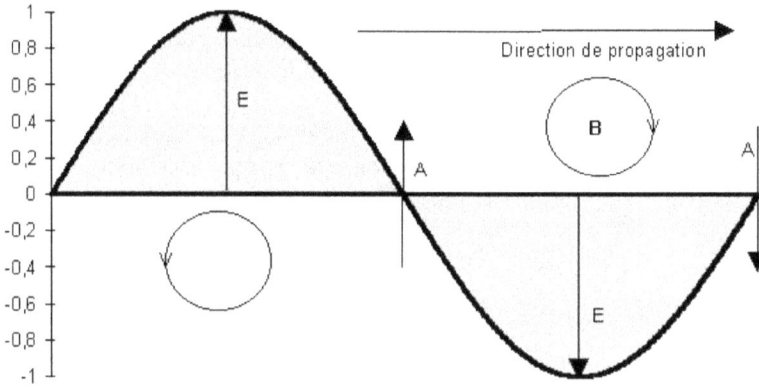

A third arrow has been drawn: the vectorial field $\vec{A}$.
It is in advance on $\vec{E}$ by a quarter of a wavelength. $\vec{A}$ is maximum when $\vec{E}$ is null. $\vec{A}$ resembles much an electric current: $\vec{A}$ which precedes, is the cause of $\vec{E}$ which follows a quarter of period later. $\vec{A}$ is called magnetic potential. In simplified terms, the electric potential V means "*There are electric charges, not far*". Similarly, the magnetic potential $\vec{A}$ means "*There are electric currents, not far*", and $\vec{A}$ has the direction and sense of the nearest and strongest current. So it is null in the center of a turn. For the most math-minded, the gyror $\check{B}$ is the rotational of $\vec{A}$.
One states that an electron, or any other electric charge, submitted to the magnetic field $\check{B}$ plus the electric field $\vec{E}$ is pushed in the direction and sense of the propagation. It is the radiation pressure.

On the next figure, let us exercise to prove it in four cases:

1. Negative charge, maximum of electric field toward the bottom of the figure.
2. Negative charge, maximum of electric field toward the top of the figure.
For that, draw with the pencil the trajectory of the charge under the influence

of the electric field, next bend it as it is bent by the magnetic field.

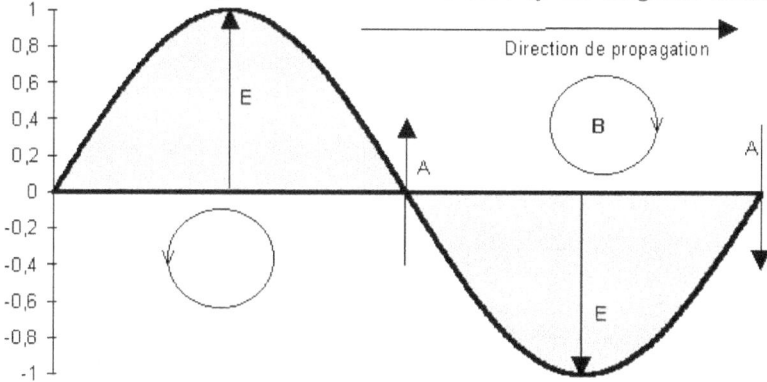

3. Negative charge, maximum of electric field toward the bottom of the figure.
4. Negative charge, maximum of electric field toward the top of the figure.

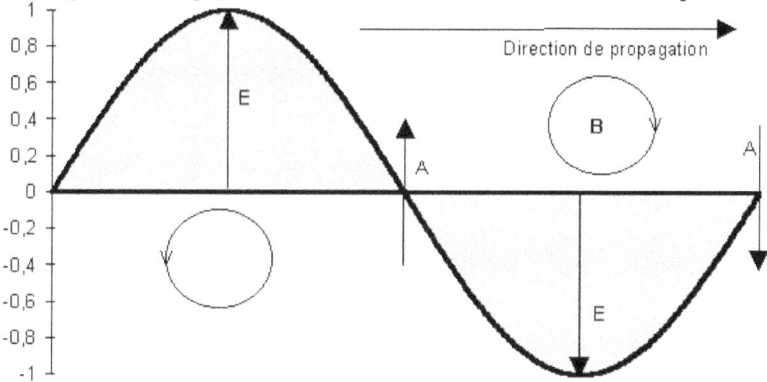

Statements:

Transversely, all the charges minus are . . .

Transversely, all the charges plus are . . .

Lengthwise, all the electric charges are . . . .

Why such a wave is called "plane polarized"?

For answering, check whether a plane contains three vectors and a gyror, characteristic of the wave: the propagation vector, the electric-field vector, the magnetic-potential vector, and the magnetic-field gyror.

Curious:

-

Open-Eyes:

- And in which plane will you place a Yagi antenna, for receiving the television UHF wave?

Curious:

-

Open-Eyes:
- Affirmative, in the horizontal plane, and perpendicular to the emitter azimuth. I have not touched the case of the circular polarizations (nor the intermediate cases). It is not more drawn in the plane of the paper, as in circular polarization the vectorand the vectorare permanently perpendicular to each other. However, the projection on the plane of the paper or screen is always what we have drawn above. The pace of the helix is equal to the wavelength. All the superpositions and mixing of plane and circular polarization with exactly the same frequency are possible. So quarter-wave plates are used to convert the circular polarization into plane, and conversely.

C.4. With light, polarization is more difficult to obtain, and to prove.

In a thermic source as the gas in a flame or a heated wire, when an atom emits light, it emits at random, without cooperation with neighbor emitters, so a plane of polarization must be at random, too. Let us calculate how long lasts the polarized emission from one atom: about 10,000 periods at the frequency 500 THz:
Is such a duration easy to measure?

Curious:
-

Open-Eyes:
- So one must use special means to observe the polarization of light. In the 19th century they had only two means:
• Some crystals such as calcite $CaCO_3$, separate and transmit differently two polarizations, according to the respective orientations of the ray and the crystal.
• The reflection and the transmission through a diopter induce a partial polarization. But several refractions or reflections in series can obtain more complete polarization.

Nowadays, we use Polaroids. These sheets are preparations of transparent high-polymers, in which are immersed conducting organic molecules (polyvinyl alcohol iodide), all aligned in the same direction. They absorb the electric field in the direction of their alignment, but let pass the perpendicular polarization.

C.5. Energy of an electromagnetic radiation.

What can do an ultraviolet radiation, an infrared cannot. The ultraviolet tans the skin, it triggers chemical reactions, it produces fluorescences, it provokes mutations in the living cells, it makes a photoconductor to conduct, etc.

In 1905, Albert Einstein expressed the law: an electromagnetic radiation interacts with matter only by **quanta**, that is the Planck constant **h**, $6.626176 \cdot 10^{-34}$ joule-second per cycle, or $1.05457267 \cdot 10^{-34}$ joule-second per radian.

These quanta are now said « **photons** ». Here is the original law by Einstein: **W = h . f.** where **f** is the frequency of the radiation, and **W** the transferred energy by one photon.

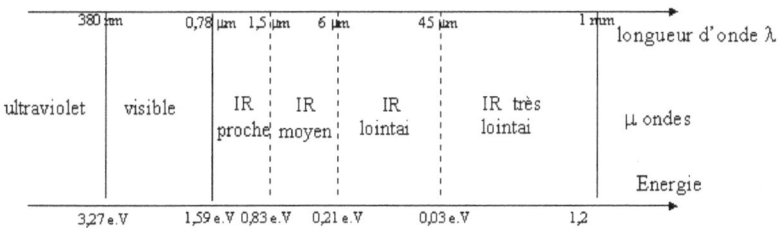

Figure 19.4.

In infrared spectroscopy, (detection of the carbon monoxide by its absorption at 4.666 µm) we have met this law in a form slightly modified by Bohr:
E_1 - E_2 = h . ν, where **E_1** is the energy of the molecule before absorption of a quantum of radiation, and **E_2** is the energy of the molecule after absorption, and **ν** is the frequency of the photon. We rewrite this law, so as the second term is a universal constant:
W T = h, or **W / ω = h / 2.π = h**. With **ω** meaning the pulsation, **T** meaning the period.

C.6. Granule?

This paragraph was written in 1995, in a time when we used floppies.
On a PC, quest what is the granule of allocation on a floppy: the smallest quantity of track that the DOS allocates to a file. Have a formatted floppy in A:, then type:

DIR A: (note the remaining place: L_0)
COPY CON A:TEST
AB   (or any short message: here 3 signs: AB^Z)
^Z
DIR A: (note the new remaining place: L_1).

The granule is the difference $L_1 - L_0$. It has varied, according to the physical medium and the BIOS of the computers.

Professor Castle-Holder:

- This trial is now obsolete, and the beginners cannot more access it. Frequently, the granule could have been 512 bytes or 1,024 bytes.

Open-Eyes:
- Now you have a familiar comparison at hand for the quantity **h**. **h** is the granule of action per cycle, and also of angular moment, given by nature. Invented by Maupertuis, the "*action*" was formalized in 1834 by W. R. Hamilton. Now we know that the constant **h** rules all the nature.

C.7. The granule of action per cycle in the molecules

The molecules may rotate and vibrate, but only at levels of half-integer values of $\hbar$. The carbon monoxide molecule CO permanently vibrates in the mode stretching-compression. It cannot de-energize under the level $W_0 = \frac{1}{2} . h.\nu_0 = \frac{1}{2} . \hbar . \omega_0$ (said energy of null).

By absorbing a photon, it may jump to level $W_1 = (1 + \frac{1}{2}) . \hbar . \omega_1$, or even higher. These pulsations ω_1 and ω_0 are almost equal to the one of the absorbed infrared radiation. W_1 - $W_0 = . .$

Curious:
- $W_1 - W_0 = \hbar . \omega = h.\nu$

Open-Eyes:
- So in infrared spectroscopy, we measure the vibration modes of the molecules. It exists much more ways of interactions between atoms or molecules and radiation. In microwave spectroscopy, we measure the rotation modes. In the visible and ultraviolet range, one measures the energy levels of the peripheric electron orbitals around the atoms. In X-rays spectroscopy one measures the energy levels of the deep electron orbitals. In gamma spectroscopy, one measures the structure of the atomic nuclei. By these means, one can know the composition of the stars: by analyzing their light, first with dispersive prims, now with lattices.

C.8. The energy of each quantum of light.

Now a graph situating the radiations in the spectrum, in energy and wavelength. And it will the end of this series of recalls.
Fig. 19.5.

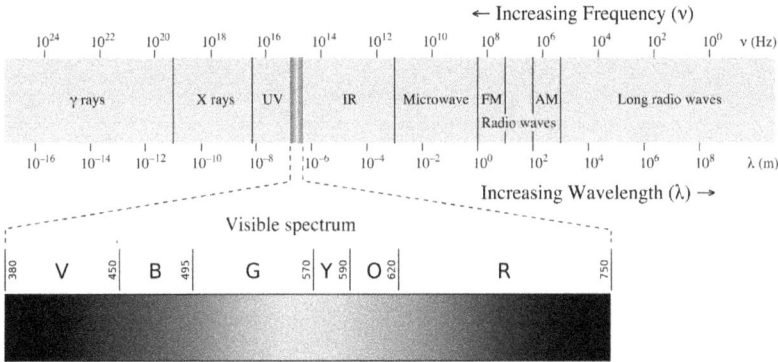

Figure 19.6.

Professor Castle-Holder:
- Our colleague strongly diverges from the traditions: under the tradition, the magnetic field $\breve{B}$ is presented as a vector, at 90° of the electric field $\vec{E}$ whith which it is in phase. Now twice per period, these two fields are simultaneously null, and their density of energy is null too! And the most awful is just in front of a mirror, when the interference of an incident ray (at normal or quasi-normal incidence) and its reflection gives two dark fringes per wavelength. Then nobody knows how the fields persist to propagate through these zones of permanent null field...

Open-Eyes:
- So, dear traditionalist colleague, *you are squelching in the mud of marsh of learned error*! You have forgotten the field $\vec{A}$ also said magnetic potential, which is a true vector, and is always in advance quadrature on. Consequently, the zones with null $\vec{E}$ and $\breve{B}$ are with maximal $\vec{A}$ as well in stationary or progressive regime. This $\vec{A}$ field persists to provide the propagation when the two others fail.

Useful analogy:
In a weight pendulum, the potential energy from gravity and the kinetic energy are exchanging twice in a period, while their total mechanic energy remains constant – but the losses in frictions and air turbulence, so decreasing in reality. In a resonating inductance-capacity circuit, the electric energy is exchanging twice a period, between the electric potential energy in the capacitor, and the kinetic (magnetic field) in the self-inductance. By considering the so despised and unpopular-in-the-clergy field $\vec{A}$, one restores the unity of the physics. Not for the birds Alexander Liapunov woke up before the others, and taught us how to represent the stability of the oscillating systems.
We still have to complete or correct the formula giving the density of energy from the fields. Stay tuned.
Figure 19.6.

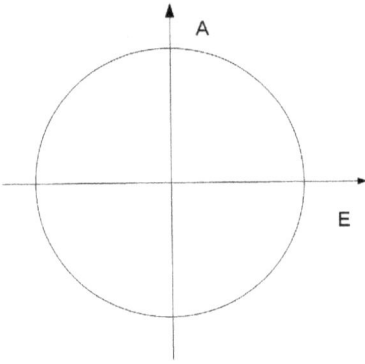

Professor Castle-Holder:
- Unless the result will be a density of power or momentum, just passing for the photons.

The Mössbauer resonance, the Pound-Rebka experiment

D.1. Conversions, formulas aide-mémoire

14.40 keV, gamma line of the ^{57}Fe, is also 0.861 Å of wavelength in the vacuum. Said in femtometers, the appropriate unit of the atomic nucleus, it is 86,100 fm. It is enormous compared to the Fe nucleus in the stable state, and even in the energized state. Therefore, to capture this photon, this ^{57}Fe nucleus must make converge the photon on it, and on very long time (here typically 3 to 10 ns) compared to the intrinsic periods in a nucleus: it is a photon of excellent definition in frequency, so of long length, about one to ten meters. Such a fact escapes to their anti-transactionist doctrine of a gunner in the 14-18 war, or a Yankee carpet-bomber: only transactional microphysics treats this. This capture is not magic but comes from frequential resonance.

Calculation detail:
14,400 electronvolts per photon $= h \cdot \nu = h \, c / \lambda$,
1 eV $= 1.60219 \cdot 10^{-19}$ C.V $= 1.60219 \cdot 10^{-19}$ joule.
Its frequency: $\nu = E \, / \, h = \frac{14400*1.60219.10^{-19}J}{6.6260755.10^{-34}J.\frac{s}{cycle}} = 3.482$ EHz (Exaherz $= 10^{18}$ Hz).
Hence the wavelength of this photon (final result in picometers):
$\lambda = hc/E = \frac{hc}{14,400*1.60219.10^{-19}J} = \frac{6.626.755*299,792,580m}{14,400*1.60219.10^{-19}J} = 86.1002$ pm
It is a hard X-ray, very ionizing, not to attend oneself... A shield is highly recommended.
We need also its momentum: $E/c = \frac{14400*1.60219.10^{-19}J}{299792580\,m/s} = 7.696 \cdot 10^{-24}$ kg.m/s.
What is the width of such a photon in a laboratory apparatus, of 80 cm length?
$2z = \frac{\sqrt{3*0.40m*86.1pm}}{2} = 5$ µm, that is one milliard of times the diameter of an iron nucleus.
What is its width at half the height of Harvard "tower" in the Pound-Rebka experiment?
$2z = \frac{\sqrt{3*10.5m*86.1pm}}{2} = = 26$ µm, that is five milliards of times the diameter of an iron nucleus.
What is the half angle of the tangent cone, near the absorber of the emitter (a nucleus of ^{57}Fe)?
$\alpha = \sqrt{\frac{3\lambda}{4a}} = \sqrt{\frac{3*86.1pm}{4*10.5m}} = \sqrt{\frac{258.3\,pm}{42\,m}} = \sqrt{6.15.10^{-12}}$rad $= 2.48 \cdot 10^{-6}$ rad.

D.2. The Mössbauer effect

The Mössbauer effect involves the resonant and recoil-free emission and absorption of gamma radiation by atomic nuclei (nuclear transition) bound in a solid. Usually, by the law of conservation of the momentum and energy, if the re-emission differs in angle of the absorption, then the momentum-energy of the photon is modified because of the recoil of the atom.

However, when the nuclei are not isolated but well held in crystal, the recoil concerns a whole lattice and so is considerably lessened, and the momentum and energy of the recoil are to rate for the entire crystal. As the mass of the crystal is considerably bigger than the one of the nucleus – in magnitude about the Avogadro constant – the speed of recoil may be negligible. With a non-negligible probability, we observe a recoil-less emission or absorption of a photon.
This is the Mössbauer effect discovered by Rudolf Ludwig Mössbauer in 1958.

So we state, in absorption as well as emission a resonance on the difference of energy between the fundamental state of the nucleus and an excited state. It gives a rich and useful Mössbauer spectroscopy. It allows investigating the environment of the atom.
Bibliography:
http://www.uni-duisburg.de/FB10/LAPH/Keune/hs/Utochkina.pdf
http://home.uni-leipzig.de/energy/pdf/freuse9.pdf
http://obelix.physik.uni-bielefeld.de/~schnack/molmag/material/
Guetlich-Moessbauer_Lectures_web.pdf
The Mössbauer spectroscopy is very sensitive, resolving the energy up to 10^{-11} in precision. So are studied the hyperfine structure of the energy levels of a nucleus, their perturbation by the chemical, electrical or magnetic environment of the atom, the oxidation degree, or different ligands, etc.
A limitation, however: this spectroscopy is only usable on solid samples, and with a limited number of chemical elements and isotopes. Among them, iron is by far the most studied. This element is abundant and important in many diverse materials, so it opens to the Mössbauer spectroscopy a large field of investigations in materials science, geology, biology, etc.

D.3. Principle of the measurements

The same way a pistol recoils when shooting, an atomic nucleus recoils when emitting or receiving a gamma photon, because of the physical principle of the conservation of the momentum. Therefore, when an atom emits a gamma photon, the energy of this photon is slightly less than the total transition energy. Conversely, for a nucleus to absorb a gamma photon, it needs the energy of the photon is slightly more than the energy of the involved transition. In both cases, a part of the energy is lost in the recoil effect, that is transformed into kinetic energy of the nucleus. Therefore, the nuclear gamma resonance, that is the emission and the absorption, of two identical nuclei resonating together via a gamma photon, is not observed if the atoms are free in space. The loss of energy is too big compared to the spectral width, and the spectra of emission and absorption do not overlap enough.

In a solid, however, the nuclei are not free but are anchored to the crystalline lattice. The recoil energy cannot take any value, but depends on emission or absorption of a phonon, converging on this atom; and in a non-negligible proportion, no phonon is involved (5.7 % in the historic case of the iridium 191, with a gamma ray at 129 keV). When no phonon is emitted nor absorbed, it is said a recoil-less event. In this case, the conservation of the momentum is fulfilled by taking the momentum of the whole crystal or solid, so the loss of energy is very small, and the shift in frequency also.

Rudolf Mössbauer discovered that a significant proportion of the emission and absorption events could be recoil-less if the temperature is low enough; this proportion is called the Lamb-Mössbauer factor. It implies that a gamma photon emitted by a nucleus may be absorbed in resonance by another nucleus of the same isotope. Measuring this absorption is the principle of the Mössbauer spectroscopy. This discovery awarded Rudolf Mössbauer the Nobel prize in 1961. He used iridium 191 and had to cool down to the liquid-helium temperature.

If the emitting and absorbing nuclei are in the same chemical environment, the transition energies are equal, and the resonant absorption is observed when both are at rest, or just very slowly converging. But differences in chemical environment bring slight changes in the transition energies. These variations may be comparable or even higher than the spectral width of the gamma rays, so may bring large variations in absorbance. To have exactly the resonance and the maximal absorption, one has to slightly modulate the energy of the gamma rays, by handling the radial speed of the emitter relative to the absorber. The variations of energy spotted by the absorbance peaks can characterize the chemical state of the absorbing atom in the studied sample.

D.4. Experimental device

In its most used form, the Mössbauer spectroscopy is by absorption: a solid sample is exposed to gamma rays, and a detector measures the transmitted intensity. The atoms of the sample must be of the same isotope as in the emitter. The source emits a radiation of constant energy. This energy is modulated by placing the source on an oscillating support, so the Doppler effect modulates this energy in the frame of the receiving sensor. The detection is synchronized with the phase of the displacement of the source, and the received intensity is plotted as a function of the relative speed. As an example, the speeds used for iron 57 (^{57}Fe) are in the magnitude of 11 mm/s, which corresponds to 48 neV

$(4.8 \cdot 10^{-8}$ eV$)$.

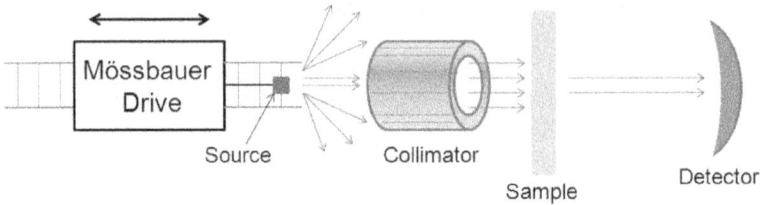

Figure 20.1.

The source for producing the useful isotope is a parent radioactive element which decays in the wanted isotope, in an excited state. For the ^{57}Fe, one uses a source of ^{57}Co, which by electronic capture decays into a ^{57}Fe nucleus, but in an energized state, which in turn decays into the fundamental state by emitting the gamma photon at 14.40 keV. Usually, the radioactive cobalt is prepared on a sheet of rhodium.

Electronic capture: one of the least frequent transmutation. The nucleus grabs a deep electron from the layer 1s, with the reaction:

$^A_Z X + ^0_{-1} e^- \rightarrow ^A_{Z-1} Y + ^0_0 \nu_e$

here expressed at the nuclear scale. Or at the nucleon scale: $\mathbf{p}^+ + \mathbf{e}^- \rightarrow \mathbf{n} + \nu_e$.

Proton plus electron gives neutron plus a neutrino, and leaves the new nucleus in an excited state.

This reaction lowers the atomic number by one, but leaves the electrons cortege in a high energy state: one electron lacks in the inner layer 1s, though the electric neutrality is preserved. This lower orbital will capture a peripheral electron, and the reorganization of the electronic cortege will emit a whole spectrum of photons, which do not concern us now.

Ideally, the parent isotope must have a half-life suitable for the duration of the experiment. Moreover, the energy of the gamma photon must be moderate, unless the proportion of recoil-less emissions will be too low, implying a poor signal/noise ratio and a long duration of acquiring. Among the elements having a suitable isotope for the Mössbauer spectroscopy, iron 57 is by far the most used, but iodine 119, tin 129, and antimony 121 are also used and studied. For a usable gamma photon in Mössbauer spectroscopy, the excited state must have a half-life time in the range 10^{-11} to 10^{-6} s.

Analysis and interpretation of the Mössbauer spectra. The Mössbauer spectroscopy gives an exceptionally good resolution in energy; it allows to detect very tiny differences in the environment of the scrutinized atoms. Typically three nuclear interactions are observed: the isomeric or chemical displacement, the quadrupolar coupling, and the hyperfine structure (the Zeeman effect).

Quadrupolar coupling. The quadrupolar coupling means the interaction between the quadrupolar moment of the nucleus and a gradient of the electric field. This coupling produces a supplementary term in the energy and a splitting of levels into a doublet in the Mössbauer spectrum.

This splitting occurs only when the nucleus has a not-null quadrupolar momentum, or in other words, a not-spherical density of electric charges. This condition is fulfilled in nuclei whose spin $\mathbf{I}$ is strictly greater than $\frac{1}{2}$, for instance, the spin $\mathbf{I} = 3/2$ of the isotopes ^{57}Fe and ^{119}Sn. When in a gradient of the electric field, the energy of such excited state splits into two sublevels corresponding to $m_I = \pm 1/2$ and $m_I = \pm 3/2$. This yields two distinct peaks in the spectrum. The quadrupolar coupling is measured as the separation between these two peaks.

The gradient of the electric field, provoking the quadrupolar splitting of levels comes from the density of electron charge in the neighboring of the atom, either the valence electrons of the atom itself, either from the near atoms.

The end of Appendix D.

Transactional (Quantum) Microphysics, Principles and Applications

Delayed choice

Basic transactional reference:
http://arxiv.org/abs/1501.00970 by Heidi Fearn, in the Fullerton University in
California.
http://arxiv.org/pdf/1501.00970v5
But in their minds of stubborn corpuscularists, alas:

Professor Marmot:

E.1. Experiment of the quantic eraser with a delayed choice.

*A delayed choice quantum eraser experiment, first performed by Yoon-Ho Kim,
R. Yu, S. P. Kulik, Y. H. Shih and Marlan O. Scully, and reported in early
1999, is an elaboration on the quantum eraser experiment that incorporates
concepts considered in Wheeler's delayed choice experiment. The experiment
was designed to investigate peculiar consequences of the well-known double-slit
experiment in quantum mechanics, as well as the consequences of quantum
entanglement.*

*The delayed choice quantum eraser experiment investigates a paradox. If a
photon manifests itself as though it had come by a single path to the detector,
then "common sense" (which Wheeler and others challenge) says it must have
entered the double-slit device as a **particle**. If a photon manifests itself as
though it had come by two indistinguishable paths, then it must have entered
the double-slit device as a wave. If the experimental apparatus is changed **while**
the photon is in midflight, then the photon should reverse its original "decision"
as to whether to be a wave or a **particle**. Wheeler pointed out that when these
assumptions are applied to a device of interstellar dimensions, a last-minute decision made on Earth on how to observe a photon could alter a decision made
millions or even billions of years ago.*

*In the basic double slit experiment, a beam of light (usually from a laser) is
directed perpendicularly towards a wall pierced by two parallel slit apertures. If
a detection screen (anything from a sheet of white paper to a CCD) is put on
the other side of the double slit wall, a pattern of light and dark fringes will be
observed, a pattern that is called an interference pattern. Other atomic-scale entities such as electrons are found to exhibit the same behavior when fired toward
a double slit. By decreasing the brightness of the source sufficiently, individual
particles that form the interference pattern are detectable. The emergence of*

an interference pattern suggests that each **particle** passing through the slits interferes with itself, and that therefore in some sense the **particles** are going through both slits at once. This is an idea that contradicts our everyday experience of discrete objects.

A well-known thought experiment, which played a vital role in the history of quantum mechanics (for example, see the discussion on Einstein's version of this experiment), demonstrated that if **particle** detectors are positioned at the slits, showing through **which slit** a photon goes, the interference pattern will disappear. This **which-way** experiment illustrates the **complementarity** principle that photons can behave as either **particles** or waves, but not both at the same time. However, technically feasible realizations of this experiment were not proposed until the 1970s.

Which-path information and the visibility of interference fringes are hence **complementary** quantities. In the double-slit experiment, conventional wisdom held that observing the **particles** inevitably disturbed them enough to destroy the interference pattern as a result of the Heisenberg uncertainty principle.

However, in 1982, Scully and Drühl found a loophole around this interpretation. They proposed a "quantum eraser" to obtain **which-path** information without scattering the **particles** or otherwise introducing uncontrolled phase factors to them. Rather than attempting to observe which photon was entering each slit (thus disturbing them), they proposed to "mark" them with information that, in principle at least, would allow the photons to be **distinguished after** passing through the slits. Lest there be any misunderstanding, the interference pattern does disappear when the photons are so marked. However, the interference pattern reappears if the **which-path** information is further manipulated after the marked photons have passed through the double slits to obscure the which-path markings. Since 1982, multiple experiments have demonstrated the validity of the so-called quantum "eraser."

Open-Eyes:
- I have emphasized above the terms proving they reason in the wording of corpuscles and Newtonian time: *"after, which slit, which-path"*. Now let us see their experimental apparatus.

Professor Marmot:
Figure 21.1.

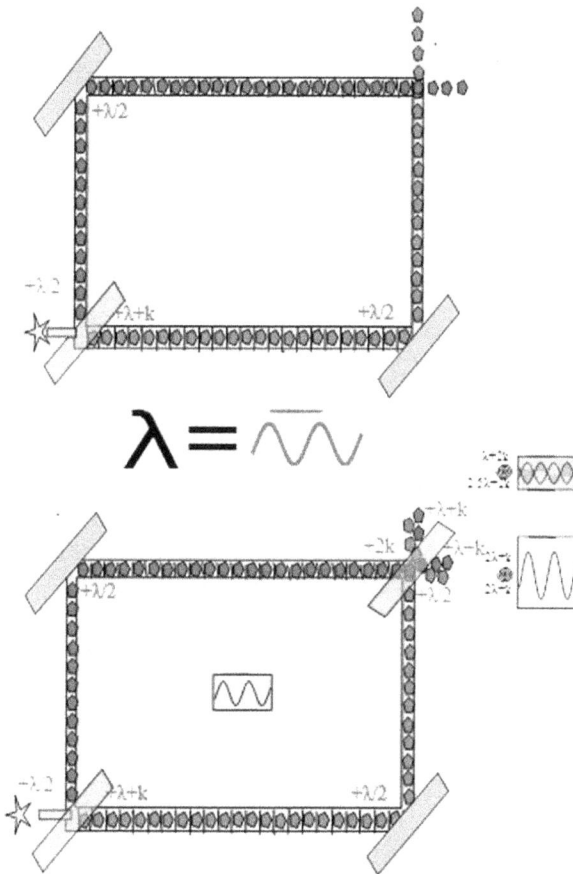

A simple version of the quantum eraser can be described as follows: Rather than splitting one photon or its probability wave between two slits, the photon is subjected to a beam splitter. If one thinks in terms of a stream of photons being randomly directed by such a beam splitter to go down two paths that are kept from interaction, it would seem that no photon can then interfere with any other or with itself.

However, if the rate of photon production is reduced so that only one photon is entering the apparatus at any one time, it becomes impossible to understand the photon as only moving through one path, because when the path outputs are redirected so that they coincide on a common detector or detectors, interference phenomena appear.

In the two diagrams in Fig. 21.1, photons are emitted one at a time from a laser symbolized by a yellow star. They pass through a 50% beam splitter (green block) that reflects or transmits 1/2 of the photons. The reflected or transmitted photons travel along two possible paths depicted by the red or blue lines.

In the top diagram, the trajectories of the photons are clearly known: If a photon emerges from the top of the apparatus, it had to have come by way of the blue

*path, and if it emerges from the side of the apparatus, it had to have come by
way of the red path.*

*In the bottom diagram, a second beam splitter is introduced at the top right. It
can direct either beam toward either exit port. Thus, photons emerging from
each exit port may have come by way of either path. By introducing the second
beam splitter, the path information has been "erased". Erasing the path information results in interference phenomena at detection screens positioned just
beyond each exit port. What issues to the right side displays reinforcement, and
what issues toward the top displays cancellation.*

Delayed choice

*Elementary precursors to current quantum eraser experiments such as the "sim-
ple quantum eraser" described above have straightforward classical-wave explanations. Indeed, it could be argued that there is nothing particularly quantum
about this experiment. Nevertheless, Jordan has argued on the basis of the
correspondence principle, that despite the existence of classical explanations,
first-order interference experiments such as the above can be interpreted as true
quantum erasers.*

*These precursors use single-photon interference. Versions of the quantum eraser
using entangled photons, however, are intrinsically non-classical. Because of
that, in order to avoid any possible ambiguity concerning the quantum versus classical interpretation, most experimenters have opted to use nonclassical
entangled-photon light sources to demonstrate quantum erasers with no classical
analog.*

*Furthermore, use of entangled photons enables the design and implementation
of versions of the quantum eraser that are impossible to achieve with single-
photon interference, such as the delayed choice quantum eraser which is the
topic of this article.*

Open-Eyes:
- Etc. I cut it there, as the further developments (like the Kim experiment) add
nothing the transactional theory cannot easily deal with. We do not recognize
anything magic in their *"entanglement"*, as the laws of the transactions do not
dwell in the Newtonian macro-time of the laboratory, and do not bother with
our macro-time.

E.2. Transactional re-reading

Open-Eyes:
- So you have well read: The professor Marmot took for hypotheses that *"the
particle"*, here a photon, is a small corpuscle, without length nor width, framed
in a macrophysical, simultaneous, and universal Newtonian time, macroscopi-
cally irreversible, THEREFORE also microscopically irreversible: *"If a photon*

emerges from the top of the apparatus, it had to have come by way of the blue path, and if it emerges from the side of the apparatus, it had to have come by way of the red path.".

So in his brain-damaged herding, the pack-follower has cut himself from any way to understand the reality.

We will not let the reader die as idiot as the professor Marmot; we give him or her the abstract from Heidi Fearn:
http://arxiv.org/abs/1501.00970.
The reading by Mrs. Fearn differs from ours, however.

E.2.1. A delayed choice quantum eraser explained by the transactional interpretation of quantum mechanics.

H. Fearn

(Submitted on 5 Jan 2015 (v1), last revised 9 Sep 2015 (this version, v5))
This paper explains the delayed choice quantum eraser of Kim et al. in terms of the transactional interpretation of quantum mechanics by John Cramer. It is kept deliberately mathematically simple to help explain the transactional technique. The emphasis is on a clear understanding of how the instantaneous "collapse" of the wave function due to a measurement at a specific time and place may be reinterpreted as a gradual collapse over the entire path of the photon and over the entire transit time from slit to detector. This is made possible by the use of a retarded offer wave, which is thought to travel from the slits (or rather the small region within the parametric crystal where down-conversion takes place) to the detector and an advanced counter wave traveling backward in time from the detector to the slits. The point here is to make clear how simple the Cramer transactional picture is and how much more intuitive the collapse of the wave function becomes if viewed in this way. Also, any confusion about possible retro-causal signaling is put to rest. A delayed choice quantum eraser does not require any sort of backward in time communication. This paper makes the point that it is preferable to use the Transactional Interpretation (TI) over the usual Copenhagen Interpretation (CI) for a more intuitive understanding of the quantum eraser delayed choice experiment. Both methods give exactly the same end results and can be used interchangeably.
Comments: 24 pages 4 figures, fifth draft
Subjects: Quantum Physics (quant-ph)
DOI: 10.1007/s10701-015-9956-8
Cite as: arXiv:1501.00970 [quant-ph]
(or arXiv:1501.00970v5 [quant-ph] for this version)

Professor Castle-Holder:
- Do you agree with the exposé from Mrs. Heidi Fearn?

Open-Eyes:
- In the details? Surely not. Yes, John Cramer published in 1986, so well before I began my own works, but I knew of him only in August 2003, thanks to the

webmaster of **amasci.com**, William J. Beaty. My works took their own direction in the years 1997-1998, and their development owes nothing to Cramer. The horrified sentences of Fearn against retro-causality leave me as cold as ice; in my sense, they have only a diplomatic interest: we do not more share the same notions of time. Those whose horizon is limited to Cramer never elaborated a theory of the Broglian ground noise; the handshaking remains a mystery for them, and they never voted their independence about conceiving the time in microphysics. They have not yet integrated the fact that for 1928, the Dirac wave equation for the electron implies two components in backward-time, the two ones the mainstream teachers hide to the students. Consequently to Dirac results, the lapping of de-Broglie-Dirac waves in which the transactions emerge is made of advanced waves as much of retarded waves.

Glossary, or micro-dictionary

The scientists are listed in chronological order (not alphabetical).

Galilean relativity, in the early 17th century. As long as the speeds are low compared to speed of light, the speeds compose as vectors in Euclidian space. But as soon the speeds exceed 5 % of the speed of light, or some electromagnetism is involved, you must pass to the Einstein Relativity, 1905.

Marin Mersenne, 1588-1648. French mathematician: pioneer of the dimensional analysis, that he applied to the weight pendulum.

Dimensional equation and dimensional analysis. After Mersenne, the main works were performed by Joseph Fourier (1822), François Daviet de Foncenex (1734–1799), and James Clerk Maxwell (1831-1879). Here we apply it to the period of the weight pendulum, simplified as having a punctual mass **m** held by a wire of length **a**, in a field of acceleration **g**, whose dimensions are respectively M, L, and L.T^{-2}. To obtain the period, dimension T, the only combination is $T^2 = \frac{L}{LT^{-2}}$, so T = coefficient x$\sqrt{\frac{a}{g}}$ where the mass eliminated by itself. Next, you need to derive a periodic movement to know that the coefficient is 2.π.

Willebrord Snellius (Snell van Royen), 1591-1626. Dutch mathematician, discoverer of the laws of refraction.

René Descartes, 1596-1650. A former mercenary, pioneer of the analytical geometry.

Pierre de Fermat, 1601-1665. Mathematician, inventor of the differential calculus. In optics, discoverer of the principle of the extremum path – most often a minimal path.

Fermat's principle: Any path really transferring light is extremum in length, compared to its neighbors. Otherwise said: this geometrical path only differs by a second order infinitesimal from its near neighbors. So no optical path is of null width, but at the time of Fermat, nobody had the wavelength of the visible light, so that no one could compute the effective width.

Fermat spindle: Application of the Fermat's Principle, answering to the question: *on which width, indeed?* Answer: less than a quarter of period late at the absorber imply the evolving width of the spindle. Beyond the Fermat spindle, no power passes. In some configuration (interferences on several slits, or several mirrors, or lattices), several spindles give each an integer number of periods delay, so a periodic admittance on the area of the potential absorbers.

Christiaan Huyghens, 1629-1695. Dutch astronomer and mathematician: observations of Saturn, improvements to clocks, mechanics of the rigid body, moment of inertia, wave theory of light, explanation of the double refraction...

Isaac Newton, 1642-1727. English mathematician: discoverer of the laws of gravitation and mechanics. Believed in a god *who sees all instantaneously*, so believed in a universal time.

Jean le Rond d'Alembert, 1717-1783. Co-directed the *Encyclopédie* with Denis Diderot up to 1757. The equation of vibrating strings in 1747. Created the operator now said the d'Alembert operator, explicit in § 6.4; in Minkowski quadridimensional space, the d'Alembertian may be considered as the extension of the Laplacian operator.

Pierre-Simon Laplace (de Laplace for 1808), 1749-1827. French mathematician, astronomer, physicist, and zealous courtier. In his *Traité de Mécanique céleste* (1799-1825), in five volumes, Laplace transformed the geometrical approach as developed by Newton into an analytical-mechanical approach. The Laplacian operator is a second-order differential operator. Here we express it in Cartesian and orthonormal coordinates to a function Ψ: $\Delta\Psi = \frac{\partial^2\Psi}{\partial x^2} + \frac{\partial^2\Psi}{\partial y^2} + \frac{\partial^2\Psi}{\partial z^2}$.

Thomas Young, 1773-1829. English physician and optical physicist: as experimentist, is the author of several interferential devices, so proved the wavelengths of the visible light.

William Hyde Wollaston, 1768-1826. English chemist and mineralogist: many discoveries, including the dark lines in the solar spectrum.

Joseph von Fraunhofer, 1787-1826. Bavarian optician and physicist: perfected the mechanical, optical and spectroscopic means of astronomy.

Augustin Fresnel, 1788, 1827. French physicist: founder of the undulatory optics.

Joseph Fourier, 1768-1830. French geometrician and physicist: theory of the Propagation of Heat in 1807, resolution of differential equations, and Fourier Transform into periodical terms (sines and cosines), or spectrum, 1822.

Michael Faraday, 1791-1867. English experimenter: several apparatus and laws in electricity, optical glass, chlorides of carbon and benzene...

William Rowan Hamilton, 1805-1865. Irish mathematician: theory of the quaternions; unification of the laws of mechanics and optics.

James Clerk Maxwell, 1831-1879. Scottish physicist: kinetic theory of gases, equations of electromagnetism.

William Kingdon Clifford, 1845-1879. English mathematician. Dead in the age of 34, leaving a tragically unfinished work.

Gustav Robert Kirchhoff, 1824-1887. German physicist: spectral analysis, a law of the black body, Kirchhoff's circuit laws in electrokinetics.

Robert Wilhelm Bunsen, 1811-1899. German chemist: photometry, spectroscopy, calorimetry, carbon-zinc battery.

Woldemar Voigt, 1850-1919. German physicist, crystallographer, and thermodynamician, author of Lehrbuch der Kristallphysik, issued in 1910. He was the first to apply in physics the **absolute differential calculus** (1892) from **Gregorio Ricci-Curbastro** and **Tullio Levi-Civita**, and especially to piezoelectricity, birefringence, and elasticity of the crystals. He created in 1898 the word "tensor", which is widely re-used since, out of the field of elasticity.

Max Planck, 1858-1947. German physicist and thermodynamician; he invented in December 1900 the constant **h**, needed for a law of the radiation of the Black Body: one can buy or sell electromagnetic interactions only by integer multiples of quantum **h** of looping (or action per cycle).

The **Black Body** is an intermediate mathematical fiction: it takes over all the dependence on temperature of the variation in spectrum and power. Next one has to multiply by the spectral luminance of the emitting body to have the complete spectral characteristic of the emission, according to the body and its temperature. For absorbance, nor the Black Body, nor the temperature intervene.

The **universal gas constant R**: 8.314,41 joule per mole and kelvin.

Boltzmann constant, k: 1.380,662 . 10^{-23} J/K (joule per kelvin). The product **kT** is the mean of the individual energies, considered at the individual level in microphysics.

Absolute Temperature: measured in kelvin, it is the quotient by the Boltzmann constant of the mean energy of each degree of freedom, that can participate in the thermal exchanges.

Planck's law of the radiation of the Black Body (December 1900): With
L_A = radiated power.
ν: frequency of the photon, in herz.
h: Planck's constant.
c: speed of light.
k: Boltzmann's constant.
T: absolute temperature.
$$L_A = \frac{2h\nu^3}{c^2} \cdot \frac{1}{e^{\frac{h.\nu}{kT}} - 1}$$
Based on fine considerations on the entropy of the radiation in a cavity, the Planck's law invalidates not only the copenhaguist mythology of 1927, but also the first version of the transactional physics, the one of John Cramer (1986), who imagined only two partners in a transaction.

This Planck's law was preceded empirically by two of its theoretical consequences:
Stefan-Boltzmann Law: The power emitted by a unit of area of the Black Body is proportional to the fourth power of its absolute temperature.
$J = \sigma T^4$.
With $\sigma = 5.67 \times 10^{-8}$ W. m^{-2}.K^{-4}
Hence comes that the stars radiate much energy by electromagnetic means, and receive little, in comparison.
Hence the **astronomers curse**: to focus on the shining properties of the emitters assemblies only (mainly stars), and neglect the too discrete properties of the absorbers, as they are much more many and more dispersed than the emitters.

Wien's Law: the frequency of the maximum of radiated emission is proportional to the absolute temperature. $\nu_{max} = $ T x 58.8 GHz / K

Nernst's Principle, or Third Principle of the Thermodynamics. The original statement by Nernst: *"The entropy of a perfect crystal at 0 K is null."*
More generally: the number of degrees of freedom (or instance, phonons) able to participate to the thermal exchanges severely decreases with temperature, so the calorific capacity of any condensed body tends towards zero when the temperature decreases.

Definition of the **entropy** according to the statistical physics:
Ω is the number of microscopic configurations giving the same macroscopic state.
S = k$_B$. Ln (Ω)
where **k$_B$** is the Boltzmann's constant. **Ln** denotes the Neperian logarithm.
The entropy is an extensive quantity: if you double the quantity of matter in

the same state, the entropy doubles also.

Action: a quantity invented by Pierre Moreau de Maupertuis (1698-1759), next largely precised by Euler (1707-1783), Lagrange (1736-1813), then Jacobi (1804-1851) and Hamilton; it has *almost* the physical dimension of an angular moment. *Almost!* By these authors, the action is a **scalar** quantity, the sum on the path of the scalar product of the momentum by the elementary displacement. While an angular moment is a **gyratorial** quantity, **outer** product of the momentum by a lever arm.

In 1924, Louis de Broglie proved that the Maupertuis *Principle of least action* and the Fermat's principle are physically identical: for any object with mass, the path is everywhere orthogonal to the isophase surfaces of the Broglian wave. Now the relationship between the three kinds of "action" quantities, and notably its quantum **h**, scalar or gyratorial, or (not yet defined), is seriously complicated since, in the development of quantum mechanics. Nobody yet has proposed a satisfying solution. The action is a relativist invariant, and lies in the core of the Emmy Noether's Theorem (1882-1935).

Hendrick Antoon Lorentz, 1853-1928. Dutch physicist: theory of light and electron. The Lorentz Transform links a material frame to another material frame. Inapplicable to the photon: it is massless.

Albert Einstein, 1879-1955. Theoretical physics. Author of the first application of the Planck's quanta to the photoelectric effect. Inventor of the Special Relativity (1905) and the General Relativity (1916).

Max von Laue, 1879 – 1960. Radiocristallography. Laureate Nobel in 1914.

Peter Joseph Wilhelm Debye (born Petrus Josephus Wilhelmus Debije) 24 March 1884 in Maastricht - 2 November 1966 in Ithaca, New York, U.S.A.) was a Dutch physicist and chemist. Laureate Nobel in 1936. Solid state physics at low temperatures, radiocrystallography.

William Lawrence Bragg, 1890-1971. Australian physicist. Laureate Nobel in 1915, jointly with his father William Henry, *"for their services in the analysis of crystal structure by means of X-rays"*, that is the foundation of the radiocrystallography. Knowing the wavelength of the X radiation, the Bragg equation links the angle of diffraction with the equidistance of the crystallographic planes in the lattice.

Bragg equation: $\mathbf{2d\ sin(\vartheta) = n.\lambda}$

Figure 22.1.

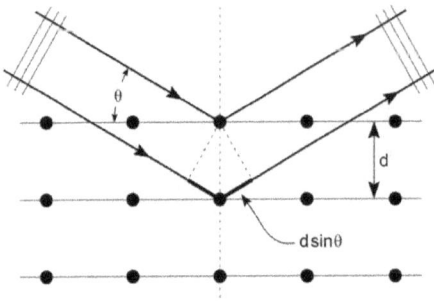

Arthur Holy Compton, 1892-1962. American physicist: discovered in 1923 the scattering of X-rays by loosely bound electrons. Laureate Nobel in 1927. Compton scattering: an incident gamma is deflected, and its frequency is lowered. The deficit in frequency corresponds to the momentum given to the ejected electron. See in the text, § 10.8.

Erwin Schrödinger, 1887-1961. Austrian physicist. Won the Nobel prize in 1933 for the (non-relativist) wave equation of the electron (1926), based on the (relativist) electron wave invented in 1924 by Louis de Broglie. Next, for 1930, worked on the Dirac's relativist equation, and obtained the decisive result of the Zitterbewegung, due to the interference between the positive-energy and the negative-energy components of the electron wave.

Louis de Broglie, 1892-1987. French physicist, discoverer of the intrinsic frequency of the particles, notably the electron, proportional to their mass: $mc2/h$. Thesis in 1924, Nobel laureate in 1929. Unified the Fermat's Principle with the Hamiltonian formalism.

Paul Adrien Maurice Dirac, 1902-1984. English physicist, with proverbial conciseness. Major contributions to quantum physics, Nobel prize in 1933, jointly with Schrödinger.

Enrico Fermi, 1901-1954. Italian physicist, many developments in particle and nuclear physics, explanation of the beta radioactivity with the emission of a neutrino (1934). Nobel 1938.

Electron: the smallest quantity of electricity, invented in 1891 and proved in 1897 as *cathodic rays*. Its charge is negative; it has a magnetic moment. The electrons are indistinguishable. You can band birds, not the electrons.

Positron: antiparticle of the electron. Its charge is positive.

Electron-volt, eV: energy taken by an electron under a difference of potential of 1 V.

Photon: smallest quantity of electromagnetic radiation, which transfers from its emitter to its absorber a Planck's quantum h of looping.

Proton: Heavy particle or *hadron*, composing the atomic nucleus. The proton has a positive electric charge, opposed to the one of the electron. This hadron is composed of three quarks: u u d. It has a magnetic moment.

Neutron: Heavy particle or *hadron*, composing the atomic nucleus, but without any global electric charge. However, owing to the quarks, it has a magnetic moment. This hadron is composed of three quarks: u d d.

Atom: only one nucleus is escorted of as many electrons as it has protons in the nucleus. The electrons balance the charge of the nucleus.

Molecule: several nuclei bound together by peripheral electrons disposed in covalent bondings (sharing of a pair of electrons of opposed spins, in the outermost orbitals), plus eventually a minority of mixed bonds or hydrogen bonds.

Neutrino: ghostly and ultra-light particle, but still carrying a spin $\frac{1}{2}$ and energy. Its helicity is only on the left, while the antineutrino twists on the right.

Fermion: any particle with the spin $\frac{1}{2}$, electron, proton, neutron, neutrino... They are ruled by the **Fermi-Dirac** statistiboson: any particle with integer spin, 0 or 1. Are bosons: photons, atoms ^{4}He, the pions (composed of two quarks), and also the Cooper's pairs of electrons in superconductivity.

Muon: unstable and short-lived *"heavy electron"*. 2.2 μs mean life, decays into an electron and two neutrinos. Weights 207 times an electron. The muons reaching the ground come from the collision of cosmic rays (usually high-energy protons) on some atom of the high atmosphere. The tauon is a heavier and again more short-lived *"heavy electron"*.

Atomic nucleus: Composed of Z protons and Y neutrons, it can retain Z electrons to form an atom.

Isotope: Though only the number of protons fixes the *Atomic Number*, therefore the chemical properties, a nucleus often remains possible with different numbers of neutrons; then the atom keeps the same place in the Mendeleiev periodic table. If the number of neutrons deviates too much from an ideal rule, this isotope is unstable. Lead and bismuth are the heaviest of the elements with a stable isotope.

Ion: Atom or molecule whose number of electrons does not balance the number of protons.

Anion: a negative ion, too many electrons for the protons. Examples: OH^-, $SO4^-$, HCO_3^-, Cl^-...

Cation: a positive ion, not enough electrons for balancing the protons. Examples: Na^+, Ca^{++}, H_3O^+.

Electrolyte: a solution containing simultaneously anions and cations, able to conduct the electric current. The most usual electrolytes are aqueous solutions, but other solvents exist, and the molten salts are electrolytes too.

Electrolysis: action involving the passage of electric current through an electrolyte, where the anions move to the anode $(+)$ and the cations to the cathode $(-)$, either spontaneously or by an external generator. The corrosion of metals in wet ambiance, and *a fortiori* in a marine environment is always electrolysis.

Mole: Quantity of chemical species containing **N** specimens of the species, atoms, or molecules, or ions, or electric charges. **N** is the Avogadro-Ampere constant: six hundred and two thousand, two hundred and fourteen milliards of milliards specimens. So the same writing of a chemical equation means as well the individual reaction, as the mass involved, rated in moles.
$CaCO_3 + heat \rightarrow CaO + CO_2$
100.09 g $\rightarrow$ 56.08 g + 44.01 g for one mole of carbonate.
But no molecule CaO nor $CaCO_3$ exists; these solids are in a more or less perfect crystalline state, with mixed ionic-covalent bonds.
Williamson synthesis of the ethylic ether:
$C_2H_5\text{-}O\text{-}Na + I\text{-}C_2H_5 \rightarrow C_2H_5 - O - C_2H_5 + NaI$
68.0514 g + 155.9666 g$\rightarrow$ 74.1238 g + 149.8942 g for one mole of each reagent.

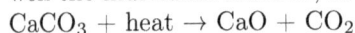

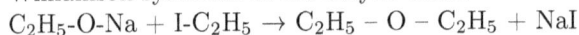

Anode: where the anions go; or the electrons in a vacuum or gas tube.

Cathode: where the cations go, or what emits the electrons in a vacuum or gas tube.
Variant, in X-ray tubes: the **anticathode** is the anode, in a refractory metal, which receives the highly energized electrons, and emits X-rays. In metallurgy, the most used anticathode is in molybdenum; in the mineralogy of the silicates, the copper and cobalt anticathode are the most used. An optimization is on the ratio of the wavelength K_α of the anticathode metal on the largest equidistances of crystal to investigate. When one has to deal with the most difficult to determine phyllites, one may prefer the iron anticathode, giving a longer wavelength.

Fermi energy in a metal: the energy of the highest electron level, at 0 K, based on the energy of the most bound electron.

Rudolf Mössbauer, 1929-2011. German physicist: has studied the gamma rays and the nuclear transitions. Nobel 1961.

Mössbauer effect. Very sharp resoning absorption of a gamma ray emitted by another nucleus of the same isotope. It requires no recoil in the absorber and the emitter. See Appendix D.

Gamma radioactivity: Photons of high energy (very penetrating) are emitted.

Beta radioactivity: emission of electrons or positrons. The β^- are soon stopped, but the β^+ will meet electrons in the matter, and the pair e^+ - e^- will annihilate into two gamma rays. In medical imaging, it is the basis of the Positron Emission Tomography (TEP): by recording the two emitted γ, the apparatus deducts where both come from.

Alpha radioactivity: emission of helium nuclei $^4_2He^{++}$.

Spectroscopy: indirect measurement of the frequencies or wavelengths of a radiation to analyze. It is a major method of analytical chemistry, and it remains usable at astronomical distances. Without spectroscopy, no astrophysics.

Dye: in medical use, you know the methylene blue and the eosin. Chlorophyll is essential to all plants. In schoolroom, you had learned the indigo and the alizarin, in the lesson on the aniline. They are molecules of moderate size, mostly those able to be bond to textile fibers. Their secret is a delocalized electron, oscillating between two extreme positions.

F-center (Farbe): What makes the biotite to be dark, and the amethyst to be violet. A lacuna in the crystal, occupied by one or more unappaired electrons. They attract and absorb light selectively.
http://www.webexhibits.org/causesofcolor/12.html

Crystalline state: perfectly ordered state of matter, whose geometry is the repetition of a parallelipipedic cell, in 3D.

Crystalline cell. In the main case, it is the minimal parallelepiped, which can be repeated to give the entire crystal. In some cases, it is preferred to consider a multiple of this cell, to profit of better symmetry; examples: the body-centered cubic and the face-centered cubic, most hexagonal and rhombohedral lattices too. Even when elementary, it happens that a primitive cell contains several atoms.

Crystalline parameters: lengths and angles of the three vectors describing the primitive cell; most often, the higher-symmetric cell is used.

Dislocation: a linear defect of the crystal. Either it is the limit of an incomplete plane, either the axis of a winding staircase.

Gaseous state: the molecules are independent, and spend more time free than in collision.

Condensed states: gather the solid and liquid state, and the pasty intermediates.

Glass and glassy states: Some liquids can set in cooling, without letting grow a crystalline germ. This metastable state may last enough to yield technical uses. Our usual glasses are made of silica skeleton, whose plasticity in the molten state was given by sodium and calcium cations. Some lavas can contain a notable proportion of silicate glass. In ceramic industry, the glazes and enamels are glasses.

Professor Marmot:
- A scandal! Not a word on Werner Heisenberg, on Wolfgang Pauli, on Pascual Jordan, on Max Born, on Niels Bohr, on Eugen Wigner!

Open-Eyes:
- This popularization book teaches semantics, and not the formalism. So I have listed the essential authors who contributed to the semantics, and not those who accumulated the blunders. Also, we had no use of the Bohr magneton.
The Bohr magneton is a physical constant and the natural unit for expressing an electron magnetic dipole moment; it appears naturally when quantifying the angular moments in atoms. It links the magnetic moment $\check{M}$ to the angular moment $\check{L}$ of the electron: $\check{M} = \gamma_e . \check{L}$
where γ_e is the gyromagnetic ratio of the electron: $\gamma_e = -\frac{q}{2e}$
In the model of the Bohr atom, the angular moment $\check{L}$ is quantified and its norm is: $\left\| \check{L} \right\|$. $n\hbar$.
Therefore the norm of the magnetic moment of the electron may be written:
$\left\| \check{M} \right\| = -n\frac{q.\hbar}{2m_e} = $ -n. μ_B where μ_B is named Bohr magneton and plays the role of the quantum of magnetic moment for the electron.

However, the proton and the neutron have surprising magnetic moments, coming from their structure by quarks. But it is another story, out of the scope of this book.

One could also reproach not mentioning André-Marie Ampère in this small glossary, nor Joseph-Louis Lagrange, George Green, Lord Kelvin (William Thomson), Hermann Grassmann, Ludwig Boltzmann, Dmitri Mendeleiev, Pierre Curie, Henry Moseley... Not mentioning the many forgotten craftsmen who made marine compasses, marine (or astronomical) chronometers, and who did so much to improve our knowledge of the laws of magnetism and mechanics.

APPENDIX G

Foiling some fool-traps in the usual folkloric mechanics

In mechanics, be sure to breakneck yourself if you do not master the vectorial tool. You draw it in arrowed sticks. The lesson in line is at:
http://deontologic.org/geom_syntax_gyr/index.php?
title=Vecteurs,_d%C3%A9finition,_propri%C3%A9t%C3%A9s

G.1. Stability of a ladder against a wall

Figure 23.1.

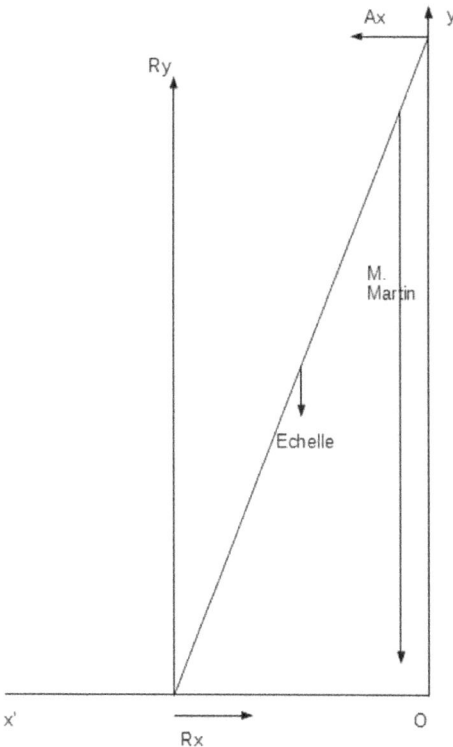

To simplify the calculation, assume that the wall is vertical, and the ground is horizontal, that the ladder is 3 m long ($= L$), that the nine rungs are equidistant at 30 cm, with the first at 15 cm, that the ladder weights 10 daN, and Mr.

Martin weights 90 daN (in mass, about 91.8 kg). Assume that the support on the wall is without friction, so produces only a horizontal reaction: $\overrightarrow{A} = \overrightarrow{A_x}$. As the figure is oriented, the weights have negative coordinates, and $\mathbf{A_x}$ also. We neglect the thickness of the ladder and its rungs. The only remaining variables are the leaning of the ladder $\mathbf{L.sin(\alpha)}$ where α is the angle from the vertical, the horizontal components of the ground reaction $\mathbf{R_x}$, and the position of Mr. Martin (and his weight) on the ladder, $\mathbf{p}$ rated from the ground contact. It is a static problem, so the resulting force is null.

Horizontal component: $\overrightarrow{A_x} + \overrightarrow{R_x} = 0$, so in coordinates, $R_x = - A_x$. R_x is positive.

Coordinate of the vertical component: $R_y = 100$ daN.

The third necessary equation comes from the null of the sum of the moments of the forces. The plane of the drawing is oriented in the counter-clockwise sense. We choose the foot of the wall as the origin of the coordinates.

Moment of $R_x = 0$.

Now, we need to name to name the lever arms of $\overrightarrow{A_x}$ and $\overrightarrow{R_y}$ that are the height of the scale and its footing.

Height: $L_y = L. \cos \alpha$. Vectorially, $\overrightarrow{L_y}$ is upwards.

Footing: $Lx = L. \sin \alpha$. Vectorially, $\overrightarrow{L_x}$ is to the left.

The symbol caret ^ denotes the outer product.

Moment of $A_x = \overrightarrow{L_y} \wedge \overrightarrow{A_x}$ (sense of rotation +, according to the orientation of the figure).

Moment of $R_y = \overrightarrow{L_x} \wedge \overrightarrow{R_y}$ (sense of rotation -, with the norm $R_y = 100$ daN). Moment of the weight of the ladder: 1.5 m . sin α . 10 daN (positive). Moment of the weight of Mr. Martin: (3 m - p) . sin α . 90 daN (positive, quasi-null when Mr. Martin is upstairs, at 2.75 m from the ground support).

Now the matter is to compare the arctangent of the quotient R_x / R_y to the limit-angle of friction of the contact ladder/ground.

Now for resolving algebraic equations, we reuse the same letters, but now meaning mere coordinates: $\mathbf{p}$ becomes a number of meters, and $\mathbf{R_x}$ the coordinate of $\overrightarrow{R_x}$ (horizontal component of the contact force from the ground to the ladder).

3. cos α . R_x - 3. sin α . 100 + 15 sin α + (3 - p). sin α . 90 = 0

R_x = tan α (100 - 5 - 90 +30 p) = tan α (5 + 30 p)

Bad luck! The instability is maximal when Mr. Martin in on the upmost rung: R_x maxi = tan α . 87.5 daN.

In the final result, the length L does not intervene, only the ratio $\mathbf{p/L}$, and the leaning α. The most the ladder is leaning, the most the friction at the ground is critical.

Critics: one could decompose the problem in two, more simple.

When does the ladder slip under its own weight $\mathbf{P}$?

$$\frac{R_x}{P} = \frac{1}{2} \tan (\alpha)$$

And simplify by neglecting the weight of the ladder, only consider the weight M of Mr. Martin: $\frac{R_x}{M} = \frac{P}{L} \tan \alpha$.

The worst is when Mr. Martin is upstairs: $\frac{R_x}{M} = \tan\alpha$.

Conclusion: Always keep the leaning of a ladder well under the limit friction angle of the contact ladder/ground. The simulation by the empty ladder is only valid if the worker does not climb more than the middle of the ladder.

G.2. The "Roman balance" and the moment of a force

Above, I have handed the moments out of a hat, and I did not draw them. For thousand years, our ancestors use levers, so the laws of the moments were empirical, until Archimedes and few other scientists theorize them.

Here we view the Roman scale. When I was a kid, it was used by the green-grocers on the market at Grenoble.

Figure 23.2.

Figure 23.3.

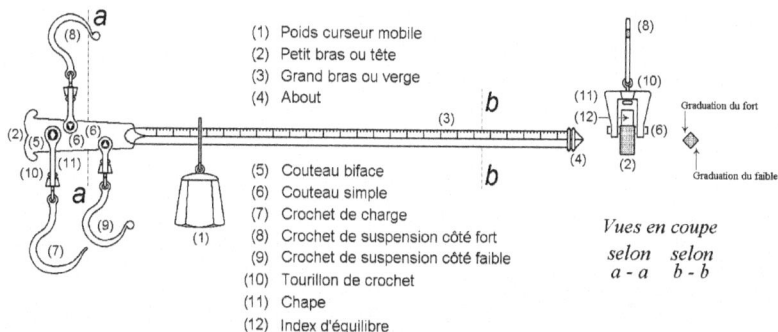

(1) Poids curseur mobile
(2) Petit bras ou tête
(3) Grand bras ou verge
(4) About
(5) Couteau biface
(6) Couteau simple
(7) Crochet de charge
(8) Crochet de suspension côté fort
(9) Crochet de suspension côté faible
(10) Tourillon de crochet
(11) Chape
(12) Index d'équilibre

Graduation du fort

Graduation du faible

Vues en coupe
selon   selon
a - a   b - b

Now we simplify the scheme. Figure 23.4.

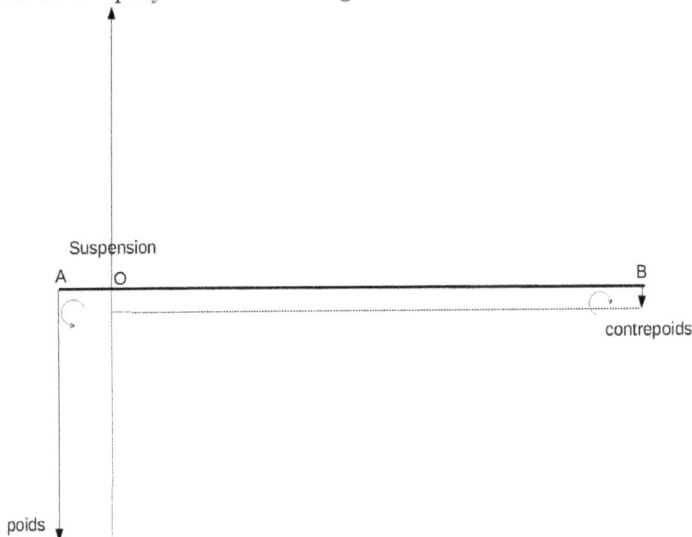

Suspension

A   O   B

contrepoids

poids

Now the dimensions:

the beam is 55 cm long, 5 cm for the segment OA, 50 cm for OB. The weight on A is 10 N, in B it is 1 N. The moment in B, as seen from the observer of the page, turns clockwise, tends to lower the point B, and has for norm 0.5 N.m. The moment in A turns anti-clockwise, and also has 0.5 N.m as the norm. The moments are represented by the areas of the parallelograms in dotted lines - areas oriented in rotation: the moments are **gyrors** - drawn on the arms of the beam and on the arrows representing the weights. There is equilibrium; the scale does not slope on one side, it can at most slowly oscillate around an equilibrium position; the manufacturer has designed a slight amount of vertical offset between the fulcrums to have a stable balance, not drawn on the simplified scheme.

Up to now, we have just mathematized an empirical knowledge; already confirmed for three to four thousand years. But now, we will prove by a **general physical principle** that it cannot be else. Suppose the beam begins to slope, letting B down for 10 mm. So the counterweight in B has worked, and it positive work is 1 N x 0.01 m = 0.01 J. This counterweight has lost potential energy in the field of gravity.

But in A, the weight went up, the work by this weight is negative, with the same norm: 10 N x (-0.001 m) = - 0.01 J. The weight in A has won potential energy, exactly as much as B has lost.

The sum of these virtual works is null ==> equilibrium of the beam.

Going beyond the simplified scheme and returning to the real scale, sure when sloping, the beam slightly gained in potential energy, as it globally wends higher than the central fulcrum; otherwise, the balance would not have any stability and would always be on the edge of a catastrophe.

G.3. The sling that bent a master-beam of the Sèvres bridge

The crane boom was not high enough? Were the lanyards fasten on too far points? The resulting compression stress went higher than the limit of bending of the beam... And the beam is ruined, not any more mountable. They had to make another one.

Scheme of the forces. Alas, the real dimensions and angles of the real bridge-building incident are not documented. It suffices to simulate by the general law: Be P the total weight of the beam, D the distance between the points of suspension, L the length of each leg of the lanyard, T the tension one each sling-leg (supposing the sling is symmetrical), C the force of compression on the beam, α the angle between the beam and each sling-leg.

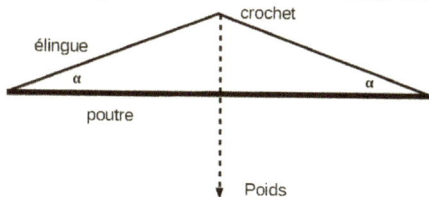

Figure 23.5.

Then $\alpha = \text{Arccos } \frac{D}{L}$, $\frac{C}{2T} = \cos \alpha = \frac{D}{2L}$
and $\frac{C}{P} = \text{cotg } \alpha$
$\frac{T}{P} = \frac{1}{2 \sin \alpha}$

Numerical applications:
The limit of the reasonable is T/P = 1, soit $\sin \alpha = 1/2$, $\alpha = 30°$, $\frac{D}{2L} = 0.866$, $\frac{C}{P} = 1.732$.
Surely unreasonable: T/P = 4, so $\sin \alpha = 1/8$, $\alpha = 7.18°$, $\frac{D}{2L} = 0.9921$, then $\frac{T}{P} = 7.94$.

G.4. The hammock suspensions; will they break?

We do not have the exact geometry of a hammock with the sleeper in, but we can study the worst case, for security purpose. The worst case is when the hiker is sitting at the center, that is when he/she climbs into the hammock (supposing not aiding with the ridge-rope which supports the roofing tarp), or when he/she will step down; this concentration of weight in the center gives lowest slopes of the suspending ropes, so the higher tension on them. The most favorable case would be with a hammock curve as a parabolic arc; we will not treat it.

We study only a case on the edge of the unreasonable, with T/P = 2 (tension of 200 daN for a sleeper weighing 100 daN), so $\sin \alpha = 1/4$, $\alpha = 14.47°$. With 3 m between trees, it gives a maximal sag of 38 cm, or 50 cm between trees distant of 4 m. Also, the ridge-rope, which supports the roofing tarp which shields the hammock from rain, must have a non-negligible sag; the hiker relies on the elasticity of the ridge-rope to accommodate the worst case where the hiker adds to the wind forces on the tarp.

Figure 23.6.

α

flèche

P/2

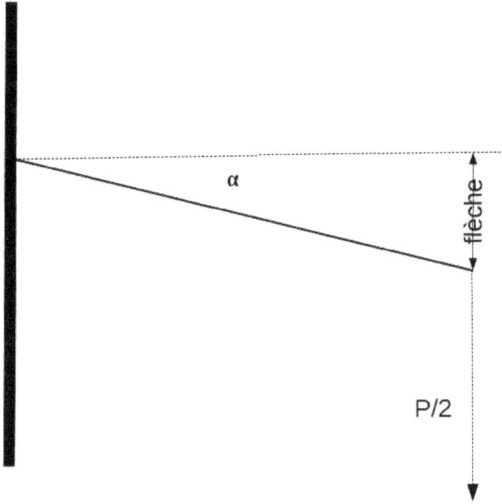

We study only a case on the edge of the unreasonable, with $T/P = 2$ (tension of 200 daN for a sleeper weighing 100 daN), so sin $\alpha = 1/4$, $\alpha = 14.47°$. With 3 m between trees, it gives a maximal sag of 38 cm, or 50 cm between trees distant of 4 m. Also, the ridge-rope, which supports the roofing tarp which shields the hammock from rain, must have a non-negligible sag; the hiker relies on the elasticity of the ridge-rope to accommodate the worst case where the hiker adds to the wind forces on the tarp.

G.5. Inertia

Throwing weapons are in the equipment of humankind for millions of years. As a proof, the savannah chimpanzees: they can tear in pieces a stuffed leopard, by throwing stones and sticks of wood. Less trained, the forest chimpanzees throw sticks as strongly, but far less accurately. Paleontologists could estimate that several Paleolithic civilizations could enhance tenfold their efficiency in hunting, and so their headcounts, when they invented and developed either arcs and arrows, either the spear-thrower arm. The Inuits are harpoon-thrower hunters: the harpoon-thrower arm increases the kinetic energy of the harpoon. The empirical knowledge of the kinetic energy of the throwing weapons are here for long, but its mathematization waited until Isaac Newton, end of 17th, beginning 18$_{th}$ centuries. And before, we had to wait for the development of the firing artillery, and the permanent armies of the 16th century, the contributions by engineers of the Renaissance period to the art of siege, and the decisive progress from the algebraists, next of the analytical geometry. The apprentice physicists in Ancient Greece could not liberate themselves from the laws of friction, omnipresent on the roads of Attica or Minor Asia, omnipresent on the seas; only the progress of the position astronomy by Tycho Brahé, next Kepler, and Galileo, could give a knowledge on movements without friction, purely inertial: the celestial movements of the planets and their satellites.

As a general rule, it is not yet assimilated, four hundred years after the Galileo relativity, three hundred and thirty years after Newton's mechanics. An extreme case in the ante-Galilean absurdity is a retired pilot of the Air Force, pretending that his plane had to constantly *push forward* the most primitive instrument of flight, the ball in the turn indicator. He had just forgotten that the drag from air acts on the fuselage, wings and governs, but not on the inside, and **at a constant speed**, the push from the reactor only serves to compensate this drag. Hyper-proud pack, imbued of their *superiority over the remaining of the world*, these retired from the AF find very hard to understand that speed is a vectorial quantity, and changes in every turn, even when its Euclidean norm as shown by the Badin indicator does not change.

Let us return to the historic didactic test, which proved that two-thirds of the students in the second year in University still have not yet assimilated neither the laws of inertia, neither the changes of frame.
Assume a liner, which takes a flat turn, to the left. A flat turn is a very bad turn, with dangerous aerodynamical effects, but we will accept such an abstraction. Air is supposed to still, no wind. The norm of the speed of the plane is supposed constant, ignoring the aerodynamical braking by the flat turn. The passenger experiments by laying an ice cube on a horizontal table in the plane, it may slide without friction. Now the question to the students was to draw the movement of the ice cube in two frames: the frame of the plane, and the frame of an observer motionless on the ground.

Correct answer: seen from the ground, the ice cube continued in a straight line, without following the turn of the plane; seen from the plane, the ice cube moved on the right, *"pushed by the centrifuge force"*.

Now make explicit the mechanical connections on this ice cube: it is borne by the table, which compensates the gravitational attraction, and that is all. Not any horizontal bond in the frame of the plane, no restriction to its inertia on the horizontal plane. So the only law for the ice cube on the horizontal plane is its inertia: to continue moving in straight line, as seen from the ground. But turning, be a flat turning or a correct turning, the plane is no more an inertial frame, but an accelerated frame, an improper frame. The term *"centrifugal force"* was invented for continuing to reason as in the familiar static case, even in an improper frame, just for describing the inertia seen from a wrong frame. If the plane undertook a correctly coordinated turn, the ice cube would not have slid on the table.
Above, I have simplified, doing as if the ground would be an inertial frame, though in fact it is submitted to the Earth rotation, whose effects are negligible

at the speeds of a jet, but major in meteorology and physical oceanography.
http://deontologic.org/geom_syntax_gyr/index.php?
title=Acc%C3%A9l%C3%A9ration_de_Coriolis_et_inertie

But what does the ball in its bent tube? In a flat turn, it starts to the right, and it is the task of the pilot to coordinate the roll with the turn to keep the ball in the center (in the frame of the plane), so he preserves the flow of air to remain from forward toward the tail, and not transversally from right to left of the plane. Even so, it remains a dissymmetry in the flow of air: it is faster on the outer side of the turn, so lifts better, and slower on the inner side. In extreme cases, the wing on the inner side stalls, and the plane pitches nose-down, or enters in spin. So crashed a B52 bomber (1994, Fairchild airbase) whose pilot showed acrobatics to the public. So crashed also a liner (Boeing 737) after taking off from Charm el Cheikh in January 2004, because of a defect in a leading edge flap; Power and corruption, only the dead pilots were accused, that extincts any criminal suit.
http://www.dailymotion.com/video/x3wk4x_crash-b-52_news
https://www.youtube.com/watch?v=GCLD6bwGq1I

G.6. Mass ≠ weight

The mass is another way to count « *how much matter?* ». It is a nonoriented quantity, a scalar. At the surface of a massive celestial body, like the Earth, this mass is attracted toward the center of the center of Earth: the law of universal attraction. This force of attraction, depending on the place and the altitude is called the weight. Weight is a vectorial quantity; as Earth rotates at a moderate angular speed, the inertia centrifugal reaction is moderate compared to the gravity attraction, and the weight points very nearly to the center of Earth.
$$\vec{P} = \vec{g} \cdot M$$
where $\vec{g}$ is the gravity acceleration, usually oriented toward the center of Earth, $\vec{P}$ is the weight, and **M** is the mass of the test body. You have all seen that on the comet **67P/Tchourioumov-Guerassimenko**, a very little massive body, the gravity is extremely small, so small it never could oblige this aggregate to take a round form, far from it.

A twin-pan balance, with the two pans submitted to the same gravitational field measures masses indeed, while a spring steelyard measures the norm of the weight vector. So does also a kitchen electronic scale, also essentially a spring with strain gauges. Laboratory electronic scales must be calibrated with accurately calibrated masses, to obtain their graduation in grams be valid.

G.7. Energy

The word "*energy*" is a pet theme for millions of adepts of magic, and magical "*medicines*", they stuff in it all the nearly anything they want. In contrast, for the scientists, "*energy*" has a precise and non-negotiable meaning: it is a

universal currency of exchange, which is measured or calculated upon rigorous and non-negotiable protocols. Physical dimension: 1 joule = 1 kg . m^2 / s
Let take the weight pendulum, mentioned when we evoked Marin Mersenne, its mass **m**, the gravity acceleration **g**, and **h** the elevation of its inertial center at the ends of its run. At the ends of runs, it has taken a potential energy **mgh**. At its low passages with the speed **v**, it has converted its potential energy into kinetic energy $\frac{1}{2}mv^2$.
Does this exchange continue forever? Surely not, as the pendulum slowly dissipates its mechanic energy into heat, by friction in the air and at its suspension point. Dissipation is inherent to macrophysics.
However, some permanent oscillatory *"movements"* exist, without dissipation, but at the microphysical scale. We have already seen the example of the longitudinal vibration of the CO molecule: it may absorb one quantum **h**, or two, even more, but cannot step down under a half-quantum, a perpetual oscillation it can never lose.

Curious:
- You say the perpetual oscillation of the diatomic molecule CO, or others, exists. But if I ask you to prove it? Will you shirk your task?

Professor Castle-Holder:
- Admit that! For you, an admissible proof would be a video, like in our macrophysical world? But light is far too big to give you the slightest image of a gas molecule. The CO molecule is 3 Å on small axes and 4.7 Å on the long axis, and these diameters are fuzzy. Thereon, you demand a resolution of the image better than 0.003 Å to have maybe a chance of detecting the longitudinal oscillation. And what will be the sensor of the movement, at the frequency 64.250 THz? Your hand? Or light? Visible light has wavelengths in the range 0.5 μm, or 5,000 Å. In microphysics all our proofs are indirect; our models are able, or are not, to draw correct predictions. As long as they succeed, we do not worry.

Dissipation into heat is inherent to macrophysics and its statistical laws. Conversely, end 18th and mainly in the 19th century, were invented thermal machines, able to convert a part of the heat energy into work, mechanical energy. Here, we will not describe the kinetic theory of the gas, achieved in the late 19th century; it describes accurately how and why one may partly convert a flux of thermal energy into mechanical work.

Let return to mechanical energy.
You are entering to port, with the motor, to take your place at quay. But at which speed? Your boat with the crew has a total mass of 1,000 kg. At 3.6 km/h (about two knots), that is 1 m/s, her kinetic energy is $\frac{1}{2}mv^2$ here 500 J. Are you sure you will damp 500 J just with your boot? 500 joules will do damages! Better to divide the speed by three, so you divide the energy by nine: 55.5 J is much more manageable to stop. And anyway, you have rather to completely stop your boat by reversing the propeller before any contact with

the quay and the other boats, and just terminate the approach with the gaff.

Beware: any energy is a quantity depending on the frame. And so is also the motion amount, we will see now.

G.8. Motion amount or momentum

In history, the notion of motion amount was even slowlier to elaborate than the kinetic energy. Isaac Newton began by expressing the fundamental law of mechanics in terms of variation of the motion amount, but alas he changed his mind, and published in terms of acceleration – the form in which it is still taught.

$\vec{F} = \frac{d\overrightarrow{mv}}{dt}$ to compare to: $\vec{F} = m\frac{d^2\vec{x}}{dt^2}$ where $\vec{X}$ is the position vector. The two forms are equivalent as long as the masses remain constant. In aerodynamics or hydrodynamics, the expression by the motion amounts is by far the most fruitful. The principle of conservation of the **global conservation of the momentum** has no exception. To change the momentum of a material object (or fluid), you must exert a force on it.

A simple formulation of the momentum (without electromagnetics, without charges, without magnetic field):

$\vec{p} = m.\vec{v}$

It is a vectorial quantity, with three coordinates on our ordinary space. Now it is no more taught, you are no more aware that it is an essential physical quantity.

Physical dimension: 1 kg.m/s.

No international name for this unit, alas. Are our students allowed to think something which has not even a name?

For the cartoonists (and even for a mock anthropologist but real feminazi activist: Françoise Héritier), not a doubt: the ancestors of humankind lived in caves and used cudgel for hunting... Did you try to feed yourself by hunting with a cudgel? Are your on-feet-menu so agreeing on letting you come so close and waits for your cudgel blow? Only two exceptions: the dodo at Mauritius (extinct), and the baby-seals waiting for their mother on the strand. However, all our real game have either superior olfaction and ear if they are mammals, either a very superior sight if they are birds. No one is willing to let a predator approach (except the dodo, then ignoring what a predator is).

Our ancestors hunted using traps or using projectile weapons. Spear-thrower arms, blowpipes, bows, and later crossbows, all these arms avail throwing the projectile farther and faster, to give it more devastating kinetic energy, as the motion amount is necessarily limited.

For a given crossbow, with a given acceptable recoil on the shoulder of the shooter, what is the best choice? To shoot a bolt of 60 g, or of 120 g? You are given the initial speed of the 60 g bolt: 360 km/h = 100 m/s. Hence its kinetic

energy is 0.06 kg / 2 x 10,000 m^2/s^2 = 300 J.
With equal recoil, the 120 g bolt carries 150 joules, at the speed of 50 m/s, but
it will be less braked by air, if correctly streamlined; so on long distance, their
differences lessen. Both carried the same motion amount: 6 kg. m/s. And
what recoil energy receives a crossbow of 3 kg?

6 kg.m/s / 3 kg = 2 m/s. And 3 kg / 2 x (2 m/s)^2 = 6 J. It is far less devastating than the 300 J of the 60 g bolt, able to pierce the armor of a knight.

Figure 23.7.

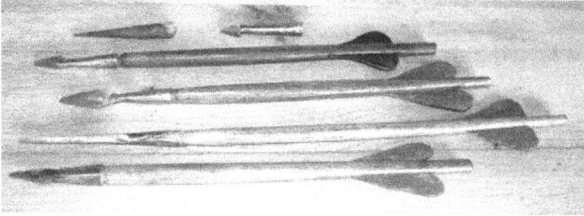

This is why our ancestors perfected the throwing arms: wildebeests and boars
run fast, and have a strong leather.

On a Tommy gun, or an assault rifle, the recoil of the gun has a disastrous
effect on the precision of the second and third bullets: the gun pitches up. So
NATO had to revise their notions on the american cartridge .30-06, too heavy
to be transferred from the long-range sniper rifle without automatic repeating,
to an assault rifle for shorter distances. So they decreased both the weight of
the bullet and the propelling charge, that decreased the recoil and the pitching
up.

Compare the recoil-energy absorbed by the shooter, to the impact energy of
the projectile.
The visual support is a video taken by ANNA agency, from 1' 06" to 1' 11",
of a precision gunner who uses an antique bolt rifle, remaining from the world
wars.
Vidéo https://www.youtube.com/watch?v=R2AEU3jTL7w
extracted from
https://www.almasdarnews.com/article/video-syrian-army-advances-
northern-aleppo/
The recoil is about 6 cm, maybe more. Before knowing the exact identification
of the rifle, let admit it is a Springfield 1903A4 of 1942, but with a new rifle
scope. It weighs 4 kg.
It shot the .30-06 (also said 7.62 x 63 mm), whose bullet weighted 150 *grains*,
about 9.72 g. About 820 m/s at the muzzle (at least, as announced nowadays
at 890m/s for hunting cartridges).
That is 7.97 m.kg/s of momentum and 3270 J at the muzzle.
Hence the speed of recoil: 820 m/s * 7.97 g / 3,980 g = 1.64 m/s.

Hence the recoil energy of 5.35 J. Here the calculation is exact, as the breech was fixed.

Admit the recoil is 6 cm. On this recoil, the work of the holding forces absorbs the kinetic energy.

Mean braking force: 5.35 J /0.06 m = 89 N. In the spread opinion in the hunters, this ammunition is at maximum the shoulder of a hunter can endure.

Now, an expert has identified the rifle in the video: a Mosin-Nagant M91-30 PU, manufactured in 1932 and later, whose design goes back to 1891. The mag before the trigger is characteristic. Its munition is a 7.62 × 54 mm R. In this version, it weighs 4 kg.

Figure 23.8.

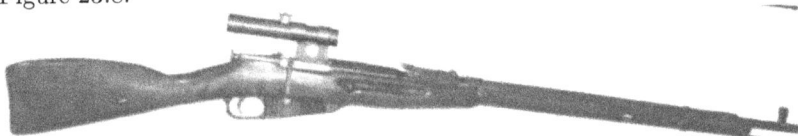

Mass of the projectile: 9.6 g. Initial speed: 870 m/s. Initial energy: 3,500 J. Momentum: 8.35 kg.m/s. Speed of recoil of the weapon: 2.09 m/s Recoil energy: 8.72 J. Mean braking force on 6 cm: 145 N.

The shooter of an assault rifle, automated, endures very different conditions: the breech recoils helically and ejects the case, charges the next cartridge, and comes back in firing position. So the shooter has far less recoil energy to absorb himself. Take the AK-47, weighing about 5,100 g with a full mag, which shoots M43 ammunition, of 7.62 mm, weighing 6.5 to 7.8 g. Initial speed: 720 m/s; initial energy: 1,991 J. For the remaining of the calculation and its coherence, we choose the mass of the projectile to be 7.68 g, almost the usual war bullet.

Figure 23.9.

Momentum: 0.00768 kg x 720 m/s = 5.53 kg.m/s. It is very near the value taken by the crossbow seen above. Initial kinetic energy with these values: 1,991 J.

What should be the recoil speed of this weapon if it had a fixed breech?
5.53 kg.m/s / 5.100 kg = 1.08 m/s. It would give a recoil energy of
5.1 kg / 2 x 1.176 m^2/s^2 = 3 J. This calculation is invalid.

As the breech is far less heavy than the whole weapon, if borrows more energy to the propellant gas, more than 10 J, maybe 20 J.

See now the firing rate, about 100 shot/minute. That is a period of 0.6 s, whose the third, 0.2 s is for absorbing the momentum of 5.53 kg.m/s. Hence a recoil force absorbed by the gunman: 5.53 kg.m/s / 0.2 s = 27.7 N, for each bullet, so a mean force of 9.2 N over the burst. Approximately, the energy taken by the gunman is in the magnitude of 0.5 J per shot bullet.

Better avoiding to try to calculate the pitching up, according to the geometry of the rifle and the morphology of the gunman.

In Relativity, the energy and the motion amount are inseparable: together they form a vector with four coordinates, **energy-momentum**; the transformation of these coordinates in a frame change is regular. One must know that light carries some motion amount (a very little one): a photon of frequency ν, transfers an energy **hν** and a motion amount **hν/c**. Hence a recoil of the emitting or absorbing atom. To observe a Mössbauer resonance, one has to block the emitter as well as the absorber in massive solids, and often cool the to very low temperature, to limit the perturbation by phonons.

G.9. Force

In the classroom, when you began to study the statics, the force was what is transmitted by a thread. We need a little more to deal with the lift on a plane wing, or to accelerate an electron in a synchrotron.
Physical dimension: 1 newton = 1 kg.m/s^2 = momentum / time = energy / length.

Taking the recoil of the Mosin-Nagant rifle: M91-30 PU. Recoil energy: 8.72 J. Mean stopping force on 6 cm: 145 N. On 3 cm: 290 N. On 1.5 cm: 580 N. Good morning the bruises on the shoulder! The magnitudes of the stopping forces are much more grave in a road accident or even worse in a plane accident. Only *Superman*, etc., but only in the cinema.

Why is a **fire hose** a so difficult to master gear, even dangerous if it escapes to the hand or the firemen? Because it throws a high flow of motion amount. A magnitude: Let's consider a "*big hose*", flowing 500 l/min, and whose final nozzle is 18 mm wide, that means a section of 2.54 mm^2. It is not the most powerful: a 1,000 l/min hose exists.
Converting the flow: 500 l/min = 8.33 l/s = 8,333 cm^3/s.
Hence we deduct the initial speed just out of the nozzle: 8,333 cm^3/s / 2.54 cm^2 = 3,281 cm/s = 32.81 m/s.
Hence the force exerted by the flow on the hose: 8.33 kg/s x 32.81 m/s = 273.3 N (it is a flow of motion amount). One has to be a strong and trained man to master it with the required accuracy. And if the hose escapes? It beats back snaking, and may cause damages to the crew.
Just for curiosity, what could be the height the spurt could reach, if air friction

did not exist? At 32.81 m/s, each kilogram of water has a kinetic energy of 538 J, to convert into potential energy **mgh**. ==> **h** = 538 J / (9.81 kg.m/s^2) = 54.8 m. More than a 15 floors building, in real conditions.

Now switch our attention to aeronautics and the mechanics of the flight.
A plane weighing 50,000 newtons (mass approx 5 tons, about a Piaggio P180) must throw how much air downwards, and with which speed, to keep flying?
As a provisional guess, we suppose the vertical speed of air, in the frame of the plane, is 10% of the horizontal speed, so in the range of 66 km/h = 18.3 m/s. In the Beaufort scale, such a speed is the force 8. At ten Beaufort, it would be a tempest. Now, the plane needs how much of such a vertical wind?

50,000 N / 18.3 m/s = 2,732 kg/s
At low altitude, assume that the density of air is 1.2 kg/m^3, so the volumetric flow is 2,277 m^3/s.
The wingspan is 14 m, roughly guess that only 13.5 m is efficient (because of the fuselage and vortex at wing tips), it comes 169.6 m^2/s for the product speed x thickness of the deviated air flow. Now this plane flies at about 183 m/s at low altitude, which yields a deviated equivalent thickness about 0.92 m (92 cm). Compared to the chord of the wing, about one meter or little more, it seems coherent. Of course, the real deviated flow has a completely fuzzy thickness, it is less and less influenced by the wing with increasing distance.

In this rough guess, I do not know whether I have under- or over-estimated the vertical deviation. If a reader has reliable data coming from a wind-tunnel, I will welcome. A control of plausibility is the configuration tanker-aircraft / refueled-fighter: the fed plane must fly under the downwash of the tanker.

Some photos to guide the reader:

Here a Cessna : Figure 23.10.

And another twin engines:

Figure 23.11.

Now a bomber, the Tu 95, with a thick wing profile, very lifty:
Figure 23.12.

For charity, in the English version, I cut the sarcasm from a retired military pilot.
https://groups.google.com/forum/?hl=fr#!searchin/fr.rec.aviation/
ravitaillement\$20vent\$20de\$20travers/fr.rec.aviation/
w780uxVCMk4/GIXyEJ9QCQAJ

While reliable data lack on the aeronautic side on the matter of vertical speed and equivalent thickness of the fluid deviated stream, many incidents, and even accidents demonstrate the enormity of the wingtips vortex, lasting several minutes after a heavy liner has taken off or landed. The practical differences between thick or thin wings are well known; For instance, refer to the French pilot Pierre Clostermann, about the thick profile of the Typhoon, with surprising reserves of lift, compared to the thin wings of the Spitfire or the Tempest. In other words, the thickness of the influenced-by-the-wing air depends on the chord, the thickness, and its eventual bending of chord.

Second plausibility control, on a helicopter, here an Alouette III: 11 m for the diameter of the rotor, max mass 2200 kg. About 85 m^2 the useful area of the rotor. To lift 21,580 N, the rotor must flow 21,580 kg.m^2/s^2 of difference of motion amount per second. For air density 1.2 kg/m^3 at mean altitude, it means

$17,980 \text{ m}^4\text{s}^{-2}$.

On 85 m², the square of the average vertical speed must be 216 m²/s², that is the square of 14.7 m/s. Indeed, in aerodynamics (also in hydrodynamics of fins, blades, or hydrofoils), the speed of the fluid intervenes twice: once in the motion amount, and also in the (temporal) rate of renewing the mass of fluid near the profile, multiplying the (fuzzy) thickness of the influenced stream.

Now we simplify all, we forget the frequency of the rotor and the thickness of the stream: we admit a uniform vertical speed under the rotor (it is false, our ears say it loudly, but let us continue nevertheless): 14.7 m/s it is the degree 7 in the Beaufort scale; "*All the trees rock. Walking against the wind is difficult*". It is confirmed that the magnitude is right. I did not consider any acceleration of the chopper, nor ascensional speed, so I say that 15 m/s is a minoration.

Among the sailing yachtsmen, we are fully aware of the practical differences between flat or full sails, but it is severely prohibited to reason it in terms of motion amount. It would be contrary to the fashion of the tribe, even *politically incorrect*, even *heretic...* Do you remember the Shadoks? Well, I am just saying that the official doctrine of the tribe, violently against any knowledge of the motion amounts, has some drawbacks for the Shadoks who sail on the sea.

Les devises Shadok

QUAND ON NE SAIT PAS OÙ L'ON VA,
IL FAUT Y ALLER !!...
... ET LE PLUS VITE POSSIBLE.

G.9.1. In the howlers collection: the exam for the driver's license.

Quote:
In the new driving license, there is the following question: a pack weighing 1 kg is laid on the rear platform of the car, and the car runs at 100 km/h. When the vehicle is stopped by a brutal shock, this pack becomes a projectile. Now, what is the **weight** *of this projectile?*
Official answer: 40 kg.

Can somebody find a justification?
It is too enormous! We let you correct yourselves. Solution in Appendix H.
In the same kind, when I was a kid, a radio hostess preaching road prevention, pretended that "*Our blood reaches the density of mercury*"...

G.10. Changing of frame

Let's begin with the most simple: the two inertial frames are in translation to each other, with a null or constant mutual speed. They remain parallel to each other, and their metrics are the same. So is the basic case of the Galilean perspective. If the position of a point M is represented by the vector $\overrightarrow{OM}$ in the first frame, of origin O, $\overrightarrow{O'M}$ in the second frame of origin O', then $\overrightarrow{O'M}$ $= \overrightarrow{O'O} + \overrightarrow{OM}$
Now divide the positions by a duration, you obtain the speed of the point M. Divide again by duration, and you obtain an acceleration. Under the Galilean restrictions written above, the summing ways are the same for speeds and acceleration, under the restriction of no mutual rotation of the frames.

G.10.1. The balloon in the wind. Jules Verne reproduced a standard question at the beginning of "**Five Weeks in a Balloon**": "*But could any balloon withstand the wear and tear of such a velocity?*", then answers with over-simplification: "*The balloon is always motionless with reference to the air that surrounds it. What moves is the mass of the atmosphere itself: for instance, one may light a taper in the car, and the flame will not even waver.*". Over-simplified indeed, as in a tempest the wind is irregular and inhomogeneous, unlike a blackboard wind, and at worst, it may contain tornadoes or waterspouts which are severely destructive. Now accept the simplification: the speed of the balloon seen from the ground is the speed of the wind where the balloon is. With reference to the surrounding air, the speed of the balloon is null.
So was in 1863 a first example of the of the usual misunderstanding of the general public: they usually confuse force with speed. We will see other samples.

G.10.2. The fall of a non-directional parachute. Provided that the wind is homogenous with altitude, the parachutist (or the container) under an old-fashioned circular parachute has no horizontal speed in reference to the surrounding air. But he has a vertical speed. In the frame of the parachutist,

the wind comes from below. The *chutes* for the pieces of equipment – say a bulldozer - remain non-directional.

G.10.3. The aircraft in the wind.
It is a mere changing of frame:
$$\vec{V}_{aircraft/ground} = \vec{V}_{aircraft/air} + \vec{V}_{air/ground}$$
So simple that the retired pilots from Air Force yell their rage to the scandal, that "*Air refueling with cross-wind, it is bloody damned complicated, you must be awful skillful, you do not suspect how much the basket is sensible to cross-wind, and you are not even pilots, only laymen!*", etc. etc. The proofs by video are abundant, but they do not tolerate them. So are the sects, packs, and tribes: obsessed by their pack-narcissism, or superiority complex of "*We, the in-pack*" over all the "*they, the out-pack*". So after dozens of years of practice, they are still unable to state that an air refueling is in the air, without mechanical bounding to the ground. Sure, the flight of the tanker must avoid to ram into the Mont-Blanc, and avoid to come in the range of enemy artillery, but this the navigator's of the tanker concern, not of the pilot who is in refueling. Even when he despises the navigators, as "*not even pilots !*". So, even covered with stripes, even stars, they confuse speed with "*forces*", kinematics with dynamics. So are they, the proud packs.

G.10.4. The boat in a flow.
The same mere changing of frame: $\vec{V}_{boat/ground}$
$$= \vec{V}_{boat/water} + \vec{V}_{water/ground}$$
$$\vec{V}_{air/water} = \vec{V}_{air/ground} - \vec{V}_{water/ground}$$

So simple, there is nothing to add. Only kinematics, with speeds. However, when examining the most enormous enormities written or said by the highest lights in sailing, it is obvious that their terminology has damageable lacks, which make them produce some serious mistakes in reasoning.

$\vec{V}_{air/ground}$ or geographic wind; we can also note it V_g. The sailing yachtsmen preferred to call it "*real wind*", implying that the other is *unreal*.

$\vec{V}_{air/boat}$ that we can note V_b, is the wind perceived in the local frame of the boat or the ship. Alas, the sailing yachtsmen disqualify it in "*apparent wind*". This local wind fills your sails, propels you, sometimes tears your sails, or even capsizes you. *Unreal*? A fallacious *appearance*?

Following the idea, maybe the wind pushed downwards by the blades of a helicopter is just an unreal *appearance*, and even the flight of the chopper itself could be an unreal *appearance*...

Now, the local wind caused by the flow, by the motion of the water $\vec{V}_{air/water}$ we note it V_c, as perceived on a drifting buoy, is absent from the allowed reasonings, and did not even receive a name. Only once I have read it as *surface-wind*. We will see on written proofs, how disastrous is this hole in their reasoning tools: they invented mysterious and magical *forces*, instead of the simple kinematics. We need two vectorial relations more:

$$\vec{V}_{air/boat} = \vec{V}_{air/water} - \vec{V}_{boat/water} = \vec{V}_{air/ground} - \vec{V}_{boat/ground}$$

An enormity in a maths handbook, under the sponsorship of some inspectors, for the Brevet d'Etudes Professionnelles: "*The river flow pushes the boat **with a force** so, and the motor pushes the boat **with a force** so, where the boat will reach the other shore of the river?*"

Total confusion of speeds with forces. You use a dynamometer to measure forces. A speed is measured with a land surveyor chain and a chronometer, plus a cork to measure the speed of the flow.

But the horizons of a maths teacher are limited to his/her blackboard: "*The Delta line is the one which is drawn in dark white*". For him/her, displacements, speeds, accelerations, motion amounts, forces, electric fields, and much worse, any gyratorial quantity, all that are mere sticks with arrowheads on the blackboard, all that is interchangeable for his/her eyes...

Well, as I am an experimenter and unbeliever, so I ask what the experimental protocol the calculus-minded had in the head is? Where will he/she hook their dynamometer? In towing? From the shore, or drifting on the implicit river? And how will they measure the **force** of the stream which **pushes** so? Where will they hook the dynamometer? From the shore or from a wharf, of course. Oh! Surprise! When we hook the dynamometer to the boat in a realistic way for crossing the river, and when the flow is strong, the canoe capsizes and sinks! Conclusion: this author of a handbook for the professional classes never understood the matters he/she is paid to teach. This calculus-minded tribe hates all that is concrete, and scorns the experimentation and the experimenters. They are not even able to distinguish kinematics from dynamics.

And on the side of the **sailing** boaters?

Often worse. They lack the basics in mechanics, so they built extravagant theories with mysterious (and inexistent) *forces*, and their contradictory properties, especially about navigation in the tides, because they confuse speeds with forces. A sailboat is a simple machine, composed of a hydrodynamic profile and an aerodynamic profile, and she exploits the relative speed of these two fluids. And only this difference of speeds, which both are vectorial. What may be the speed of the water in reference to land changes the strategy of the skipper, and the work of the navigator, but does not change anything to the maneuvering nor the sails and their trimming. Example of a change of strategy: in the Channel, where the tides are strong, and the bottom not so far, by a faint wind a moored sailboat may "*run*" quicker on water, as anchored to ground, than another, doing its best with only the wind and the water, without help from the ground...

If you are beating windward, for the sails trimming only matters the speed of the wind in reference to water. Where the tide drives you is another worry, distinct, concerning the navigator and the skipper: they are in charge of driving you on the aim, and out of the dangers, so they give instructions to the crew and the helmsman.

The dramas exist.

By ebb tide, a sailing cruiser is approaching the channel out the Morbihan Gulf. The geographical wind is faint from the north-east, seven knots. The ebb tide is favorable, south-west-flowing. However, in the channel near Port Navalo, the stream accelerates, here to seven knots[1] (it may run at ten knots). So it remains not any wind/water V_c , and the sailboat can no more maneuver in any way. The ebb-flow carries to a reef with a beacon, no auxiliary motor serviceable, the cruiser is thrown on the reef and stay torn there. At low tide slack, ones rescue the crew, evacuate them. A tide later, the wreck disappeared with the flow and ebb, and was never seen again. Maybe, if they had the reflex to moor the anchor soon enough, and if the anchor set without breaking, maybe they could save their yacht.

Exercise (energy and forces, about this kind of accident):

What was the energy of the shock of the yacht on the reef? Mass of the yacht: 2,500 kg. Speed of the stream: 3.6 m/s.

16 210 joules.

It is also the energy your anchor must absorb to stop you in emergency before the crash on the reef, when it hooks the bottom. Now the big question: will the stopping-forces be held by all the dynamic chain [bottom + anchor + shackle + chain + shackle + eye + rope + bow roller + mooring cleat or bitt or the end knot], without that something breaks, or the anchor drags. Personally, I have broken a CQR anchor, and I have seen another one to break in critical conditions. So I do not feel the question as too academic.

Be optimistic, and suppose their rope is a stranded 16 mm, with a breaking force of 55 kN, and strain of 11 % at 50 % of the rupture force. Your mission is to compute:

1° What is the minimum stopping distance which permits to the stopping energy to be elastically absorbed by the rope, with a compatible tensile force?

2° In this condition (11% strain), what is the necessary length of rope to let run, before braking?

3° Is this compatible with the withstands of the apparatus on board, bow roller, mooring cleat or bitt? Or unrealistic?

You will be terrified by the results, and the consequences on your fittings and maneuvers.

The solution in Appendix H. In alpinism also, we are concerned by vital calculations on what a climbing rope must absorb, in case of fall of the leader.

This is the end of the digression on stopping forces and energies of shock.

In the tides, you have access only to the local wind-on-water V_c, not to the geographical wind V_g. Even when you prefer to confuse V_g and V_c together under the name of "real wind", you still do not have access to the geographical

---

[1]The report is published by Alain Grée, in **Navigation, le cap, la route, le point** ; édition Gallimard. Coefficient tide 101 (the maximum is 120), the ebb-flow was 7 knots, and the faint wind was west (the only point I have changed). The reef was the Grand Mouton. Guessing the mass of the cruiser: This Aloa 25 is 7.80 meters long and 2.70 m wide, weighs empty 1,800 kg, without motor, water, equipment, supplies, nor crew. Guess 2,500 kg as the total mass in time of the accident.

wind, unless you moor at anchor.

Enormities by William F. Crosby (1890-1953): The text I have is a french translation at Editions Maritimes et Coloniales. Crosby never caught that unless the water is motionless, the wind he has on water is not the geographical wind. So he invents *"tidal **forces** on the keel or centerboard"* of his Snipe, instead of merely considering changes of frames for the **speeds**. Consequently, close haul to the wind against the tide, he recommends to hoist the centerboard *"not to let too much hand to the tide"*. Etc. etc.

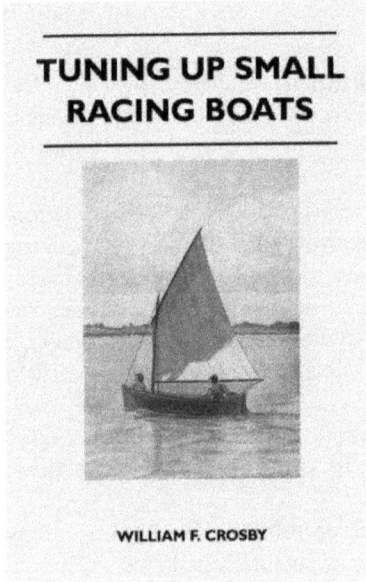

**TUNING UP SMALL
RACING BOATS**

WILLIAM F. CROSBY

The enormities of Yves-Louis Pinaud (*directeur technique national*, years sixties): *"Wind against tide make this Dragon very weather helm. Look how much the helmsman keeps the helm on weather!"*

Figure 23.15.

Courant contre vent rendant ce dragon très ardent. Remarquez comme le barreur maintient sa barre au vent.

And what is the experimental protocol of Yves-Louis Pinault to prove that the weather-helm character if this racing yacht depends on the tide in reference to the ground, and not of the wind in reference on the water? Nothing at all, of course, YLP had a great mouth, and it was enough for him, for all practical purposes. Indeed, the general system of waves and the acuity of their crests depends on the shallows and on the tidal streams, and an experimented sailor is trained to distinguish the two effects, as the profiles of the waves differ.

Another proof that he confused the geographical wind with the wind-on-water, exactly as Crosby did thirty years sooner:

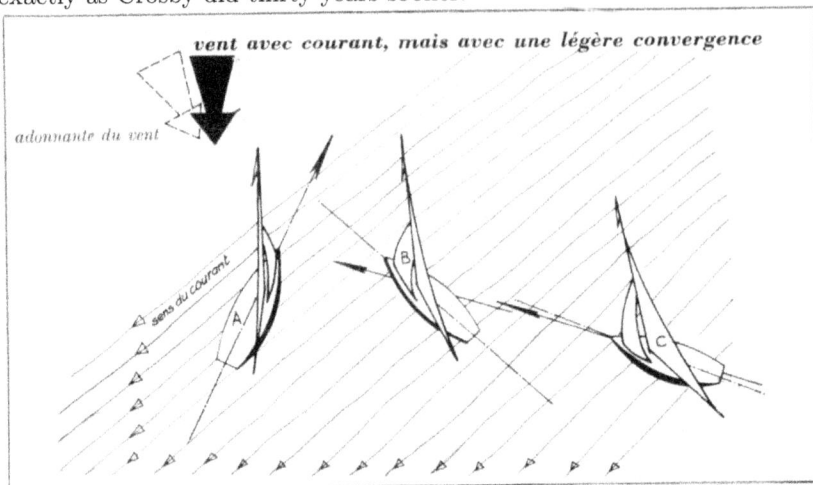

vent avec courant, mais avec une légère convergence

adonnante du vent

sens du courant

Le voilier A bâbord amures profite des évolutions adonnantes du vent réel et fait une route lui permettant de trouver appui contre le courant. Les voiliers B et C sont ramenés aux cas du premier dessin. Le voilier B serre trop le vent. Le voilier C fait courir de manière à bénéficier d'une grande résistance à la dérive.

Figure 23.14.

English translation: *Wind with the current, but with a slight angle. Winglift. Direction of the current. The sailboat A is on port tack and benefits from winglifts from the real wind; her road provides support againts the tide. The sailboats B and C are in the cases of the first drawing. The sailboat B sails too close to the wind. The sailboat C lets running, so benefits of better support against the drift.*

Here again, he had replaced the elementary kinematics by some magic and mysterious dynamics. And I fell for that, fifty-two years ago, to his argument from authority...

Did I succeed in inciting you to revise your basics in kinematics and mechanics? I keep a doubt.

Curious:
- Objection: « *adonnantes* » in the legend you quote? What does it mean?

Open-Eyes:
- More favorable when you sail windward, close hauled, depending on your tack. Sure this notion is familiar to the sailing boatsmen, but unknown from the landlubbers. Supposing that the geographical wind comes from the north and you will route to the north. Supposing that on a run with starboard tack, you had the best close hauled heading on NW, on the heading 315° (the compass rose is divided into 360 degrees). If the wind turns 10° about in the North-quarter-East, you can do a better heading on 325° with the same speed on water, leading you more quickly on your goal. This windshift is favorable for you, it is a gift, and the french sailors say the wind has "*adonné*", in english you say the windshift gave you "*a lift*". If on the contrary, the windshift blows

from the 350, it is "*a header*" for you starboard-tacker, it obliges you to head 335°, or to tack over, as now the port tack would be more favorable, heading at 35° instead of 45° previously.

Professor Castle-Holder:
- Someones flagged to me also other enormities, again under the confusion between kinematics and dynamics, and no-assimilation of the Galilean relativity: "*An instructor of gliding, teaching that after a taking off against the wind, you should not turn at 180° too quickly, risking the penalty of stalling, **because the inertia of the glider hinders the wind to re-give enough speed to the glider in the other direction**". Another one who has not yet assimilated that when a glider is in the air, it is bound to air, not to the ground. On the contrary of the observer on the ground...

G.11. "The wind rises of 4°"...

This hoax seems due to Manfred Curry, in the thirties, and was propagated enthusiastically by Jean Merrien in the fifties. If it were true, soon all the lower layers of the atmosphere would be empty of air, to the profit of the highest layers, even the intersidereal space... "*And the birds would fall*", adds Jean-Claude Van Damme.
Origin: an erroneous protocol, then a very wrong interpretation.

G.12. "Centrifugal force"

Recalling the lesson for the centripetal acceleration in a circular motion:
http://deontologic.org/geom_syntax_gyr/index.php?
title=Les_quart-de-tours_entre_vecteurs_:_gyreurs
Let take a « material point » M in uniform rotation around a center O, at the fixed distance R from the center of rotation O. The plane of rotation is fixed.

Figure 23.15.

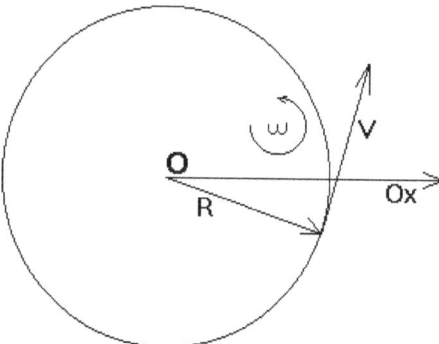

The operator "angular speed" (which contains a quarter-turn operator), applied

to the vector radius (from the axis to the point M) gives the peripheral speed:
$$\vec{V} = \breve{\omega}.\vec{R}$$
The orientations of $\vec{R}$ and $\vec{V}$ always change, but their norms do not change. And their quotient the angular speed $\breve{\omega}$ is also a constant, in a uniform circular motion.

The centripetal acceleration is the product of the peripheral speed by the angular speed: $\vec{\gamma} = \breve{\omega}.\vec{V}$. In both cases, $\breve{\omega}$ links two orthogonal vectors. In the plane of rotation we draw these two geometrical relations: $\vec{V} = \breve{\omega}.\vec{R}$ so :

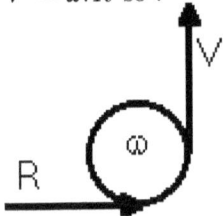

Figure 23.16.

Unit of $\breve{\omega}$: the radian per second.

The centripetal acceleration $\vec{\gamma} = \breve{\omega}.\vec{V}$ by this connector:

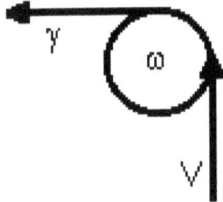

Figure 23.17.

And we can unite these two connectors, to directly link the centripetal acceleration to the vector radius: $\vec{\gamma} = -\left|\breve{\omega^2}\right|.\vec{R}$

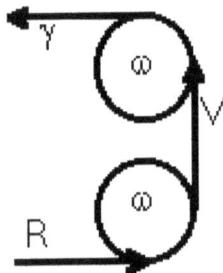

Figure 23.18.
Two quarter-turns do a half turn...

Paradox: What is this radian whose square is –1? It is the quotient of two perpendicular lengths (and equal in norm). So when you are in an improper frame, in rotation, and are willing to forget it is an improper frame, call *centrifugal force* on an object the product of the mass of this object, by the opposite of the centripetal acceleration, calculated above. It is not a fictitious force: it can explode a badly made grinder, or explode an alternator in case of a wrong manoeuver, with serious damages around.

G.13. Speed: covered distance / duration

In the course of the dialog, we have seen that for measuring a speed, you need a basis of length, and a mean to measure the time spent on the journey. In Galilean relativity, the bases of length and the bases of time are transportable, and are invariable so. For 1905 and Albert Einstein, we know the limits of this approximation, and we know that we have to specify in which frame we pretend to describe things. Moreover, there are conditions to the measure of a speed: it needs some time, and it needs some space. The less you have, the less your measure is valid. In Relativity, considering that the speeds form a vectorial space is over; Now only the "**swiftness**" maintain this habit, inherited from the times of Galileo and Newton. The link between their norms: Speed = c.tanh(Swiftness). The directions are preserved.

G.14. Acceleration

Acceleration is the variation of speed per unit of time. In our macroscopical and non-relativist domain, it is a vectorial quantity. However, a convenience is lost: In a frame change, with a mutual rotation of the frames, the simple additivity which existed for speeds, is lost. The correction was given by Gaspard de Coriolis; the lesson is at:
http://deontologic.org/geom_syntax_gyr/index.php?
title=Acc%C3%A9l%C3%A9ration_de_Coriolis_et_inertie

The notion of acceleration is dependent on the macrophysical world; it may be extrapolated in microphysics only within tight and non-relativist limits. Yes, you can accelerate an electron in an electric field, and all the vacuum tubes, the cathodic tubes, and the X-rays tubes work so. For ionic implantation in the making of microchips, the concept of acceleration by an electric field works also. Only the ion or the electron are not macroscopic in a completely macroscopic apparatus. In these conditions, the undulatory character of these individual waves is not or little perceived by our apparatus.

But for the Compton scattering of X or gamma photons by a free electron in a metal or in graphite, it is a new affair: here it is the reflexion of a wave train (the photon) on the temporarily stationary Dirac-Schrödinger electron wave of the electron re-bouncing on the photon, which for a limited time exists in both directions, the ancient and the new direction of propagation.

An amazing **fool trap**: in what direction, the gravity acceleration?
For anybody, it seems evident that it is downward: any dropped object falls.

Nevertheless, when you brake hard in your car, the lucky charms or rosary you hang at your inside mirror hang forward; though the braking acceleration is backward. And nobody can contest the direction this kinematic acceleration. But why, in the absence of kinematic acceleration, the pendulum hangs downward? And how to make those pilots, stubborn and rigid in their Squadron Lounge, understand how in this simplified accelerometer, the "*ball*", turn indicator, the accelerations compose? At rest on the ground, the ball is low, at the center of the bent glass tube; and flying in steady straight flight, with the wings flat, the ball is in the center also.

The solution exists only in the frame of the **General Relativity**: to be standing on the ground is disobeying to the gravity, and the valid reference frame is now a geodesic of the Riemann space, that is free fall. "Valid" here means validity regarding the GR theory; it does not imply anything regarding the ease nor even the opportunity to put a metrological central in free fall... When you stand motionless on the ground, you are resisting to gravity, and your weight pendulum or the ball accuse a counter-gravity acceleration; the acceleration is upward. *Et voilà!* A weight pendulum or ball or any other accelerometer indicates the sum of the kinematic acceleration plus the counter-gravity acceleration.

Like it or hate it.

Exercise: A stuntman wishes a free fall on 20 meters in front of the camera. You have to calculate a mean to damp his fall, not imposing a deceleration above 2 **g** (taken as constant, for simplicity). This mean will absorb all the kinetic energy of the stuntman. No crushing of the bones of the stuntman is allowed, no mashing of his flesh either.
We will admit the equation of the move is: $x = a\,t^2 + bt + x_0$.
Where **b** is the speed at the date t = 0.
x_0 is the abscissa at the date t = 0.
a is the half of the constant acceleration. Take
g = 9.8 m/s^2.

Professor Castle-Holder:
- I have heard another enormity, from a retired fighter pilot: He explained that *in a nosedive, the pilot could not eject because the downward speed in m/s is higher than the upward acceleration of the ejector seat in m/s².*

Curious:
- If I understand well, he compared a speed to an acceleration. Absurd!

Professor Castle-Holder:
- And worst: he compares a number of meters per second, a number, to another number, a number of meters per squared second. Just change the unit of time, and the numbers change, but not in the same way. The third blunder made

by this retired man is that he confuses two different directions: the vertical in reference to Earth, and the direction of ejection of the seat, which is perpendicular to the axis of the fighter, to clear out of the course of the tail-plane, whatever this axis is oriented in reference to Earth. When nose diving at 90°, this speed of ejection in the frame of the plane is horizontal. But not in the frame of the Earth.

Open-Eyes:
- I will add two exercises, for reviewing. The elevator broke, and Mrs. Smith must climb by the stair. All harnessed, she weighs 60 kg. She climbs each stage in twenty seconds, and the height between stages is 271 cm. What is the mean mechanic power developed by Mrs. Smith? Take g = 9.8 m/s².

A fighter pilot also owns a personal plane with a propeller, which takes off shortly: from 0 to 100 km/h in 100 m.
He deducts that his plane with a propeller accelerates more than a Mirage III, as for accelerating from 0 to 300 km/h, this one needs much more than 300 m. Your mission is to prove his error.

The same calculation also applies to the braking course on a dry roadway, for a car as well as for a plane. The braking course at 10 m/s or 36 km/h is multiplied by four at 72 km/h, by nine at 108 km/h, and by sixteen at 144 km/h, as the deceleration is limited by the adherence of the tires. In case of aquaplaning on a thin layer of water, the news is even worse. The calculations are the same for the depth of damping by the vehicle in case of frontal shock, and the practical results are terrifying.

G.15. Angular speed Ω, in radians per second

The angular speed describes how much a solid turns, in one unit of time; and by extension the speed of change of phase for a periodic *thing*. It may be expressed in radian per second, revolution per minute, or cycle per second.
About the rotation of a solid, its angular speed $\check{\Omega}$ has all the characteristics of a **gyror:** invariant plane, sense of rotation, and norm of this angular speed. This geometrization has no more interest when the phase of a non-geometric quantity is considered.
At what is the link with the rotational of the field of speeds?

This scheme sketches a wheel, rolling without sliding.

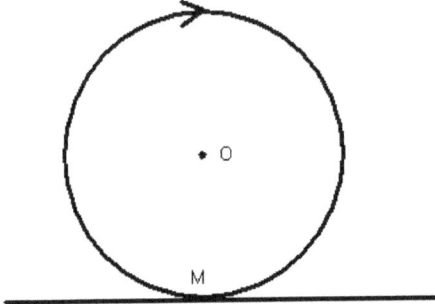

Figure 23.19.

For a rotating solid, the rotational is exactly $2\check{\Omega}$. This factor 2 is somehow puzzling. We had seen in the chapter on the Coriolis complementary acceleration – in case of mechanical bonds – or the contrary, the inertial deviation anti-Coriolis of the marine streams or the winds around an anticyclone or a low, that the inertial free movement (anti-Coriolis) is very similar to a block-rotation as a solid.
http://deontologic.org/geom_syntax_gyr/index.php?
title=Acc%C3%A9l%C3%A9ration_de_Coriolis_et_inertie

This in contrast to a vortex with central aspiration, where the movement is irrotational but the central hole (like the famous Mælstrøm described by Edgar Poe): the speed increases near the center in $1/\mathbf{r}$, where $\mathbf{r}$ is the distance to the center. A tornado presents the same irrotational scheme, but the central hole. Somehow similar to the tornados and waterspouts, but at much larger scale, is a tropical cyclone. An intermediate case, however, because of the friction of the wind on the see, a mechanical damping. The central evacuation to the stratosphere of the converging air is the eye of the cyclone.

G.16. Angular moment

The moment of a force in reference to an axis is the outer product of the lever arm by the force: $\check{M}_{yx} = \overrightarrow{L_y} \char`\^ \overrightarrow{A_x}$. We stated that the moment of two opposing forces, or couple, does not depend on any frame nor axis.
The angular moment of the motion amount of a small material object, respective to an axis, is defined as the outer product of the vector radius to the object by the motion amount of this object:
$\check{L} = \overrightarrow{R} \char`\^ \overrightarrow{p}$. This moment is also a gyror.
Spins are angular moments, but independent of the frame. A gyror has a direction of plane, and a sense of rotation in this equiplane. The gyror angular moment may be considered as the contracted product of two second-order tensors: the inertia tensor, which is symmetric, by the angular-speed tensor, which is anti-symmetric.
$\check{L} = \overline{\overline{I}} . \check{\Omega}$

The theorem of conservation is very similar to the one of the momenta: the angular moment is conservative. To alter it, you must do something, to apply a couple of forces, or a moment of a force. Therefore, the angular moment has all its place in the Noether's theorem, which is fundamental in all the theoretical physics.

Noether's Theorem: If a system has a continuous symmetry property, then there are corresponding quantities whose values are conserved in time.

No absolute spatial position	Invariance by translation in space (the laws are the same everywhere)	Conservation of the momentum
No absolute time	Invariance by translation in time (the laws are always the same)	Conservation of the energy
No absolute orientation	Invariance by rotation in space (no privileged direction in space)	Conservation of the angular moment
No absolute speed regarding this of light (Special Relativity)	Invariance of the light-cone passing by a point of space-time. Transforms of the Lorentz group.	Conservation of the space-time interval
No absolute acceleration (General Relativity)	Diffeomorphisms (General covariance)	Action of Einstein-Hilbert
No individual identity of particles	Permutation of identic particles	Fermi-Dirac Statistics, Bose-Einstein Statistics.
No absolute reference for the phase of charged particles	Invariance by a change of phase	Conservation of the electric charge

Applications of the conservation of the angular moment:
spins, flywheels, gyroscopes, gyroscopic compasses, gyroscopic turn indicators, inertial navigators, anti-roll stabilizers for ships, shells shot from rifled barrels, Foucault's pendulum...

Lastly, the spin of elementary particles with mass (electron, proton, neutron, etc.) has the properties of an elementary unit of angular moment, plus some other properties without equivalent in our macrophysical world.

Avoiding to distract the reader, we waive developing astronomical applications. Some despotic maths-minded fools play the clever in the sandpit, by pretending that "*the Sun turns around Earth*" is not more false than "*Earth turns around the Sun*", and heavily laugh at these *clumsy* physicists "*who do not even master the notion of changing of frame*". I encourage the curious reader to calculate and compare the angular moments in the two cases. One is wise and coherent with the astronomical data (included the *elliptic aberration of the stars*), the other is insane. Next the same exercise for the energies. The funniest is to try to apply the Kepler and Newton laws to the Sun "*orbiting around Earth*": the enormity of the howler *jumps to the eyes*.

G.17. Moment of inertia

For a sensorial approach, the best step to take is to watch the video
http://www.nfb.ca/film/aviron_qui_nous_mene_-_double_elementaire
with Bill Mason and his son, canoeing at sea. In the whitewater river, they sat
near both ends, the son at the front, and the father on the aft, so each has the
most powerful lever arm for the evolutions. When they continue at sea, with
swell, from 22 min 28 s up to 24 min 20 s, on the contrary, both pull together
at the center of the canoe. Goal: to reduce the moment of inertia in pitch, to
let the front and aft lift and drop as easily as possible according to the waves,
without shipping sea water.

Sooner, we have seen the dimensional analysis. We have seen the kinetic energy
in translation: $K_T = \frac{1}{2}mv^2$ and now we ask for a similar form in rotation, on
the style $K_R = \frac{1}{2}I\omega^2$ where ω is the angular speed in radians per second, and $\mathbf{I}$
the inertia moment in reference to a plane direction of rotation and the axis of
this rotation. K_R is of dimension M.L². T⁻², so I is of dimension M.L². Now for
the question of the numerical coefficient, we return the angular moment: per
elements, it is the sum of moments of momenta around the axis of rotation. No
fuss: the elementary mass multiplied by the square of the lever arm, coefficient
1 at this scale; just to sum on the geometry of the object.

A beginning of the mathematization for Bill Mason and his son: At the beginning, each one was at 150 cm from the center of the canoe, next each is at 50
cm of the center, and we decide that man each weighs 70 kg.
Before: $I_{paddlers}$ = 140 kg x 1.5 m x 1.5 m = 315 kg·m².
After: $I_{paddlers}$ = 140 kg x 0.5 m x 0.5 m = 35 kg·m².
Such a difference is very efficient. Now we have to add the moment of inertia
of the canoe, in pitching. This is more difficult to obtain, and worse with some
water inside.

Now we can mathematize. It is easier if you already know the center of gravity
or the center of inertia of the object. Let G this center, and M the current
scanning point. The constraint is that for a model of the inertia in rotation you
must specify the direction of this axis of rotation, or equivalent, the direction
of the equiplane of rotation, perpendicular to the axis.

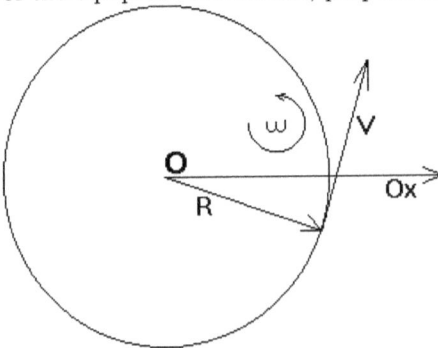

Figure 23.20.

The complete result for all possible directions is a symmetrical tensor of second order. You are not familiar with the mathematical object; here it contains squares of distances, distance GM for instance.

Imagine a thin hoop, of radius r, and mass m, turning in its plane around its center. Then the moment of inertia in this plane is $m.r^2$, and you have just multiply by an angular speed to obtain an angular moment.

Mechanically more credible, more realistic, the next object is a full disk, again turning in its plane around its center (a music disk, a video disk or a brake disk, it works the same). Now one has to sum all these virtual hoops, each according to its radius. For the calculation, we need the areal mass: ρ, in kg/m^2 . $\rho = m/(\pi.R^2)$ where R is the radius of the disk. Hence the moment of inertia:

$$I = \int_0^R 2\pi\rho r^2 dr = \rho\frac{2\pi R^4}{4} = \tfrac{1}{2}mR^2$$

And if on the contrary, we want to minimize the moments of inertia for the same solids, supposed unalterable? The case of the hoop, turning around a diametral axis:

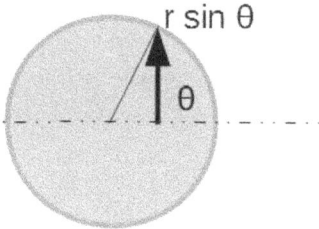

Figure 23.21.

The parameter of integration is the angle at center, ϑ, varying from 0 to 2π. We need the lineic mass of the hoop: $\sigma = m/2\,\pi\,r$. Hence the mass element is $\sigma\,r\,d\vartheta$, at the distance $r\,|\sin(\vartheta)\,|$ of the axis.

$$I_T = \int_0^{2\pi} r^2 \sin^2(\theta)\,\sigma r d\theta = \sigma r^3 \int_0^{2\pi} \sin^2(\theta)\,d\theta = \pi\sigma\,r^3 = \tfrac{mr^2}{2}$$

(the half of the maximal moment).

We proceed to the same integration on the previous disk, and obtain its transverse moment: $\tfrac{1}{4}m.R^2$ (again the half of the maximal moment).

We approximate roughly the moment of inertia in pitch of Mason's canoe. We replace it by a beam 5 m long, of constant cross-section, weighing 50 kg, that is 10 kg/m.

$$I_{canoe} \approx \sigma \int_{-2.5m}^{+2.5m} x^2 dx = \tfrac{2}{3}\,10 \text{ kg/m} * (2.5 \text{ m})^3 = 104 \text{ kg.m}^2$$

It is an expeditious approximation, but its magnitude is compatible with observational facts. The real canoe is heavier, and slightly longer (about 5.5 m), but her masses are more concentrated than in the above approximation.

Curious:
- I would have experimental measures. May I have?

Professor Castle-Holder:

- Sure! But the experiment is not toil-free and not without some dangers for the precious structure of the canoe. First, you weigh the canoe: m. It allows you to use the theorem of parallel axes: for any axis at the distance d of the center of inertia, the moment of inertia is increased by this one of the simple weight pendulum md^2. Next, you make a seesaw from the canoe: you suspend her successively by three horizontal fixed axes in the section containing the center of inertia, vertically spaced, and you measure the period of rocking. You know the acceleration of gravity in your location. The period would be infinite if the axis passed just by the center of gravity. You put in drawing the inverse of squares of the periods as a function of the distances di of the experimental axes. Where the experimental straight line meets the null ordinate, the abscissa is the position of the center of inertia. Now you know the real distances d_1, d_2, d_3 for the center of gravity, so you deduce the three added moments of inertia md_i^2 and by substraction, you deduce the irreducible moment of inertia, this one by the axis of pitch passing by the center of inertia.

Curious:

- One or two days of work to build this tunable suspension of the seesaw, and then two hours to proceed the measures and the calculations...

Open-Eyes:

- Please decide what you want. The experimental approach is fundamental to any science, but we never pretend that it is gratis. Sometimes, but rarely. Another way for experimenting asks less material but more calculations: if you have a handle at each end you can suspend, each in turn, the aft end, next the fore-end to a spring, soft enough so that you can time the oscillation period around the other end, the fixed pivot end. Then, by subtractions of the calculated md_i^2 moments, you can calculate from the two measured moments of inertia, the irreducible moment. Subtractions imply poor final precision, alas. Another *caveat* is that the springs do not age during the measurement; and the bungees age by working hard.

Up to now, we have only taken the simplest cases, when one of the principal axes of the inertia tensor is parallel or identic to the axis of rotation. Things become far less pleasant when these axes do not match in direction: the two gyrors, angular moment, and angular speed do not more have the same plane, and the holonomic constraints must *do something* to maintain the rotating piece in a geometry which is not the one of the angular moment. So the machine vibrates. Nowadays, the garages have the right machinery for the dynamic equilibration of the wheels of your car. When I was a kid, and my parents drove, it was not so: when approaching 100 km/h, the wheels vibrated, the steering wheel vibrated, it was terrifying, even dangerous.

We stop here now: the reader will easily find inline the complete mathematization.

APPENDIX H

Solving the exercises

Chapter 3: **The atomic limit, its fundamental quantities.**

Broglian frequency of the proton: $= m_p c^2/h$
$= 1.67265 \cdot 10^{-27}$ kg $\cdot$ 89.87552 $\cdot$ 10^{15} m^2/s^2 / 662.6076 $\cdot$ 10^{-36} kg$\cdot$m^2/s $=$ **2.26873 $\cdot$ 10^{23} Hz**

Under 500 V it takes a kinetic energy of 500 eV $= 80.109 \cdot 10^{-18}$ J, that is a speed of $\sqrt{\frac{2*80.109 \cdot 10^{-18}J}{1.67265 \cdot 10^{-27}kg}} = 309,740$ m/s. A non-relativist speed.

Wavelength: $\frac{h}{mv} = \frac{662.6076 \cdot 10^{-36}kg.m^2/s}{1.67265 \cdot 10^{-27}kg*309,740m/s} = 1.279$ pm. It is 447 times smaller than an interatomic distance in aluminum. An experiment of interferometry in diffraction could only succeed at very grazing incidence.
We store the intermediate and universal result: $c^2/h = 1.35639 \cdot 10^{50}$ kg^{-1} $\cdot$ s^{-1}.

Broglian frequency of a 5 g bullet: $= 5 \cdot 10^{-3}$ kg $\cdot$ 1.35639 $\cdot$ 10^{50} kg^{-1} $\cdot$ s^{-1} $=$ 6.781 $\cdot$ 10^{47} Hz.
Wavelength at 800 m/s: $= \frac{662.6076 \cdot 10^{-36}kg.m^2/s}{5 \cdot 10^{-3}kg*800 \, m/s} = 1.6565 \cdot 10^{-34}$ m.
It is up to you to figure how you could measure such kind of "*wavelength*".
We physicists, we gave up for evident reasons... Just look at the result of the ballistics of a real bullet.
Practically, the mechanics are undulatory only for very small mass: electrons, protons, neutrons, so on.

Chapter 4: **Crystals, theoretical density of the pure iron.**
Volume of a cell: V $= a^3 = 2.32956 \cdot 10^{-29}$ m^3
Mass of a cell: m $= 2$ (atoms per cell) $* 55.845.10^{-3}$ kg / $6.022.10^{23} = 1.8547.10^{-25}$ kg.
Theoretical volume mass: $\rho = 7,962$ kg.m^{-3}, which is more than the experimental value, 7,874 kg.m^{-3} at room temperature; a difference to make us perk up.

Chapter 7: **Afshar's experiment**, dressed in Elitzur and Vaidman dressing.
Frequency ν of the photon: 565 THz. Transmitted momentum: $= 1.249 .10^{-27}$ kg.m/s.
Assume it will be absorbed by a molecule N_2, of mass 28/N g $= 46.5 .10^{-27}$ kg, it will give it the speed of 0.027 m/s from the rest (at 0 K).
And on a milligram of a thread? A speed of $1.249 .10^{-21}$ m/s. So, to travel a

distance equivalent to an atomic nucleus, it will take a million of seconds, or twelve days.

Good gracious! What a *bloody blow*! It makes you so wobbly!

Appendix G: **The 100 kg stuntman, falling on 20 m** (in the field of the camera).

The statement was ambiguous: Was the constraint on the kinematic acceleration only? Or on the total acceleration, including the withstand to the Earth gravitational attraction?

In 20 m of free fall, the stuntman lost a potential energy: 100 kg * 9.8 m/s² * 20 m, equals 19,600 joules.

The most simple reasoning, in the first reading of the statement: during the remaining of his fall, the braked part, of length **x**, the stuntman still loses **mgx** of potential energy. The sum of these potential energies must be compensated by the work of a braking force: **mg + m(2g) = 3 mg**. That is a braking work of **3 mgx**. Difference: **2 mgx** = 19600 joules. Hence **x** = 10 m. On this braking length, the stuntman is submitted to a braking force triple of his weight: 2,940 N.

For the second reading of the statement, to be braked with only twice his weight, the stuntman would need a height of braking equal to the height of filmed height, that is 20 m. The kinematic acceleration would be 1 **g**.

Reasoning by the equation of motion at constant acceleration and force:

His speed at the end of the unbraked free fall is:

v² = 2 * 19,600 joules / 100 kg = 392 m²/s² (= 2 * 9.8 m/s² * 20 m).

Hence **v = 19.80 m/s** at the end of the filmed fall, beginning of the braked fall.

Some stuntmen use a heap of cardboard boxes as the damping material for their falls.

With the acceleration of **2 g**, what time does it take to stop the falling stuntman?

v = 2g.t + v₀. Hence the time of deceleration: **t** = 19.80 m/s / 19,6 m/s² = 1.0102 s.

In this time, height run: **x = g .t² = 10 m**.

It is the same kinematic result, obtained with less physics and more maths.

The gain of potential energy by **Mrs. Smith**:

2.71 m x 9.80 m/s² x 60 kg = 1,593 joules per floor.

Average mechanical power spent: 1,593 joules in 20 seconds = 80 watts.

The **military pilot** who owns a small plane (0 to 100 km/h in 100 m) and compares its acceleration with a military fighter. Let **a** denote the acceleration. The speed at origin v_0 is null. The origin of abscissa is where the pilot releases the brakes.

Distance: 100 m $= 1/2$ a. t^2

Speed at 100 m $= 100$ km/h $= 27.78$ m/s $=$ a.t

Dividing the first equation by the second one gives the time at 100 m:

$t = 2$ x $\frac{100\,m}{27.78\,m/s} = 7.2$ s.

Hence we deduct the acceleration: a $= \frac{27.78\,m/s}{7.2\,s} = 3.86$ m/s^2.

With the same acceleration, supposed constant, the military jet will use thrice the time to reach 300 km/h, that is 21.6 s, and will run:

1.93 m/s^2 * 21.6 s * 21.6 s $= 900$ m.

With the same acceleration, it will run three times longer than 300 m.

The two tons and half sailing cruiser, becalmed and not-maneuvring, driven by a 3.6 m/s tide. It may be worse: in a perigean neap tide, the stream reaches 10 knots in front of Port Navalo.

Energy of the stopping: (2,500 kg / 2) x (3.6 m/s)$^2 = 16,210$ joules.

We want that the stopping force does not exceed 55 kN / 2 $= 27.5$ kN ($= 2,750$ daN $= 2,800$ kilogram-force).

Theoretically, the rope could not break at his tension if it is new. However, such a tension, laterally on the bow-davits could pry it, and tear the bow.

Multiplying this force by the run of its work gives the stopping energy; hence this run: 16,210 J / 27,500 N $= 59$ cm. Surely it is a minimum!

Hence the minimal length of the elastic working rope: 0.59 m x 100 / 11 $= 5.4$ m. It does not fit with the practical experience; surely we have forgotten some data.

The forgotten fact: the usual anchor of this cruising yacht, this one which should be ready to be let go mooring in ten seconds, probably weighs no more than ten kilograms, and cannot supply such a stopping force excepted if it hooks on a rock (or a wreck)... and does not break.

First correction: we multiply by ten the run of the stopping braking, up to 5.9 m (you must have purchased and installed a 50 m or better 60 m mooring rope), so the stopping force decreases to 2.75 kN (275 daN, about 280 kgf). The elastic strain of the rope does not change, still about 60 cm for 54 m of useful mooring rope, so the most of the braking effect along 530 cm should be supplied by dry friction: your boot on the rope running in the davits. Not mentioning the unwinding equipment you have not (but should have), to let this mad rope orderly run at 3.6 m/s. Which one is the anchor which will provide in five seconds the required stopping force? Some tests are published in nautical reviews:

https://www.hisse-et-oh.com/articles/367-divers-bancs-d-essais-d-ancres

During these five seconds of hooking in the ground, the becalmed boat has already drifted 18 m. Add the distance of the hooking in the ground, about 2 m if things turn well.

Another bad news: on laminaria seaweed, flattened by the rapid stream, or on Posidonia in Mediterranea, no anchor can hook if it already has a horizontal speed, or not enough vertical diving speed. In the offshore industry practice, they core drill the bottoms before ordering the anchors for the offshore drilling platform.

For an anchor, a ratio of efficiency traction/weight of 10 is pretty good, and 20 is exceptional; The real ratio depends heavily on the nature and consistency of the ground, and of the length of the mooring line – for inescapable trigonometric reasons. Let admit ten, that is 1,000 N, for the anchor weighing 100 N (10 kg) taken as the hypothesis. We see that we must multiply again by three the stopping run of the sailing cruiser, up to 16.20 m, twice the length of the boat, to not overload the traction the anchor and the sea bottom can hold. The major part of the braking work will be provided by the boot of the crew on the davits, to not rely too much on the elasticity of the hawser – it would be too short and to brutal.

Another interesting question is to convert the 15,000 joules of braking, into burnt centimeters of the sole of the boot.

Reminding the trigonometry of a mooring.
Vertical lifting force on the anchor = traction force on the rope * [depth of water + height of the bow-davits on the water] / length of the rope between the anchor and the davits.
A length of chain, between the anchor and the rope, much helps to decrease, if possible to nullify the so dangerous lifting force from the rope to the anchor; instead, the anchor should bury as soon and deep as possible, to gain a withstand against dragging. Here we have neglected in the calculations the effect of about 10 meters of 8 or 10 mm. The only good news is that in the channel of Port Navalo and its strong tide, the depth is moderate in comparison to the calculated length of mooring rope, so the trigonometry does not much play against us.

Conclusions:
In a strong tides zone, for a sailing boat of two tons and a half, supply and mount a 15 kg anchor, not less, already waiting in the davits, ready to dive, and of excellent quality!
And be extremely cautious in your navigations and maneuvers, anticipate! We have seen a case where 70 to 80 m are run between the moment the anchor reaches the bottom, and the complete stopping of the boat. Not counting the distance run between the decision to moor, and when the dropped anchor reaches the sea-bottom. A hair-raising thinking.

And, and, and... do not approach a narrows or a shallow, like the channel of the Gulf of the Morbihan unless the tide is slowed down, near a slack-tide!

Appendix G: **the new driving license**.
In full absurdity, the authors of the official statement blinded themselves about the distance of stopping and the time of stopping of the object. Moreover, they use the word "*weight*" where inapplicable. Not all forces are weights, only the gravity forces may be weights, in static cases.
We retro-engineer from their weird assertion to the implicit hypotheses. They pretend that the acceleration is 40 g = 392.4 m/s² (no sign is written: no necessity to orient the axis of abscissas). We neglect the gravity, and we liken the cosine to 1. We know the initial speed, v_0 = 100 km/h = 27.78 m/s, we suppose that the deceleration is uniform, the time for stopping is 0.0707 s:
v_0 = γ t <==> t = v_0 / γ
So they supposed that the stopping distance was 0.98 m: x = 1/2 γ t²
It is more or less coherent with the norms of shortening of the front hoods of the cars, but totally incoherent with the trajectory of an object initially laid on the rear, when the shortening of the fore part has ended. Rear objects are much more dangerous than postulated by the officials of the driving license.
To estimate the real danger of the rear object of 1 kg mass, calculate its kinetic energy: $\frac{1}{2}m\left(v_0\right)^2 = \frac{1}{2}1\,kg\,.\,(27.78\,m/s)^2$ = 386 joules. Worse than a crossbow bolt.

But the failure of our science teaching is so disastrous, that no one in the general public accepts to know what a joule is.

Transactional (Quantum) Microphysics, Principles and Applications

The physical quantities

This refresher course is inline for several years at
http://deontologic.org/geom_syntax_gyr/index.php?
title=Les_grandeurs_physiques

I.1. Many more properties than the numbers, more constraints, more safeguards, too.

I.1.1. Abstracting?

TO ABSTRACT is to exempt oneself from considering lots of properties and features (etymology: *trahere* pull, *ab* out of). In words of the theory of sets, an abstraction is the quotient-set by the set of equivalence classes. It is up to the abstractor to write black-on-white what is his/her equivalence relation, and to prove his/her operation is reliable.

Corollaries: "abstract" does not imply "*better*". It is the task of the analyst to verify and prove that with the process of abstraction he chose, he has not made professional faults, that he has not discarded some vital properties. Next, he has to prove that his abstraction is sensible in productivity, and, if possible, optimal.

The master abstraction in the history of humankind is surely the invention of the numbers.

At three weeks, our babies clearly discern one babar (a doll representing the elephant Babar) from two babars, and two babars from three babars. Adult magpies and pigeons do as well. But about three, four, five, their arithmetic abilities loose foot, are prone to be deceived. A magpie who has seen four men (looking alike in their aspect) entering in a hut, and only three outgoing later, does not smell the misled: if the clothing and look (such as the voices, or rhythms of gestures) are similar enough, the magpie did not notice the differences.

The numbers are all-abstract.
The number *four* is not "*four donkeys*", nor "*four pebbles*", it is only what is common in them by the relation "*One can establish a bijection* (a one by one correspondence) *between these two collections*". So it is an equivalence class: the relation is symmetric, reflexive, transitive. Our Magdalenian ancestors lived

the long and slow transition from hunters of wild reindeers to stockbreeders of reindeers. They could chase herds towards corrals, where they could choose and capture some, next distribute this property between several families in several clans. So they had to invent ways to count. To transfer a gravel stone from one heap to another for each reindeer who enters the corral, or is transferred to a sub-corral, is an easy solution. A more portable solution is to cut a notch in a stick, one per animal in the collection. The Roman numerals are the direct offspring of this shepherds practice: I, II, III, V, X, L...

But a number, just a number at all, has properties differing from those of *"a number of something"*. A herd of four geese has not the properties of a herd of four adult reindeers, which has not the properties of a herd of four mammoths. A grove of four pine-trees has not the properties of a grove of four birches. The number *"four"* is exempted of all that. The additions which are free between pure numbers are seriously restrained in numbers-of-something.
First, the two collections must be disjointed, without any common elements, and next, their union must be still a valid set with sense and use, where the elements have some common equivalence.

I.1.2. Disjointed sets. It was an ancient box-card plant repurposed into low-cost housing. A few flats had a second floor above the roof-terrace, built in light Siporex blocks, rock-wool and iron sheets. My daughter was soon seduced by a charming kitten who came visiting and lapped all the good things to drink or to eat that we gave him. The kitten was very serious in learning how the hunt the pigeons. However, from a neighboring islet, a young boy called the kitten, for eating and coming in for the night. If you reunite the two collections, the cat of my daughter and the cat of the neighbor boy, does it make a collection of two cats, or one cat?
Similarly, by the end of crossing the Atlantic, aboard Kurun, Jacques-Yves Le Toumelin announces to his crew Fargue:
"The first to see land will win two glassfuls of rum. The second, only one glassful, and the last will win a kick in the ass.
- What will win the second last?"
If you reunite the first, the second, the second last and the last, how many sailors do you gather? Le Toumelin disembarked Fargue at La Martinique and continued alone his round-the-world cruise.

Remember: adding requires that the elements of the united collection are distinct; no overlapping.

I.1.3. The union set must have a logical coherence. Here is another gag due to René Goscinny.
With bumped helmets and armors, a Roman patrol of four legionaries reports:
"- Ave centurion! We were attacked and outnumbered by a Gauls band.
- Give me their description!

- Err, a big one and a small one, wearing each a wild boar on his back, and a little dog.
- They were five, indeed."
A little more reflection. OK for counting the two Gauls as one set. With many "if" and "but", the dog, here *Dogmatix*, could team with his human masters in a fight. In that case, one could count three fighters. With some mind acrobatics, one could build a set with the two slaughtered boars and the little dog: they are all Laurasiatherians, and maybe smell the beast? Or the all three could be a meal for a family of bears or a pack of wolves? But to count all the five as, as what? As *"a very outnumbering band of Gauls"*? Err, are you sure it is a coherent set? With a relation of equivalence between the five bodies?
However, they could be all the five buried by a landslide or an avalanche, and their bones could be sorted out by a future team of paleontologists, centuries later.

I.1.4. Fuzziness and irreducible errors.

It is frequent that a collection may be defined only within an irreducible fuzziness. Unthinkable for a pure number.

Let us take the population of France: all the time, some die, babies are born, some enter, some others quit. Some disappear, some enter surreptitiously, clandestine agents may change of identity... In the meantime, counting the people takes time to the enumerators who cannot simultaneously be everywhere. Whichever the methodological precautions are taken in the census, it remains a margin of errors and irreducible uncertainty; these margins depend on the methods and the means of the counting. Therefore, we are compelled to distinguish several levels in the abstraction process starting from the complex, and partly elusive reality of the population in a country.

I.2. Fitting into a scheme of interprofessional normalized abstractions.

The physical quantities are half-abstract, they keep a property shared with the real world, of which the numbers are exempted: **a meaning, a calibration benchmark, and the associated grammar of variance.**

The physical quantities are exempted of many very complex properties of the measures, and of the results of measures (with many kinds of uncertainty and errors), or of the storing the results on a paper or an electronic machine, with all its problems of limited width or resolution.

For the concrete quantities and the physical quantities, equality has a meaning, **provided we can define a protocol of comparison and measure.** For instance, to compare two length, two areas, or to compare two lots of box of detergent under different shrinkwrapping... This condition fulfilled, we are entitled to write an equality of quantities, even with different units, such as: 1 cycle = 2 π radians.

The physical quantity is defined as the abstraction common to, in reality, is relevant from the same family of measuring protocols, whose precision and cost may differ, but all having the purpose of measuring a same kind of quantity. For instance, a length, or a rest mass.
So it is a class of equivalence. It is far less abstract than a number, as it is **a descriptor of something**. It is bound to the syntax inherent to the semantic vocation of each physical quantity.

The need exists for characterizing these practical and concrete quantities (for instance, 500 flasks of medicine) on the one hand, and physical quantities on the other hand, their links with the unit quantity, and with the set of the real numbers. So we define the scalable quantities. We also need to distinguish the arbitrarily scalable quantities, from the numberable quantities, for which nature has provided a natural unit, independent of the human conventions.

In 1873, James Clerk Maxwell devoted the first paragraph of his **Treatise on Electricity and Magnetism** in explaining that all physical quantities are the product of a number by a sample of this quantity, called Unit.

ON THE MEASUREMENT OF QUANTITIES.

1.] EVERY expression of a Quantity consists of two factors or components. One of these is the name of a certain known quantity of the same kind as the quantity to be expressed, which is taken as a standard of reference. The other component is the number of times the standard is to be taken in order to make up the required quantity. The standard quantity is technically called the Unit, and the number is called the Numerical Value of the quantity.

There must be as many different units as there are different kinds of quantities to be measured, but in all dynamical sciences it is possible to define these units in terms of the three fundamental units of Length, Time, and Mass. Thus the units of area and of volume are defined respectively as the square and the cube whose sides are the unit of length.

Figure 25.1.
Let us take the width of a sheet of paper issued by my printer. It is in A4 format, and is 21 cm wide. Or 0.21 m, or 210 mm. But no function with real-number values, may give 21 = 0.21 = 210.
Though it is perfectly correct to write: 21 cm = 210 mm = 0.21 m.
These two equalities are not written between numbers but between physical

quantities. They denote equivalence classes, defined by the principles for measuring. Here only the principles of at least one protocol for measuring the lengths, at least one principle for building the measuring apparatus, and campaigns of measures verifying the equivalence and the right adequation of the materials and protocols, can justify the definition of the physical quantity, and the conditions of the written equality.

Figure 25.2.

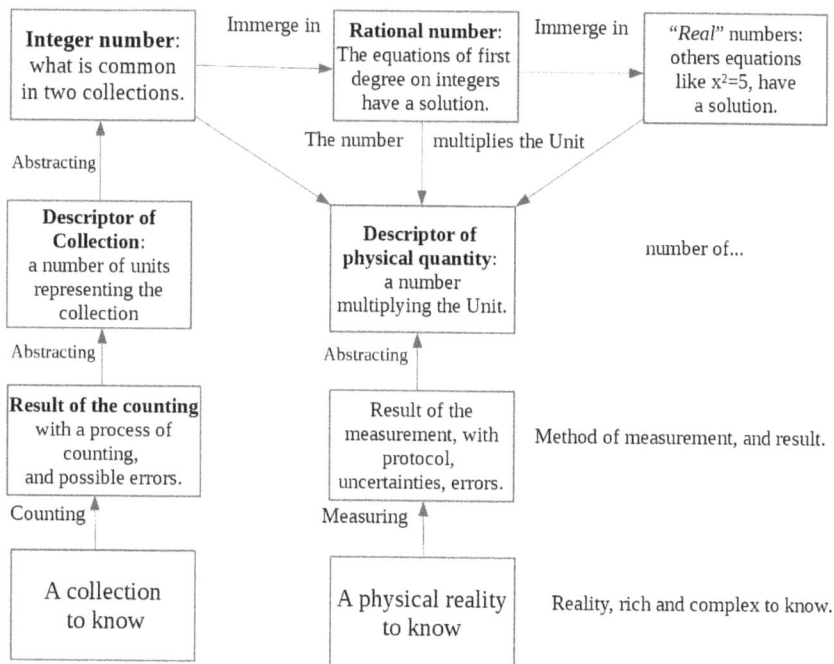

We set the distinction: A quantity is said **numerable** if naturally it is a multiple of a unit fixed by nature. The headcount of a herd of horses is, but the births and grow of the foals, numberable. An electric charge is always the multiple of an elementary charge (electrons or protons), but this multiple may be fluctuating and inaccessible in macrophysical situations. A baryonic number, or a number of nucleons in a matter, are essentially numerable quantities. In a warehouse, the numbers of cardboard boxes of a specified good, are numerable, and the pills in a medicine box also are. Such units are "natural" for the accountant or the stock-keeper: once they are made so, he cannot change the definition of the unit.

On the contrary, a quantity is **arbitrarily scalable**, or briefly said "*scalable*", if the choice of the unit is left to some human arbitrariness. So are the electric tensions, the intensities of electric current, the masses, the lengths, the durations, etc. whose standard units come from human decisions, as arbitrary as they are necessary.

So are the major part of the quantities concerning the physicist, almost all macroscopic quantities, as long as we are far from the atomic limit.

I.3. Five practical rules, usable as axioms.

1. You may add (or subtract) only quantities of same nature, provided they are extensive ones. Intensive quantities, either are not additive at all, such as the temperatures, either their addition is submitted to strong experimental conditions.
2. You may multiply a numerable quantity by an integer. You may multiply a scalable quantity by a real number.
3. In the family of the physical quantities, writing an equality requires you could define an experimental method, and instruments for comparing. Then you are entitled to write an equality between quantities, such as 210 mm = 21 cm. We express the quantity as the product of a coefficient by a unit of this quantity. You must have defined a standard unit.
4. Any physical quantity has an inverse, so it is not necessary to define a new method for measuring. Example: a decameter is graduated in 100 divisions per meter. The inverse of the meter is written $\mathbf{m^{-1}}$.
5. We are free to multiply physical quantities by any another physical quantity; or divide by any not-null quantity. This operation is external: it generates another and distinct physical quantity Whether the result is useful or not, is only the next question. One can state that the really practical results make a simple and remarkable structure.

These pleasant properties of linearity are only valid in a space comparable to the our one: weak gravity, gaseous state of moderately condensed, moderate temperatures. In a neutrons star, our familiar physics would bear lots of surprises.

I.4. The dimensional analysis, first safeguard for the physicist.

Pioneering the dimensional analysis in the 17th century, the French mathematician **Marin Mersenne**, (1588-1648), applied it to the weight pendulum.

Return to his example of the weight pendulum, and treat it with the MKSA units: **On what does depend its period?**
We have the acceleration of gravity, and it is uniform in the bulk of the pendulum, its mass, the distance from the center of gravity to the point of suspension, and the maximal angle of oscillations (if simple pendulum, the angle from vertical); for more complex pendulum we should also consider its moment of inertia.
Dimensions:
Gravity g : L. T^{-2},
Mass m : M,
Length a : L,
Period T : T.

Only one combination gives seconds: $T^2 = \frac{L}{LT^{-2}}$, so
$T = \{\text{numerical coefficient}\} * \sqrt{\frac{a}{g}}$

Only remains to decide the coefficient: 1 or 2π, or the inverse, $\frac{1}{2\pi}$?
It depends on the unit of phase. T is a full cycle.
But when you approximate the potential energy by the first order of the sine (otherwise you should need elliptical functions, which are not more taught), m.g.h = m.g.a. $\frac{\theta^2}{2}$ where the angle ϑ is in radians.
Neglecting the damping, you write that the mechanical energy is constant, so its derive in reference to time is null. Therefore the square of the pulsation is.
Hence the coefficient 2π for the period: $T = 2\pi \sqrt{\frac{a}{g}}$.
The mass was eliminated by the dimensional analysis.

I.5. The variance of the scalable quantities

By definition, they are the product of a sample called "unit" by a *real* number.

So if we change of unit of measure, the quotient of the quantity by the unit is **contravariant** to the unit: 210 mm = 21 cm. In spaces of dimension above 1, each of these quotients is called "coordinate" and each coordinate is also contravariant to the base vectors.
Examples: you may clear a debt of 1,000 F with two banknotes of 500 F, of five of 200 F, or twenty of 50 F. You may send 60 t of goods by three lorries of 20 t, or two lorries of 30 t each. **The quotient is contravariant to the divider**. When the unit is 100 times bigger, the quotient is 100 times smaller, so their product denotes the same thing.

Scheme: scalable quantity = quotient * unit-of-this-quantity.

What if the unit is more complex? If gasoline costs 5.60 F per liter, and one dollar rates 5.60 F, then this gasoline costs 1 \$/l. But it costs 3.785 \$/gallon or 21.20 F/gal as one gallon is 3.785 l: **covariance with the unit in the denominator**, here the volume.

These notions of variance are at the core of any measure and mathematization of physics. This basic syntax is neglected with damages.

I.6. In physical quantities, the conversions are straightforward.

To calculate a wavelength: wavelength = celerity * period.
The other combinations that our pupils throw at random will never yield a wavelength.
L = L/T . T
"wavelength" = 343 m/s . 0.025 s = 343 . 0.025 s.m/s = 343 . 0.025 m = 8.6 m

Then any conversion of units is straightforward. We just have to **write equalities true in physical quantities**.
For instance, we have to convert a volume mass (here a steel) from non-S.I. Units into S.I. Units, first we write the tautology:

$7.8 \cdot \frac{g}{cm^2} = 7.8 \cdot \frac{g}{cm^2}$

Up to now, we have not violated nor the physics, nor the mathematics... Next, we multiply the second hand by fractions equal to 1. They are chosen so that the unwanted terms of units may be eliminated by simplification. Here the gram in the numerator is unwanted, and the cm3 in denominator also is unwanted.

$7.8 \cdot \frac{g}{cm^2} = 7.8 \cdot \frac{g}{cm^2} \cdot \frac{1\,kg}{1000\,g} \cdot \frac{1000\,cm^3}{1\,dm^3} = 7.8 \cdot \frac{kg}{dm^3}$

(by commutativity number.Unit)

$7.8 \cdot \frac{kg}{dm^3} = 7.8 \cdot \frac{kg}{dm^3} \cdot \frac{1000\,dm^3}{1\,m^3} = 7{,}800 \cdot \frac{kg}{m^3}$

To the gifted students, the method seems heavy, and socially not selective enough. They prefer to divine the result, with flair. The less gifted students prefer this robust method, much more liable: it is unbreakable.

I.7. Putting into equation; safeguard by the units and the dimensional analysis.

Here is the scheme for action (a praxeogram if you prefer) given in the classroom, for solving any real problem with the help of algebra, through two steps of abstraction, next un-abstraction:
Figure 25.3.

Solving and posing equations.
Diagram of the actions to do.

Read the statement:
Specifications given
by the client

Write down a **dictionary**
of the identifiers
you will use.

Axis of increasing abstraction

Rephrase the statement
around this dictionary
and its entries.
All the predicates.

Domain of the physicist:
physicals equation,
physical results.

Domain of the mathematician:
numerical equation and results.

Translate this pseudo-code
into mathematical code.
Each predicates gives one equation.

Separe the physical part:
only the units.
Verify it is correct.

Is the equation
reduced to the units
correct?

No

yes

Domain of the client.
He has not to explain you
how to use your tools!

Units

Write the equation reduced
to only the numbers.

Solve this equation .

Restore the physical unit
to the numerical result.

Answer
to the question
in the words
of the client.

End of the process.

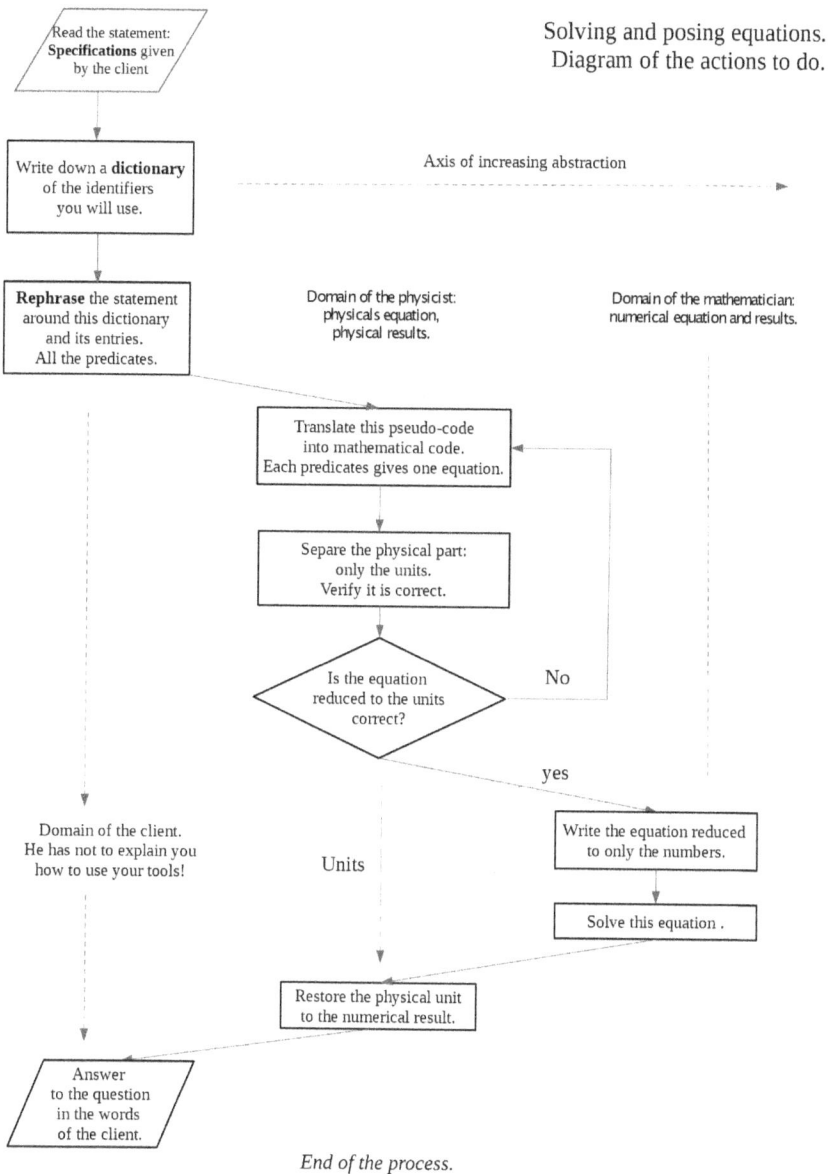

· First abstraction: from the problem as stated by the customer, toward a schematization of each sentence into irreducible predicates; writing a dictionary of the variables and their symbols (it does not matter whether the dictionary has one, two, three or twenty entries), and the transcription of the predicates into equations.

· · Second abstraction: when the dimensional verification is over and does not more reveal errors, then resolve the numerical sub-system by the algebraic methods.

· · Second un-abstraction: restore the physical units, to answer in physical quantities to the physical problem.
· First un-abstraction: answer in the language of the customer, in words he understands, and which answer to his initial problem.

I.8. Operator of proportion.

If I order fabric per meter, and that the price is 3.90 euro per meter, the operator of proportion which transforms the meters into price of the cutting is "multiply by 3.9 €/m".
It is a distance purchasing, so the vendor adds a postage fee, say 10 €.
So we consider a new operator: "add 10 €".
One may inverse the operation, to answer the question: "I have this budget. How many meters of this fabric can I order?".
So we have to reverse the previous operators:
Subtract 10 € to the budget.
Divide by the price per meter.
Only take the integer part of the previous result (the vendor does not send fractions of a meter).

I.9. Operator half-turn.

Examining the scientific literature of the 18th and 19th centuries, for instance the article written by Jean le Rond d'Alembert in the *Encyclopédie*, "*Nombres négatifs*", reveals the difficulties brought by the confusion between the numbers and the operators. The notion of number, triumphant for the Antiquity with Pythagoras, ate up the operator. D'Alembert admitted that the rules for calculating with the negative numbers were correct, while he maintained that they were embarrassed with any interpretation. Be **-1** a number, why not? But surely multiplying by **-1** is **an operator**.

Lots of our pupils lose their footing then: nobody thought to give them nor image, nor an immediate application, nor any concrete reference for this operator, and the field of application.

In charge of these lost pupils in a technical course, I had to remedy their scholar failures; So I laid the support, an oriented axis, with a unit vector; for a parenthesis, I used the image of a miniature wagon (but a bus works, too). When you return the wagon, all the passengers inside are returned too. When you remove the roof and the walls of the wagon while leaving the passengers, they remain returned. So is the minus sign in front of a parenthesis: multiply a parenthesis by **-1** is returning the wagon (including its passengers). Multiplying twice by **-1** is twice a half turn, so coming back to the initial situation.

We will return to this later, with the operator quarter-of-turn, applicable to vectors.

I.10. Links to following chapters: geometric quantities of the physics.

Vectors, definition, properties.
Metrics of the vectorial quantities in physics.
The quarters-of-turn between vectors: the gyrors.
http://deontologic.org/geom_syntax_gyr

I.11. Bibliography and references for the physical quantities.

APMEP, Commission Mots, n° 6 : **Grandeurs et mesures**. APMEP, 1982, Paris. Brochure n° 46.

André Pressiat. **Calculer avec les grandeurs**. Actes de l'Université d'été de Saint-Flour.

André Pressiat. **Quotients - Proportionnalité - Grandeurs**.

Jean-François MUGNÏER. **Recherche et rédaction de problèmes au Collège**. Feuille de Vigne n° 100 – Juin 2006.

Transactional (Quantum) Microphysics, Principles and Applications

Index